VOLUME SEVENTY THREE

ADVANCES IN PHARMACOLOGY

VOLUME SEVENTY THREE

ADVANCES IN PHARMACOLOGY

Diversity and Functions of GABA Receptors: A Tribute to Hanns Möhler, Part B

Edited by

UWE RUDOLPH

Laboratory of Genetic Neuropharmacology, McLean Hospital, Belmont, and Department of Psychiatry, Harvard Medical School, Boston, Massachusetts, USA

Serial Editor

S. J. ENNA

Department of Molecular and Integrative Physiology, Department of Pharmacology, Toxicology and Therapeutics, University of Kansas Medical Center, Kansas City, Kansas, USA

Managing Editor

LYNN LECOUNT

University of Kansas Medical Center School of Medicine, Kansas City, Kansas, USA

ELSEVIER

AMSTERDAM • BOSTON • HEIDELBERG • LONDON
NEW YORK • OXFORD • PARIS • SAN DIEGO
SAN FRANCISCO • SINGAPORE • SYDNEY • TOKYO

Academic Press is an imprint of Elsevier

Academic Press is an imprint of Elsevier
225 Wyman Street, Waltham, MA 02451, USA
525 B Street, Suite 1800, San Diego, CA 92101-4495, USA
125 London Wall, London, EC2Y 5AS, UK
The Boulevard, Langford Lane, Kidlington, Oxford OX5 1GB, UK

First edition 2015

Notices

Knowledge and best practice in this field are constantly changing. As new research and experience broaden our understanding, changes in research methods, professional practices, or medical treatment may become necessary.

Practitioners and researchers must always rely on their own experience and knowledge in evaluating and using any information, methods, compounds, or experiments described herein. In using such information or methods they should be mindful of their own safety and the safety of others, including parties for whom they have a professional responsibility.

To the fullest extent of the law, neither the Publisher nor the authors, contributors, or editors, assume any liability for any injury and/or damage to persons or property as a matter of products liability, negligence or otherwise, or from any use or operation of any methods, products, instructions, or ideas contained in the material herein.

ISBN: 978-0-12-802658-8
ISSN: 1054-3589

For information on all Academic Press publications visit our website at http://store.elsevier.com/

CONTENTS

PREFACE
A Tribute to Professor Hanns Möhler

Few scientists make such major contributions that their name is forever linked to a particular molecule, pathway, or theory. Professor Hanns Möhler is one of those individuals. Everyone working in the areas of γ-aminobutyric acid (GABA) neurochemistry and neuropharmacology is familiar with Hanns' work and its importance in characterizing the $GABA_A$ receptor. While many made seminal contributions in defining this system, none can match Hanns in terms of groundbreaking discoveries and creative insights regarding the structure, function, and clinical importance of $GABA_A$ sites. Even a cursory glance through the literature reveals Hanns' impact on neuropharmacology over the past four decades, his uncanny ability to sense the direction of the field, and his skills in adopting the techniques and technologies necessary for advancing the understanding of GABAergic transmission.

Besides his contributions to the basic neurosciences, Hanns' efforts in drug discovery led to significant clinical advances. It all began while he was working at Hoffmann-La Roche in Basel, Switzerland, where he defined the molecular mechanism of action of the benzodiazepines, drugs used extensively for the treatment of anxiety, sleep disorders, muscle spasms, seizures, and in anesthesia. Diazepam (Valium^R) is a prototype of this chemical class. Hanns' work contributed greatly to identifying the benzodiazepine target as a component of the $GABA_A$ receptor. He and others demonstrated that when bound to this site the benzodiazepines enhance allosterically the activity of this major inhibitory neurotransmitter receptor system. His subsequent discovery of flumazenil (Anexate^R), a benzodiazepine antagonist, was the direct result of his earlier findings, demonstrating the translational nature of his work and its practical clinical applications. Upon leaving Hoffmann-La Roche, Hanns established an academic laboratory at the University of Zurich, where he remains as Emeritus Professor. With his team at the University of Zurich, he established and categorized the distinct pharmacological functions of $GABA_A$ receptor subtypes. This information is proving critical for designing and developing the next generation of drugs as it makes possible the selective targeting of those $GABA_A$ receptors responsible for mediating anxiolysis while leaving untouched those responsible for sedation and effects on cognition. In addition to his work on the GABA system, Hanns developed strategies for exploiting adenosine as a treatment for epilepsy and glycine for dampening some symptoms of schizophrenia. These efforts underscore his dedication to identifying the clinical opportunities presented by discoveries in the basic neurosciences. Given these accomplishments, Hanns' work has had a significant impact on the careers of countless neuroscientists and millions of patients. His legacy will always be tied to the critical role he played in defining the properties of the $GABA_A$ receptor, with his influence lingering for decades as his findings continue to be exploited for therapeutic gain.

Hanns Möhler received his BA in Chemistry from the University of Bonn, his MA in Biochemistry from the University of Tübingen, and his Ph.D. in Biochemistry from the University of Freiburg in Germany. While pursuing his doctoral degree, he spent 18 months studying in the Department of Biochemistry at Michigan State University. His interest in the budding field of neuroscience led him to pursuing postdoctoral work at the Medical Research Council in London before accepting a research position at Hoffmann-La Roche in Basel. The Basel appointment was particularly attractive as it allowed him to pursue his passion for the neurosciences. During his time at Hoffmann-La Roche, he held academic appointments at the University of Freiburg, his alma mater. In 1988, Hanns was elected Professor of Pharmacology and Director of the Institute of Pharmacology on the Medical Faculty of the University of Zurich, with a joint appointment in the Department of Chemistry and Applied Biosciences at the Swiss Federal Institute of Technology (ETH) in Zurich. He was also the first Director of the Swiss National Center of Neuroscience Research where some 200 basic and clinical neuroscientists and bioengineers collaborate in the areas of synaptic plasticity and repair.

Hanns is the recipient of a number of major awards in recognition of his research accomplishments. Included among them are the Prize of the Foundation Forum, Engelberg, Switzerland, the Wachter Prize for Medical Research from the University of Innsbruck, The Neuroscience Award of the European College of Neuropsychopharmacology, the Theodore-Ott-Prize, the most prestigious prize in Switzerland for a neuroscientist, from the Swiss Academy of Medical Sciences, and the Golden Kraepelin Medal from the Max-Planck-Institute of Psychiatry, Munich. He is an elected member of the European Academy of Sciences and an honorary member, and a member of the board, of the Swiss Academy of Medical Sciences. He has been appointed as a fellow of the Collegium Helveticum. He serves on the editorial boards of a number of leading journals in the fields of pharmacology and neurosciences.

The impact of Hanns' work is evident from his publication record. As of September 2014, PubMed listed 249 articles authored or coauthored by Hanns Möhler. Much of this work appears in *Nature*, *Science*, *Nature Neuroscience*, and the *Proceedings of the National Academy of Sciences USA*. Data published by Thomson Reuters indicate Hanns' work has been cited 23,263 times, with an h-index of 80. He has been consistently listed by the Institute of Scientific Information among the "Highly Cited Researchers" in Neuroscience.

In addition to his scientific discoveries, Hanns is recognized for his contributions to education in the field. Together with Kurt Jungermann, he coauthored a popular German biochemistry textbook and is routinely rated as an outstanding lecturer by students of medicine, pharmacy, pharmaceutical sciences, chemistry, and biology. He has mentored dozens of individuals in his laboratory, from research group leaders to visiting scientists, postdoctoral fellows, and graduate and undergraduate students. Many have gone on to successful, independent research careers, as evidenced by the authors contributing to these volumes.

Volumes 72 and 73 of *Advances in Pharmacology* were prepared to honor Hanns Möhler and his work. Contributing authors were drawn from the universe of his scientific collaborators, colleagues, students, and fellows. All those invited eagerly agreed to participate in the creation of these volumes. While many of the chapters include sections highlighting the authors' past associations with Hanns, the vast majority of the reports are devoted to describing the latest developments in the field. Given this design, the contents of these volumes provide not only historical perspectives on the development of $GABA_A$ pharmacology but also state-of-the-art research in the field and the promise for the future. Topics covered include the behavioral and pharmacological phenotypes of $GABA_A$ receptor subunit knockout and knockin mice animals and the effects of point mutations on the allosteric properties of this site. Other subjects include the role of glutamatergic neurons in the sedative actions of benzodiazepines, the role of $GABA_A$ receptors in substance abuse, and the mechanisms responsible for the development of tolerance to the benzodiazepines. Even this partial list of volume subjects indicates the breadth of Hanns' interests and those of his associates. It also illustrates the productivity that results from working in a collegial, collaborative, and multidisciplinary environment, such as that provided by Hanns at the University of Zurich. It was an honor for us to work on the development of these volumes to recognize Hanns' contributions to our discipline. Along with the other contributing authors, we are privileged to know such an accomplished scientist and scholar, and to have him as a colleague, collaborator, and mentor. Besides showing our respect for Hanns' scientific insight, creativity, and productivity, we hope that these volumes will in some small way demonstrate our appreciation for the friendship, warmth, and encouragement he has so generously bestowed on each of us over the past 40 years.

Uwe Rudolph
Laboratory of Genetic Neuropharmacology,
McLean Hospital, Belmont, and Department of Psychiatry,
Harvard Medical School, Boston, Massachusetts, USA

S.J. Enna
Departments of Physiology and of Pharmacology,
University of Kansas Medical Center,
Kansas City, Kansas, USA

CONTRIBUTORS

Mario A. Acuña
Institute of Pharmacology and Toxicology, University of Zurich, Zurich, Switzerland

Karthik Balakrishnan
Institute of Pharmacology and Toxicology, University of Zurich, Zurich, Switzerland

Dietmar Benke
Institute of Pharmacology and Toxicology, University of Zurich, Zurich, Switzerland

Bernhard Bettler
Department of Biomedicine, Institute of Physiology, Pharmazentrum, University of Basel, Basel, Switzerland

Norman G. Bowery
Department of Pharmacology, University of Birmingham Medical School, Edgbaston, United Kingdom

R.L.M. Faull
Centre for Brain Research, and Department of Anatomy with Radiology, Faculty of Medical and Health Sciences, University of Auckland, Auckland, New Zealand

Jean-Marc Fritschy
Institute of Pharmacology and Toxicology, University of Zurich, Zurich, Switzerland

Thomas Fuchs
Department of Biology; Department of Biochemistry and Molecular Biology, and Center for Molecular Investigation of Neurological Disorders, The Huck Institutes of the Life Sciences, The Pennsylvania State University, University Park, Pennsylvania, USA

Bernhard Luscher
Department of Biology; Department of Biochemistry and Molecular Biology, and Center for Molecular Investigation of Neurological Disorders, The Huck Institutes of the Life Sciences, The Pennsylvania State University, University Park, Pennsylvania, USA

Richard W. Olsen
Department of Molecular & Medical Pharmacology, David Geffen School of Medicine at UCLA, Los Angeles, California, USA

William T. Ralvenius
Institute of Pharmacology and Toxicology, University of Zurich, Zurich, Switzerland

Adi Raveh
Department of Biomedicine, Institute of Physiology, Pharmazentrum, University of Basel, Basel, Switzerland

Rostislav Turecek
Department of Biomedicine, Institute of Physiology, Pharmazentrum, University of Basel, Basel, Switzerland, and Department of Auditory Neuroscience, Institute of Experimental Medicine, ASCR, Prague, Czech Republic

Kaspar Vogt
International Institute for Integrative Sleep Medicine (WPI-IIIS), University of Tsukuba, Tsukuba, Ibaraki, Japan

H.J. Waldvogel
Centre for Brain Research, and Department of Anatomy with Radiology, Faculty of Medical and Health Sciences, University of Auckland, Auckland, New Zealand

Hanns Ulrich Zeilhofer
Institute of Pharmacology and Toxicology, University of Zurich, and Institute of Pharmaceutical Sciences, Swiss Federal Institute of Technology (ETH) Zürich, Zurich, Switzerland

Khaled Zemoura
Institute of Pharmacology and Toxicology, University of Zurich, Zurich, Switzerland

CHAPTER ONE

Reflections on More Than 30 Years Association with Hanns

Norman G. Bowery[1]
Department of Pharmacology, University of Birmingham Medical School, Edgbaston, United Kingdom
[1]Corresponding author: e-mail address: n.g.bowery@bham.ac.uk

Contents

Abstract

I first met Hanns in 1977 and soon learnt of his extraordinary ability as a researcher. He became a friend as well as a mentor providing enthusiasm for my own research. I watched closely over the years how his research uncovered details of the association of the benzodiazepines and GABA and delineated the structural composition of the $GABA_A$ receptor associated with the action of individual drugs such as antianxiety and antiepileptic agents. His work produced many important contributions to medicine notable of which was the discovery of the first benzodiazepine antagonists, which are now routinely used in clinical practice. But for me his most important contribution was the discovery of the benzodiazepine receptor. During this time, my group uncovered a novel receptor for GABA and my progress in this work was encouraged and enhanced by discussions with Hanns.

1. INTRODUCTION

It is a great honor for me to contribute to this volume in recognition of the work of one of the most, if not the most, prolific researcher in the field of benzodiazepines, Hanns Möhler. He has been foremost in the discovery of the basis for their actions and for facilitating the introduction of different benzodiazepines into clinical medicine and there is no doubt that his contribution has been of paramount importance.

I first met Hanns in Spatind, Norway, where we were attending one of the most influential symposia in the field of amino acid neurotransmission.

Advances in Pharmacology, Volume 73
ISSN 1054-3589
http://dx.doi.org/10.1016/bs.apha.2014.11.001

It was organized by Frode Fonnum under the auspices of NATO. One focus of the presentations was the establishment of binding sites for GABA in mammalian brain tissue as exemplified by Enna, Beaumont, and Yamamura (1978), Lloyd and Dreksler (1978), and Olsen, Greenlee, Van Ness, and Ticku (1978). In addition, there were many "firsts" at this meeting. An example of which was the possibility that endogenous "inhibitors" of GABA receptor binding such as phospholipids (Johnston & Kennedy, 1978) are present in mammalian brain tissue. It was 1977 and nothing was known at that time about the structure of the GABA receptor or the exact distribution of receptor binding sites in the mammalian brain. But Curtis, Duggan, Felix, and Johnston (1970) in Australia had described the first competitive antagonist of the GABA receptor. Recognizing this, Hanns and his colleague, Okada, were able to use radiolabeled bicuculline to demonstrate binding sites on synaptic membranes for the first time at the meeting (Möhler, 1979; Mohler & Okada, 1978).

I had been working on peripheral nervous tissue, namely, the rat superior cervical ganglion, at this time and had found that GABA receptors were present on neurones in this tissue and when activated produced neuronal depolarization (Bowery & Brown, 1974). This appeared to be analogous to the depolarization of primary afferent fibers produced by GABA in mammalian spinal cord (Curtis, 1978). In fact, evidence had shown that this depolarization of nerve terminals (primary afferent depolarization) was responsible for physiological inhibition of dorsal roots. Thus, activation of GABAergic interneurones within the spinal cord can reduce sensory input.

Thus far, the response to GABA in superior cervical ganglia was detected by electrophysiological surface recording from intact isolated tissue. The advent of GABA receptor binding techniques, which were adequately described at this meeting, prompted Hanns and me to consider the possibility of detecting the presence of binding sites in homogenates of ganglia. As a consequence, we arranged to do a series of experiments in his laboratory in Basel. These experiments were performed over a period of about 6 weeks during which we obtained preliminary evidence for the existence of saturable binding sites on this tissue. This culminated in studies conducted by David Hill, in my laboratory in London (Bowery, Hill, & Möhler, 1979), showing the nature of these binding sites in bovine superior cervical ganglia.

While in Basel, Hanns and I discussed in great detail about the possibility of a novel receptor for GABA existing on neurones within the brain. This was prompted by findings that my colleague, Alan Hudson, and I had obtained in isolated atria of the rat (Bowery & Hudson, 1979). There was no evidence at that time for any other GABA receptor with distinct pharmacological properties being present within the brain or elsewhere. Binding sites with different affinities had been described by, for example, Johnston and Kennedy (1978), Olsen et al. (1981), and Guidotti, Gale, Suria, and Toffano (1979), but there was no pharmacological distinction between them. In fact, evidence indicated that any separation might be due to the removal of endogenous inhibitors when neuronal membranes were extensively washed. Washing appeared to serially uncover binding sites with higher affinity for GABA.

Our studies suggested the presence of a novel receptor for GABA in synaptic membranes, the pharmacology of which is quite distinct from the classical chloride-dependent GABA receptor. The observations that led to this discovery emanated from experiments using rat-isolated atria. We hypothesized that if GABA receptor activation on neurones of superior cervical ganglia could produce the same effect on the nerve terminals of these ganglionic neurones, this would produce terminal depolarization analogous to that occurring at primary afferent terminals in the spinal cord (Curtis, 1978).

Of course, we could not examine any depolarization produced by GABA directly but instead decided to study the effect of GABA on the release of noradrenaline from atrial tissue evoked by transmural stimulation. For this purpose, we chose to look at the release of radiolabeled noradrenaline as it had been recently established that 3H-noradrenaline taken up by isolated atria was released from transmitter stores in nerve terminals within the heart tissue in response to nerve stimulation (Iversen, 1974). The results of our experiments showed that, as predicted, GABA reduced the evoked release of 3H-noradrenaline (Bowery et al., 1981; Bowery & Hudson, 1979). This was most evident in the presence of an α1 adrenoceptor antagonist to suppress feedback inhibition of noradrenaline (Bowery et al., 1981; Kalsner, 1973). We initially assumed that this inhibition by GABA was due to nerve terminal depolarization. However, when we began to examine the pharmacology of this effect, we found that the recognized GABA receptor antagonist, bicuculline, would not prevent the action of GABA (Bowery et al., 1981), and, moreover, the GABA analogue, β-chlorophenyl GABA (baclofen) acted as an agonist mimicking the action of GABA. This compound has no effect at bicuculline-sensitive receptors (see Bowery, 1993).

I remember discussing these results in detail with Hanns, and he strongly encouraged me to pursue these findings. We subsequently published our initial findings, which prompted us to discover whether this action of GABA could be detected in brain tissue. It soon became evident that we were looking at a novel receptor for GABA, which was not chloride-dependent and had a distinct pharmacological profile (Bowery, 1993). We wanted to emulate the binding studies for GABA that by now had become firmly established but could not see a way of detecting this novel site in the presence of the established GABA receptor. Using the recognized 3H-GABA binding technique in sodium-free medium and in the presence of bicuculline, no residual saturable binding was observed. The addition of cations such as calcium and nickel had no effect on this binding (Enna & Snyder, 1977). We decided that the only way forward was to obtain some radiolabeled baclofen and to examine for any saturable binding. For this, I returned to Basel not to Hanns' laboratory at Hoffman LaRoche but to that of Helmut Bittiger at the then CIBA-Geigy laboratories.

Baclofen was first discovered by this group in an attempt to produce a GABA mimetic that might be used as a centrally active sedative/muscle relaxant. It was initially designed as a GABA analog, which, unlike GABA, would cross the blood–brain barrier (Bein, 1972; Keberle & Faigle, 1972). However, there was never any evidence to show that it acted at the chloride-dependent GABA receptor even though it produced muscle relaxation in humans. It was first marketed for the treatment of muscle rigidity in 1972 and remains the drug of choice in such conditions.

Fortunately, the CIBA-Geigy group had produced tritiated baclofen and was kind enough to provide us with a sample. They also provided us with their raw data from experiments in which they had attempted, but failed, to obtain evidence for the presence of binding sites for 3H-baclofen on synaptic membranes. All of their experiments had been conducted in Na^{+}- and Ca^{2+}-free media as had been employed for 3H-GABA binding. So David Hill and I decided that we would use the same physiological medium that we had used for our release studies in isolated atria and brain slices (Bowery et al., 1980), namely, Krebs' solution. In the first experiment using synaptic membranes prepared from rat whole brain, we obtained a small amount (~17%) of saturable binding. This indicated to us that some component(s) in Krebs was responsible. By a process of elimination removing one component at a time, we were able to show that the binding was dependent on the presence of Ca^{2+}, and Mg^{2+} was able to substitute to some extent for Ca^{+} (Hill & Bowery, 1981). This meant that if this binding was at a GABA site then GABA should displace 3H-baclofen. Also 3H-GABA binding in the presence of Ca^{2+} would be enhanced by an amount that could be displaced by excess baclofen. This was readily demonstrated and enabled us to name this novel receptor $GABA_B$ as distinct from the bicuculline-sensitive receptor which we designated $GABA_A$ (Bowery, Hill, & Hudson, 1983; Hill & Bowery, 1981).

During this time, Hanns and his colleagues had been focusing on the distribution of $GABA_A$ binding sites in mammalian brain and on the nature of the interaction ofbenzodiazepines with the GABA sites (Haefely, Kulcsár, & Möhler, 1975; Haefely, Kulcsár, Möhler, Pieri, et al., 1975; Hösli, Möhler, Richards, & Hösli, 1980). This culminated in 1977 with Hanns and "Tom" Okada publishing evidence for the presence of specific binding sites for benzodiazepines in rat brain tissue (Möhler & Okada, 1977). This coincided with similar observations made by Braestrup, Albrechsten, and Squires (1977) (see chapter in this volume by Hanns Möhler) (Fig. 1).

These binding sites are coupled to $GABA_A$ receptors, and numerous studies have since shown evidence for a functional link (see Upton & Blackburn, 1997). In support of this, Hanns and colleagues showed that the distribution of benzodiazepine binding sites in rat brain, using receptor autoradiography and immunohistochemistry, concurs with the distribution of $GABA_A$ receptors (Richards et al., 1984). It was this study, the details of which I knew in advance from discussions with Hanns prompted us to examine the distribution of $GABA_B$ sites in rat brain sections by developing an autoradiographic technique to compare with that of $GABA_A$ sites (Bowery, Price, & Hudson, 1987; Wilkin, Hill, Hudson, & Bowery, 1981). It became clear that there were distinct differences in the distributions of the two receptors although there were also similarities. One region of notable distinction was in the rat cerebellum where $GABA_A$ sites are

Figure 1 Hanns, me, and "Tom" Okada on a boat in Kyoto, in 1981 after the IUPHAR congress in Tokyo.

predominantly observed in the granule cell layer while $GABA_B$ sites are confined to the molecular layer. In general, there are more $GABA_A$ sites throughout the brain but occasionally the opposite occurs such as in the interpeduncular nucleus where $GABA_B$ sites predominate (Price, Blackburn, Hudson, & Bowery, 1984).

This was a very active period in global GABA research, and Hanns was very much a part of this. His group produced the first antagonists of the benzodiazepine receptor, which is nowadays referred to as "benzodiazepine site" (Hunkeler et al., 1981) and among the compounds that emerged were partial agonists as well as competitive antagonists. This led to the concept of "inverse" agonists which produced the opposite effect of full agonist benzodiazepines. Competitive antagonists such as flumazenil would block the effect of both full agonists such as diazepam as well as that produced by the inverse agonists, and these studies led to the introduction of this compound into clinical medicine to reverse the effects of full agonists.

It was around this time at the end of the 1980s that Hanns moved from industry to the University of Zurich as Head of Department. I remember discussing this with him when he was trying to decide whether to go or to stay at Roche. We were both attending a symposium in Japan and during the morning break he raised the matter over a cup of coffee. I had moved about 18 months earlier from Merck, Sharp, and Dohme to take the Chair at the School of Pharmacy in the University of London, and he wanted my opinion about making such a move. He had had a very successful career at Roche and was concerned that this might not continue if he went into academia. These concerns were completely unfounded, as we now know as his research continued to flourish.

He played a major role in determining the subunit composition of $GABA_A$ receptors and defining their distributions in the mammalian brain (e.g., Caruncho, Puia, Mohler, & Costa, 1995; Frischy & Mohler, 1995; Mohler et al., 1997). Moreover, he was able to show the importance of specific compositions as potential selective targets for drugs acting as, for example, antianxiety and antiepileptic agents (Möhler, 1992; Möhler, Fritschy, & Rudolph, 2002; Rudolph, Crestani, & Möhler, 2001). His research group was able to demonstrate that the intrinsic activity of benzodiazepine site ligands can vary among the $GABA_A$ receptor subtypes with the degree of $GABA_A$ receptor modulation being influenced by the type of alpha subunit in the GABA receptor (Knoflach, Drescher, Scheurer, Malherbe, & Mohler, 1993). Subtype-specific pharmacology may also be important in the alleviation of cognitive dysfunctions in schizophrenia, based on the treatment of patients with an alpha (2)/alpha (3)-$GABA_A$ receptor ligand (Möhler, 2009).

The structure of the $GABA_A$ receptor and its multiple forms was eventually discovered at the end of the 1980s and early 1990s (see Barnard et al., 1998, for a full review). Its classification as a five subunit protein controlling chloride ion channels was established at this time, details of which were given in Barnard et al. (1998), and recognized the possible existence of many forms of "the" receptor. Hanns was a member of the team that formulated this classification (Fig. 2).

A few years after the emergence of the structure of $GABA_A$ receptors came the discovery that $GABA_B$ receptors comprise two protein subunits which are dissimilar and G-protein coupled (see Bettler, Kaupmann, Mosbacher, & Gassmann, 2004). The subunits classified as B1 and B2 are both required for optimal function of the receptor, but each subunit appears to have a separate role. While B1 is important as the Ca^{2+}-dependent binding site for GABA (Galvez et al., 2000) and for the G-protein linkage, B2

Figure 2 Hanns with his wife on the left and my wife on the right at a neurochemistry symposium in Barcelona 2002.

possesses allosteric modulatory sites analogous to the allosteric sites for benzodiazepines on $GABA_A$ receptors. Interestingly, the distribution of the individual receptor subunits for $GABA_{B1}$ namely, B1a and B1b in mammalian brain can differ between pre- and postsynaptic locations (e.g., Towers et al., 2000), suggesting different functional roles. Using antisera, which recognize selectively GBR1a, GBR1b, and GBR2, GBR1a and GBR1b were both found by immunoprecipitation to occur as heterodimers associated with GBR2 in mammalian brain tissue. Furthermore, monomers of GBR1a, GBR1b, or GBR2 were not detectable, suggesting that practically all $GABA_B$ receptors are heterodimers *in situ* (Benke, Honer, Michel, Bettler, & Mohler, 1999).

By analogy with the allosteric mechanism of action of the benzodiazepines at the $GABA_A$ receptor, compounds have also been shown to allosterically modulate the action of agonists at the $GABA_B$ receptor (Gjoni, Desrayaud, Imobersteg, & Urwyler, 2006; Gjoni & Urwyler, 2008; Urwyler, 2011). The site of action appears to be on the B2 subunit, and thus far, only positive modulators have been described unlike the positive and negative modulatory actions of the benzodiazepine ligands (Mohler, 2006).

Figure 3 Lakeside in Zurich in 2010.

2. CONCLUSION

I have followed Hanns' research closely for more than 30 years and have watched the benzodiazepine area unfold to such an extent that we now have drugs with selective effects within this class. Hanns made a major contribution to this but for me he has provided more, that of scientific leadership and above all great friendship (Fig. 3).

CONFLICT OF INTEREST

The author has no conflicts of interest.

REFERENCES

Barnard, E. A., Skolnick, P., Olsen, R. W., Mohler, H., Sieghart, W., Biggio, G., et al. (1998). International Union of Pharmacology. XV. Subtypes of g-aminobutyric acid A receptors: Classification on the basis of subunit structure and receptor function. *Pharmacological Reviews*, *50*, 292–313.

Bein, H. J. (1972). Pharmacological differentiation of muscle relaxants. In J. W. Birkmayer (Ed.), *Spasticity—A topical survey* (pp. 76–82). Vienna: Hans Huber.

Benke, D., Honer, M., Michel, C., Bettler, B., & Mohler, H. (1999). Gamma-aminobutyric acid type B receptor splice variant proteins GBR1a and GBR1b are both associated with GBR2 *in situ* and display differential regional and subcellular distribution. *The Journal of Biological Chemistry*, *274*(38), 27323–27330.

Bettler, B., Kaupmann, K., Mosbacher, J., & Gassmann, M. (2004). Molecular structure and physiological functions of $GABA_B$ receptors. *Physiological Reviews*, *84*, 835–867.

Bowery, N. G. (1993). $GABA_B$ receptor pharmacology. *Annual Review of Pharmacology and Toxicology*, *33*, 109–147.

Bowery, N. G., & Brown, D. A. (1974). Depolarising actions of γ-aminobutyric acid and related compounds on rat superior cervical ganglia in vitro. *British Journal of Pharmacology*, *50*, 205–218.

Bowery, N. G., Doble, A., Hill, D. R., Hudson, A. L., Shaw, J. S., Turnbull, M. J., et al. (1981). Bicuculline-insensitive GABA receptors on peripheral autonomic nerve terminals. *European Journal of Pharmacology*, *71*, 53–70.

Bowery, N. G., Hill, D. R., & Hudson, A. L. (1983). Characteristics of $GABA_B$ receptor binding sites on rat whole brain synaptic membranes. *British Journal of Pharmacology*, *78*, 191–206.

Bowery, N. G., Hill, D. R., Hudson, A. L., Doble, A., Middlemiss, D. N., Shaw, J., et al. (1980). (-)Baclofen decreases neurotransmitter release in the mammalian CNS by an action at a novel GABA receptor. *Nature*, *283*, 92–94.

Bowery, N. G., Hill, D. R., & Möhler, H. (1979). Sodium independent GABA receptor binding in peripheral nervous tissue. *British Journal of Pharmacology*, *67*(3), 445P–446P.

Bowery, N. G., & Hudson, A. L. (1979). γ-Aminobutyric acid reduces the evoked release of 3-H noradrenaline from sympathetic nerve terminals. *British Journal of Pharmacology*, *66*, 108P.

Bowery, N. G., Price, G. W., & Hudson, A. L. (1987). $GABA_A$ and $GABA_B$ receptor site distribution in rat central nervous system. *Neuroscience*, *20*, 365–385.

Braestrup, C., Albrechsten, R., & Squires, R. F. (1977). High densities of benzodiazepine receptors in human cortical areas. *Nature*, *269*, 702–704.

Caruncho, H. J., Puia, G., Mohler, H., & Costa, E. (1995). The density and distribution of six $GABA_A$ receptor subunits in primary cultures of rat cerebellar granule cells. *Neuroscience, 67*(3), 583–593.

Curtis, D. R. (1978). Pre- and non-synaptic activities of GABA and related amino acids in the mammalian nervous system. In F. Fonnum (Ed.), *Amino acids as chemical transmitters* (pp. 55–86). New York: Plenum Press.

Curtis, D. R., Duggan, A. W., Felix, D., & Johnston, G. A. R. (1970). GABA, bicuculline and central inhibition. *Nature, 226*, 1222–1224.

Enna, S. J., Beaumont, K., & Yamamura, H. I. (1978). Comparison of 3H-muscimol 3H-GABA receptor binding in rat brain. In F. Fonnum (Ed.), *Amino acids as chemical transmitters* (pp. 487–492). New York: Plenum Press.

Enna, S. J., & Snyder, S. H. (1977). Influences of ions, enzymes detergents on γ-aminobutyric acid receptor binding in synaptic membranes of rat brain. *Molecular Pharmacology, 13*, 442–453.

Frischy, J. M., & Mohler, H. (1995). $GABA_A$ receptor heterogeneity in the adult rat brain: Differential regional and cellular distribution of seven major subunits. *The Journal of Comparative Neurology, 359*, 154–194.

Galvez, T., Urwyler, S., Prézeau, L., Mosbacher, J., Joly, C., Malitschek, B., et al. (2000). Ca(2+) requirement for high-affinity gamma-aminobutyric acid (GABA) binding at $GABA_B$ receptors: Involvement of serine 269 of the $GABA_BR1$ subunit. *Molecular Pharmacology, 57*(3), 419–426.

Gjoni, T., Desrayaud, S., Imobersteg, S., & Urwyler, S. (2006). The positive allosteric modulator GS39783 enhances $GABA_B$ receptor-mediated inhibition of cyclic AMP formation in rat striatum in vivo. *Journal of Neurochemistry, 96*(5), 1416–1422.

Gjoni, T., & Urwyler, S. (2008). Receptor activation involving positive allosteric modulation, unlike full agonism, does not result in $GABA_B$ receptor desensitization. *Neuropharmacology, 55*(8), 1293–1299.

Guidotti, A., Gale, K., Suria, A., & Toffano, G. (1979). Biochemical evidence for two classes of GABA receptors in rat brain. *Brain Research, 172*, 566–571.

Haefely, W., Kulcsár, A., & Möhler, H. (1975). Possible involvement of GABA in the central actions of benzodiazepines. *Psychopharmacological Bulletin, 11*(4), 58–59.

Haefely, W., Kulcsár, A., Möhler, H., Pieri, L., Polc, P., & Schaffner, R. (1975). Possible involvement of GABA in the central actions of benzodiazepines. *Advances in Biochemical Psychopharmacology, 14*, 131–151.

Hill, D. R., & Bowery, N. G. (1981). 3H-Baclofen and 3H-GABA bind to bicuculline-insensitive $GABA_B$ sites in rat brain. *Nature, 290*, 149–152.

Hösli, E., Möhler, H., Richards, J. G., & Hösli, L. (1980). Autoradiographic localization of binding sites for [3H]gamma-aminobutyrate, [3H]muscimol,(+)[3H]bicuculline methiodide and [3H] flunitrazepam in cultures of rat cerebellum and spinal cord. *Neuroscience, 5*(9), 1657–1665.

Hunkeler, W., Möhler, H., Pieri, L., Polc, P., Bonetti, E. P., Cumin, R., et al. (1981). Selective antagonists of benzodiazepines. *Nature, 290*(5806), 514–516.

Iversen, L. L. (1974). Uptake mechanisms for neurotransmitter amines. *Biochemical Pharmacology, 23*(14), 1927–1935.

Johnston, G. A. R., & Kennedy, S. M. E. (1978). GABA receptors and phospholipids. In F. Fonnum (Ed.), *Amino acids as chemical transmitters* (pp. 507–516). New York: Plenum Press.

Kalsner, S. (1973). Mechanism of potentiation by amines of non-equilibrium blockade of the alpha-adrenoceptor. *British Journal of Pharmacology, 47*(2), 386–397.

Keberle, H., & Faigle, J. W. (1972). Synthesis and structure-activity relationships of the γ-aminobutyric acid derivatives. In J. W. Birkmayer (Ed.), *Spasticity—A topical survey* (pp. 90–93). Vienna: Hans Huber.

Knoflach, F., Drescher, U., Scheurer, L., Malherbe, P., & Mohler, H. J. (1993). Full and partial agonism displayed by benzodiazepine receptor ligands at recombinant gamma-aminobutyric acidA receptor subtypes. *The Journal of Pharmacology and Experimental Therapeutics*, *266*(1), 385–391.

Lloyd, K. G., & Dreksler, S. (1978). 3H-GABA binding to membranes prepared from post-mortem human brain: Pharmacological and pathological investigations. In F. Fonnum (Ed.), *Amino acids as chemical transmitters* (pp. 457–466). New York: Plenum Press.

Möhler, H. (1979). GABA receptor binding with [3H]GABA and [3H](+)bicuculline-methiodide: An improved method. *Advances in Experimental Medicine and Biology*, *123*, 355–362.

Möhler, H. (1992). GABAergic synaptic transmission. Regulation by drugs. *Arzneimittel-Forschung*, *42*(2A), 211–214.

Mohler, H. (2006). Allosteric modulation of $GABA_A$ receptors. A pharmacology of sleep, anxiety, memory and consciousness. In N. G. Bowery (Ed.), *Allosteric receptor modulation in drug targeting* (pp. 79–92). London: Taylor and Francis.

Möhler, H. (2009). Role of $GABA_A$ receptors in cognition. *Biochemical Society Transactions*, *37*(Pt 6), 1328–1333.

Mohler, H., Benke, D., Benson, J., Luscher, B., Rudolph, U., & Frischy, J. M. (1997). Diversity in structure, pharmacology and regulation of $GABA_A$ receptors. In S. J. Enna & N. G. Bowery (Eds.), *The GABA receptors* (pp. 11–36). Totowa, NJ: Humana Press.

Möhler, H., Fritschy, J. M., & Rudolph, U. (2002). A new benzodiazepine pharmacology. *The Journal of Pharmacology and Experimental Therapeutics*, *300*(1), 2–8.

Möhler, H., & Okada, T. (1977). Benzodiazepine receptor: Demonstration in the central nervous system. *Science*, *198*(4319), 849–851.

Mohler, H., & Okada, T. (1978). GABA receptor in rat brain: Demonstration of an antagonist binding site. In F. Fonnum (Ed.), *Amino acids as chemical transmitters* (pp. 493–498). New York: Plenum Press.

Olsen, R. W., Bergmann, M. O., Van Ness, P. C., Lummis, S. C., Watkins, A. E., Napias, C., et al. (1981). γ-Aminobutyric acid receptor binding in mammalian brain. Heterogeneity of binding sites. *Molecular Pharmacology*, *19*, 217–227.

Olsen, R. W., Greenlee, D., Van Ness, P., & Ticku, M. K. (1978). Studies on the gamma amino acid receptor/ionophore proteins in mammalian brain. In F. Fonnum (Ed.), *Amino acids as chemical transmitters* (pp. 467–486). New York: Plenum Press.

Price, G. W., Blackburn, T. P., Hudson, A. L., & Bowery, N. G. (1984). Presynaptic $GABA_B$ sites in the interpeduncular nucleus. *Neuropharmacology*, *23*, 861–862.

Richards, J. G., Möhler, H., Schoch, P., Häring, P., Takacs, B., & Stähli, C. (1984). The visualization of neuronal benzodiazepine receptors in the brain by autoradiography and immunohistochemistry. *Journal of Receptor Research*, *4*(1–6), 657–669.

Rudolph, U., Crestani, F., & Möhler, H. (2001). $GABA_A$ receptor subtypes: Dissecting their pharmacological functions. *Trends in Pharmacological Sciences*, *22*(4), 188–194.

Towers, S., Princivalle, A., Billington, A., Edmunds, M., Bettler, B., Urban, L., et al. (2000). $GABA_B$ receptor protein and mRNA distribution in rat spinal cord and dorsal root ganglia. *The European Journal of Neuroscience*, *12*(9), 3201–3210.

Upton, N., & Blackburn, T. (1997). Pharmacology of mammalian $GABA_A$ receptors. In S. J. Enna & N. G. Bowery (Eds.), *The GABA receptors* (pp. 83–120). Totowa, NJ: Humana Press.

Urwyler, S. (2011). Allosteric modulation of family C G-protein-coupled receptors: From molecular insights to therapeutic perspectives. *Pharmacological Reviews*, *63*(1), 59–126.

Wilkin, G. P., Hill, D. R., Hudson, A. L., & Bowery, N. G. (1981). Autoradiographic localisation of $GABA_B$ receptors in rat cerebellum. *Nature*, *294*, 584–597.

CHAPTER TWO

Significance of $GABA_A$ Receptor Heterogeneity: Clues from Developing Neurons

Jean-Marc Fritschy[1]

Institute of Pharmacology and Toxicology, University of Zurich, Zurich, Switzerland

[1]Corresponding author: e-mail address: fritschy@pharma.uzh.ch

Contents

Abstract

Briefly after the landmark discovery by Hanns Möhler that $GABA_A$ receptors ($GABA_AR$) are the site of action of benzodiazepine site ligands, their distribution in the rodent CNS during development was mapped by autoradiography, demonstrating early and widespread expression of $GABA_AR$ in the brain and spinal cord. Ten years later, the first studies using subunit-specific antibodies revealed unsuspected heterogeneity in the subunit composition of $GABA_AR$ in developing brain, with striking regional specificity and rapid changes in expression and subcellular localization correlating with the maturation of neuronal circuits. These data contributed to the wealth of evidence that GABAergic neurotransmission, acting both synaptically and extrasynaptically, modulates major steps of neuronal development (proliferation, migration, differentiation, and circuit formation). In immature neurons, $GABA_AR$ activation leads to neuronal depolarization and activation of Ca^{2+} signals, which mediate many of the developmental effects of GABA. Therefore, $GABA_AR$ heterogeneity might be essential to fine-tune GABA actions in the dynamics of CNS maturation. Furthermore, since a disturbance of GABAergic function during ontogeny can potentially affect many aspects of CNS maturation and contribute to the etiology of major brain disorders, $GABA_AR$ heterogeneity provides a substrate for the development of a tailored pharmacology targeting specific

Advances in Pharmacology, Volume 73
ISSN 1054-3589
http://dx.doi.org/10.1016/bs.apha.2014.11.006

receptor subtypes. This chapter provides a brief overview of these issues with a special focus on the seminal contributions of Hanns Möhler to the emergence of these concepts of fundamental relevance in today's neuroscience research and pharmacological developments.

ABBREVIATIONS

GABA γ-aminobutyric acid
$GABA_AR$ $GABA_A$ receptor

1. INTRODUCTION

$GABA_A$ receptors ($GABA_AR$) belong to the family of Cys-loop ligand-gated ion channels. They are selectively permeable for Cl^- (and HCO_3^-) ions and mediate most of the fast inhibitory neurotransmission in the CNS. $GABA_AR$ are encoded by a family of 19 subunit genes, with multiple α, β, and γ subunits homologues (Olsen & Sieghart, 2008). As each receptor comprises between 3 and 5 different subunits, with a stoichiometry of 2α, 2β, 1γ (or δ) subunit, there is room for considerable heterogeneity in their subunit composition. Although evidence is compelling that $GABA_AR$ structural heterogeneity determines their functional properties, pharmacological profile, subcellular localization, and posttranslational/posttranscriptional regulation (as reviewed in this book), we still have little understanding why any specific $GABA_AR$ subtype, defined by its subunit composition, needs to be present in a given cell at a particular location (or developmental stage) in order to fulfill its role. Nevertheless, $GABA_AR$ are not readily interchangeable, and complex mechanisms have evolved to restrict specific $GABA_AR$ subtypes to well-defined neuronal circuits, implying that such mechanisms likely are functionally relevant. The analysis of developmental mechanisms regulated by GABAergic transmission in neurons, both during ontogeny and in adult neurogenesis, has provided the clearest evidence that specific steps of development, ranging from progenitor cell proliferation to functional integration of mature neurons, involve distinct $GABA_AR$ subtypes. It is the purpose of this chapter to review this evidence and discuss its implications for our understanding of $GABA_AR$-mediated transmission in health and disease.

Hanns Möhler, with his teams at Hoffman-La-Roche LTD and at the Institute of Pharmacology and Toxicology of the University of Zurich, has contributed enormously to our current understanding of GABA$_A$R biology and pharmacology. Among the major groups which also led this research field forward, most are contributing to this book. However, none of this would have happened, if Hanns Möhler had not first established that GABA$_A$R are the only site of action of benzodiazepines (and other molecules that share the same high-affinity binding site on neurons), which operate through allosteric modulation of the γ-aminobutyric acid (GABA)-binding site (Möhler & Okada, 1977). Hanns Möhler had the foresight to establish multidisciplinary teams, allowing him to be for decades at the forefront of research related to GABA$_A$R biology. Therefore, he has contributed significantly to many facets of our understanding of GABA$_A$R pharmacology and distribution, and upon initiation of cloning, identification of specific subunits, functional characterization using recombinant systems and *in vivo* preparations, posttranscriptional regulation, and, most importantly, relevance of GABA$_A$R heterogeneity for mediating the spectrum of action of classical benzodiazepine-site ligands.

I had the privilege to join his team in Zurich in 1991, as the morphologist in charge of analyzing the distribution and regulation of GABA$_A$R in the CNS. I quickly learned that Hanns Möhler did not compromise for quality or excellence in research and that he liked to challenge his people and to be challenged by them in order to pursue cutting-edge research projects. Engraved in my memory is the following recollection: I asked him one day what he thought of studying GABA$_A$R during brain development. He replied that I should not waste my time, it had already been done by autoradiography, showing early onset of expression, gradual increase in parallel with synaptogenesis, and no special change thereafter (Schlumpf, Richards, Lichtensteiger, & Mohler, 1983). I pondered his advice, but concluded there was nothing to lose by giving it a try with subunit-specific antibodies. Therefore, I went on and analyzed the α1 and α2 subunits at different postnatal stages in the rat and observed striking changes in expression of these two subunits, with exquisite regional and temporal specificity. When I showed this data to Hanns Möhler, he reacted enthusiastically (despite my earlier challenge of his wisdom and authority) and gave me a Ph.D. student, Jacques Paysan, and a summer student, Anne Enna, to rapidly advance the project. He contributed tremendously to the writing of the landmark manuscript that came out of this work (Paysan, Kossel, Bolz, & Fritschy, 1997), but declined to be a coauthor on it, saying "Jean-Marc, this is your work and you should take full credit for it." I wish many scientists had great mentors like him.

1.1. Early days

The first immunochemical analyses of $GABA_AR$ in rodent, bovine, and human brain were performed by Hanns Möhler and his team at Roche, using a panel of monoclonal antibodies raised against purified bovine receptors (Haring et al., 1985; Schoch et al., 1985). These landmark studies uncovered the widespread distribution of $GABA_AR$ across the CNS, containing both α and β subunit variants (which were not cloned at the time), the co-occurrence of these subunits within a receptor complex, and the presence of both the high-affinity GABA binding site and the benzodiazepine binding site on the receptor. It was concluded that, unlike evidence from photo-affinity labeling, $GABA_AR$ have a *homogeneous* subunit composition across the brain!

Prior to this pioneering work, Hanns Möhler's team had already demonstrated ultrastructurally that radiolabeled benzodiazepines bind preferentially in the vicinity of GABAergic terminals, both synaptically and non-synaptically (Mohler, Richards, & Wu, 1981). Using the monoclonal antibody bd-17, they could confirm the presence of $GABA_AR$ in cell membranes facing GABAergic terminals, the majority of the labeling being extrasynaptic (Richards, Schoch, Haring, Takacs, & Mohler, 1987). They could also show that the distribution of the $GABA_AR$ mapped with this antibody correlated very well with the density of autoradiographic labeling using benzodiazepine agonists, except in the cerebellum, where autoradiographic labeling was much weaker. Later work showed of course, that the relative low abundance of autoradiographic labeling was due to the presence of a majority of diazepam-insensitive $GABA_AR$ in the granule cell layer, containing the α6 and/or δ subunits (Laurie, Seeburg, & Wisden, 1992). A few years later, the presence of both postsynaptic and extrasynaptic $GABA_ARs$ was confirmed ultrastructurally in the cerebellum by the of Peter Somogyi's group in Oxford, still using monoclonal antibody bd-17 (Somogyi, Takagi, Richards, & Mohler, 1989). Altogether, these early ultrastructural studies highlighted three essential features of $GABA_ARs$, before their molecular composition could be determined by cloning: ubiquitous presence, probably in every neuron of the CNS; distribution at postsynaptic and extrasynaptic sites (implying a potential function upon activation by GABA in the extracellular fluid); paucity of presynaptic receptors on axons and axon terminals, despite early functional evidence of their existence, notably mediating primary afferent depolarization (Barker & Nicoll, 1972).

The molecular cloning of cDNAs encoding $GABA_AR$ subunits rapidly revealed the existence of a large gene family. The apparent homogeneity of the subunit composition of GABA$_A$Rs initially revealed by immunochemical analysis using the monoclonals bd-17 and bd-24 (Haring et al., 1985) was due to the fact that these antibodies recognize abundant and widely distributed subunits. In particular, bd-17 binds to an epitope common to both the β2 and β3 subunits (Ewert, Shivers, Luddens, Mohler, & Seeburg, 1990), which together account for >80% of GABA$_A$R (Benke, Fritschy, Trzeciak, Bannwarth, & Mohler, 1994). Strong evidence for heterogeneity nevertheless arose, in particular, from the identification of the rat δ subunit, whose distribution was shown to be largely different from the γ2 subunit, and to correspond to sites where muscimol, but not classical benzodiazepine agonists, bind with high affinity (Shivers et al., 1989).

Identification of protein sequence of each of the 19 GABA$_A$R subunits (Olsen & Sieghart, 2008) afforded the possibility to generate subunit-specific RNA probes and antibodies to investigate their relative expression patterns distribution and pharmacological properties in the adult CNS and during development. Hanns Möhler's group has played a leading role in this enterprise, providing early on a detailed biochemical and pharmacological analysis of the major GABA$_A$R subunits and determining their relative abundance. This work has contributed significantly to the emergence of the concept of "*GABA$_A$R subtype*," representing a distinct pharmacological and functional entity, with a specific localization in defined neuronal circuits (Benke, Mertens, Trzeciak, Gillessen, & Mohler, 1991; Fritschy et al., 1992; Mohler, Benke, Mertens, & Fritschy, 1992). In particular, the six α subunit variants could be shown to represent distinct receptor subtypes, paving the way to abolishing diazepam binding to specific GABA$_A$R subtypes by substituting the conserved His101 residue with an Arg residue (H/R knock-in mice), thereby allowing to differentiate GABA$_A$R mediating the multiple effects of benzodiazepine site ligands (Mohler, Fritschy, & Rudolph, 2002; Rudolph & Möhler, 2004). This approach was made possible by the seminal observation of Wieland et al. that the histidine residue at position 101 of the α1 subunit determines sensitivity to diazepam of the pentameric receptor complex (Wieland, Luddens, & Seeburg, 1992).

Almost 30 years after the first immunohistochemical analysis of GABA$_A$R in the brain by Hanns Möhler (Haring et al., 1985), it should be emphasized that despite the concerted efforts of several groups (De Blas, 1996; Fritschy & Mohler, 1995; Hörtnagl et al., 2013), it has not been possible to raise antibodies against each of the 19 GABA$_A$R subunits. Therefore, knowledge on the relative abundance, cellular, and mainly subcellular distribution patterns of these subunits (which likely correspond to multiple $GABA_AR$ subtypes) remains fragmentary at best. Much of information about the expression pattern of $GABA_AR$ subunits relies, therefore, on the mapping of their mRNA distribution, which has been largely completed by the Allen brain atlas, providing detailed regional maps of the brain of the distribution of the 19 subunits.

Furthermore, as noticed in the first ultrastructural studies of $GABA_AR$ immunohistochemical distribution in the CNS (Nusser & Somogyi, 1994; Somogyi et al., 1989), the subcellular labeling pattern and the sensitivity of antibodies are strongly dependent on tissue fixation. In particular, the detection of postsynaptic receptors, anchored at postsynaptic sites, requires a mild fixation, which might be detrimental for tissue preservation and conservation of other epitopes (Fritschy, Weinmann, Wenzel, & Benke, 1998; Notter, Panzanelli, Pfister, Mircsof, & Fritschy, 2013; Sassoè-Pognetto et al., 1995). The precise reason why postsynaptic receptors are hard to detect with antibodies directed either to the extracellular N-terminal domain, or to the intracellular loop between transmembrane segments 3–4 remains obscure (van Rijnsoever, Sidler, & Fritschy, 2005).

Finally, it should also be emphasized here that the precise repertoire of $GABA_AR$ subtypes expressed by any given neuron remains elusive. This is due to the fact that co-localization of two α (or β) subunit variants within the same postsynaptic site (Kasugai et al., 2010) does not allow distinguishing whether it reflects the presence of two receptor populations, or a single population containing two different α (or β) subunits.

1.2. $GABA_A$ receptors and GABAergic transmission in developing CNS

Hanns Möhler collaborated with his future colleagues of the Institute of Pharmacology and Toxicology to investigate the distribution of $GABA_ARs$ in fetal brain using autoradiography with radiolabeled benzodiazepine-site ligands (Schlumpf et al., 1983). This study revealed the early expression of $GABA_AR$ in fetal CNS, being detectable first in the spinal cord and lower brainstem around E14 and increasing gradually in a region- and layer-specific manner to establish the pattern seen in mature brain postnatally. Since these receptors were detected by radiolabeled benzodiazepine ligands, it implies expression of $\gamma 2$ subunit-containing receptors, similar to those found in adult CNS, thereby reinforcing the early concept of $GABA_AR$ homogeneity, at least in terms of distribution and function. Furthermore, since the expression of the receptor preceded synaptogenesis, they concluded in favor of a role of $GABA_AR$-mediated transmission prior to formation of GABAergic synapses. Recent evidence, however, points to the existence of synaptically connected networks regulating spontaneous activity of embryonic motoneurons in the E12 rodent spinal cord via $GABA_AR$-mediated mechanisms (Czarnecki et al., 2014), indicating that expression of functional $GABA_AR$ starts several days earlier as suggested 30 years ago.

Following the cloning of $GABA_ARs$, others have revisited their developmental expression pattern by *in situ* hybridization, using subunit-specific antisense mRNA, and reported early expression of multiple subunits, with age- and region-specific changes during spinal cord and brainstem maturation (Ma, Saunders, Somogyi, Poulter, & Barker, 1993), as well as ganglionic eminence, neocortex, hippocampus, and olfactory bulb (Cuzon Carlson & Yeh, 2011; Gambarana, Beattie, Rodriguez, & Siegel, 1991; Ma & Barker, 1995; Poulter, Barker, O'Carroll, Lolait, & Mahan, 1992). The unexpected structural heterogeneity unraveled by overlapping distribution patterns of several subunits opened the possibility that multiple $GABA_AR$ mediate diverse functions in developing CNS (Cuzon Carlson & Yeh, 2011; Maric et al., 1999).

These findings were in line with observations that GABA exerts trophic effects on neurons in culture and promotes cell proliferation, survival, migration, and outgrowth of neurites (Barbin, Pollard, Gaiarsa, & Ben-Ari, 1993; Behar, Schaffner, Scott, Greene, & Barker, 2000; Heck et al., 2007; Liu, Morrow, Devaud, Grayson, & Lauder, 1997; Spoerri, 1988). In particular, GABA was shown to regulate the tangential migration and cortical entry of interneurons derived from the ganglionic eminence (Cuzon, Yeh, Cheng, & Yeh, 2006; Inada et al., 2011). Furthermore, evidence that GABA exerts depolarizing effects in immature brain was emerging in parallel to these morphological studies (Cherubini, Gaiarsa, & Ben-Ari, 1991; Loturco, Owens, Heath, Davis, & Kriegstein, 1995; Serafini, Valeyev, Barker, & Poulter, 1995), in line with its role mediating primary afferent depolarization in the spinal cord (Barker & Nicoll, 1972). It is also around this time that the implications of the early role of GABA in shaping CNS development for the pathophysiology of neurodevelopmental and psychiatric disorders, notably autism and schizophrenia, started to be recognized (Lewis, 2000).

Depolarizing effects of GABA in developing neurons have been proposed to be of fundamental importance for at least three reasons: first, the depolarization is sufficient to induce opening of voltage-gated Ca^{2+} channels and, therefore, enable Ca^{2+}-dependent effects on neuronal maturation. In particular, $GABA_AR$-dependent Ca^{2+} signaling has been implicated in the control of precursor cell proliferation (Fiszman, Borodinsky, & Neale, 1999; Wang, Johnston, Ball, & Brinton, 2005) and migration (Bortone & Polleux, 2009), although the underlying signal cascades activated by Ca^{2+} to regulate these processes are not yet clear. Ca^{2+} signaling has also been involved in the control of activation of transcription factors, notably CREB (Merz, Herold, & Lie, 2011; Tao, Finkbeiner, Arnold, Shaywitz, & Greenberg, 1998), which regulate BDNF expression (Obrietan, Gao, & Van Den Pol, 2002), neuronal differentiation, and survival (Ghosh, Carnahan, & Greenberg, 1994). Second, GABA depolarization has been suggested to occur prior to the onset of excitatory glutamate transmission (Hosokawa, Sciancalepore, Stratta, Martina, & Cherubini, 1994; Tyzio et al., 1999), ensuring the possibility to firing axon potentials and contributing to the emergence of synaptically connected networks. Third, GABA depolarization allows the emergence of activity patterns unique for developing neurons, such as giant depolarizing potentials in the hippocampus (Ben-Ari, Cherubini, Corradetti, & Gaiarsa, 1989) and other CNS regions, believed to play a key role in establishment of appropriate neuronal maturation and network connectivity (Ben-Ari, Gaiarsa, Tyzio, & Khazipov, 2007; De Blas, 1996). The reality of GABA-induced depolarization *in vivo* (Ben-Ari et al., 2012; Kirmse, Witte, & Holthoff, 2010; Valeeva, Valiullina, & Khazipov, 2013) has been questioned in reports suggesting that it might be a side effect of energy-deprived cells (Rheims et al., 2009) and/or of lesions inherent to the preparation of brain slices (Bregestovski & Bernard, 2012). These controversies highlight the need for a better understanding of the ionic mechanisms underlying depolarizing GABA effects and their consequences for network activity in health and disease.

A fundamental aspect of the effects of GABA during fetal and early postnatal development is the paucity of GABAergic synaptic contacts at this stage, implying an extrasynaptic, most likely tonic, effect of GABA upon receptors distributed in the cell membrane (Vautrin, Maric, Sukhareva, Schaffner, & Barker, 2000), in particular for the regulation of progenitor cell proliferation and neuroblast migration. Hanns Möhler recognized this issue in his first autoradiographic analysis, although first direct evidence for $GABA_AR$-mediated extrasynaptic (or volume) transmission was provided much later (Mody, 2001). The important role of tonic GABAergic transmission in the regulation of neuronal networks is now well established (Farrant & Nusser, 2005). There is also strong evidence that specific interneurons release GABA-mediating tonic effects on stem and progenitor cells, as demonstrated in developing cerebral cortex, where Cajal-Retzius cells in the marginal zone provide an early source of ambient GABA (Luhmann, Kirischuk, Sinning, & Kilb, 2014). However, it has also been suggested that tonic inhibition might partly be mediated independently of GABA (Wlodarczyk et al., 2013), possibly involving spontaneous GABA$_A$R openings or the action of another endogenous agonist. The latter could be taurine, as shown in a study using mouse transgenesis to prevent GABA formation (Furukawa et al., 2014).

1.3. Switch in GABA$_A$ receptor subunit composition during development

Starting in 1993, we investigated by immunohistochemistry the expression pattern of the main α subunit variants in late fetal and early postnatal brain. Specifically, we aimed to get further insight into possible differences between GABA$_A$R being expressed prior to and after synaptogenesis, as well as to test whether the switch from depolarizing to hyperpolarizing action of GABA in developing neurons—which had just been discovered (Cherubini et al., 1991) – was dependent on GABA$_A$R subunit composition. Looking first at the α1 subunit, which is by far the most abundant α subunit variant in adult brain, we noted that its expression levels in neonatal rats (and mice) is very low (see Figs. 1–3). Furthermore, its localization during the early postnatal period exhibits a striking regional and laminar specificity, notably in the thalamocortical system. Thus, the α1 subunit immunoreactivity is first upregulated in the cortical plate and maturing neocortex selectively in layers III–IV of primary sensory areas (somatosensory, visual, auditory), delineating precisely their cytoarchitectural organization, notably individual barrels in primary somatosensory cortex (Paysan, Bolz, Mohler, & Fritschy, 1994). At later stages, a gradual increase of α1 subunit expression in the remaining layers gradually masks this pattern, although the α1 subunit immunoreactivity remains highest in layer IV of adult neocortex.

In contrast to the α1 subunit, the α2 and α5 subunits exhibit a strong immunoreactivity both in regions where they remain expressed in adult brain (e.g., striatum, hippocampus) and in regions where they are only transiently expressed (e.g., thalamus, globus pallidus, specific cortical layers) (Fritschy, Paysan, Enna, & Mohler, 1994). In particular, the α5 subunit immunoreactivity disappears rapidly from layer IV of primary sensory areas, forming a pattern opposite to that of the α1 subunit. While these results were originally obtained in rat tissue, they essentially hold true also for mice, as illustrated in Figs. 1–3, in which sections from C57Bl6/J mice at postnatal day 5 are shown.

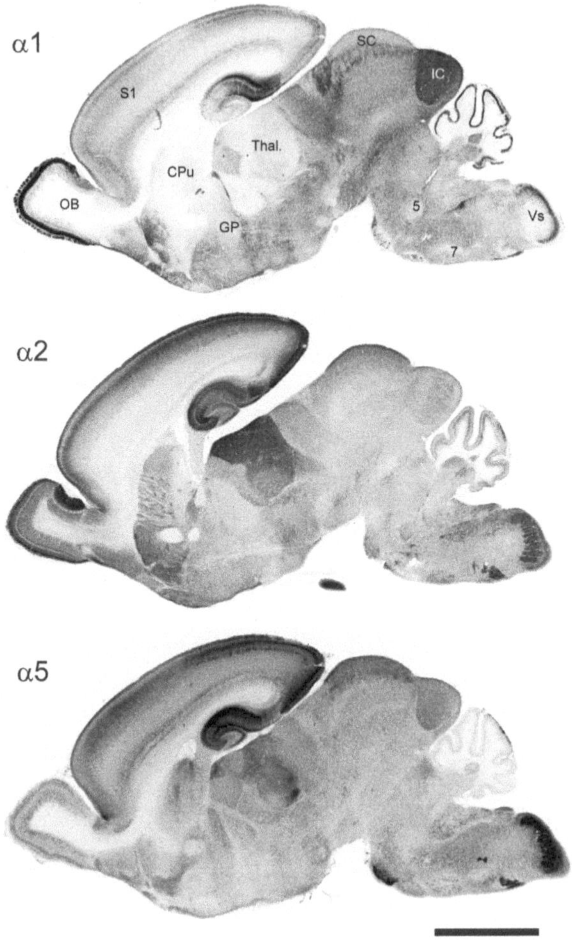

Figure 1 Differential distribution of the α1, α2, and α5 subunits in mouse brain at postnatal day 5, as detected by immunoperoxidase staining in sagittal sections. Age-dependent features include the prominent α2 and α5 subunit immunoreactivity in the thalamus, the specific outline of the whisker-to-barrel pathway, notably in the spinal trigeminal nucleus, and the low level of α1 subunit staining in cerebral cortex and thalamus. Note also the sometimes overlapping, sometimes complementary staining for the α2 and α5 subunit, and the very prominent presence of α5 in specific regions, such as the pontine nuclei (arrow). *Abbreviations*: 5, trigeminal motor nucleus; 7, facial nucleus; CPu, striatum; GP, globus pallidus; IC, inferior colliculus; OB, olfactory bulb; S1, primary somatosensory cortex; SC, superior colliculus; Thal., thalamus; Vs, spinal trigeminal nucleus. Scale bar, 1 mm.

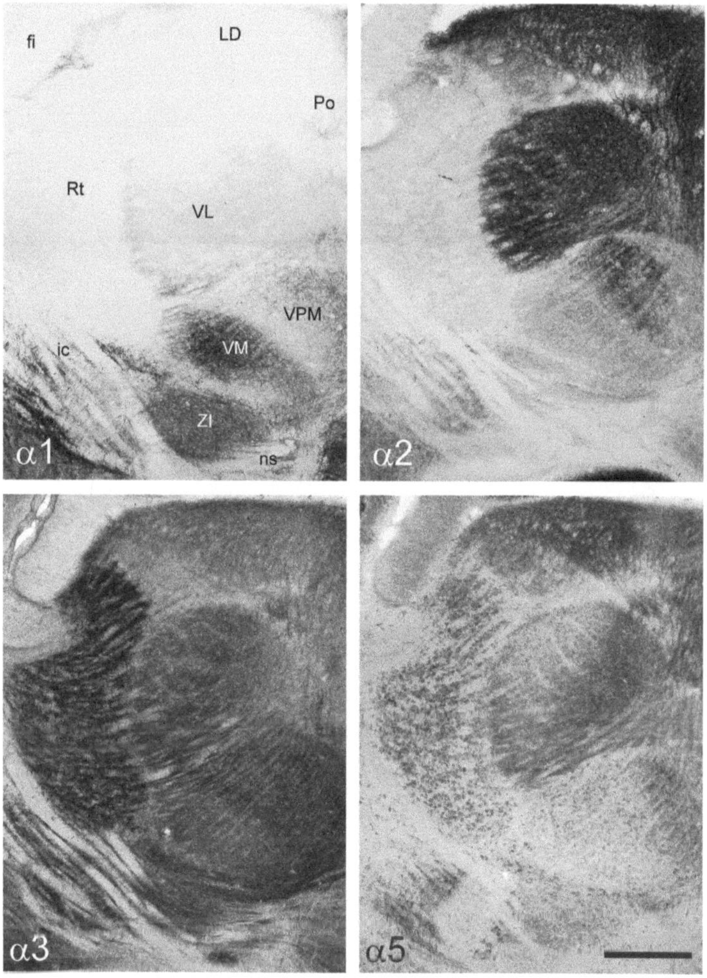

Figure 2 Higher magnification view of the differential distribution of four α subunit variants in the mouse thalamus at postnatal day 5 (sagittal sections). By comparison, in adult thalamus, the α1 subunit is prominent everywhere, except the reticular nucleus, the α2 and α5 subunits are almost undetectable, whereas the α3 subunit is restricted to the reticular nucleus. Note in these images how the complex cytoarchitecture of the thalamus is revealed by the differential staining intensity across nuclei exhibited by each of these 4 subunits. The barrelettes in the ventrolateral nucleus are outlined by the α2, α3, and α5 subunits. These findings point to the coexistence of multiple GABA$_A$R subtypes in most thalamic nuclei during postnatal development. *Abbreviations*: fi, fimbria; ic, internal capsule; LD, laterodorsal nucleus; ns, nigrostriatal bundle; Po, posterior thalamus; Rt, reticular nucleus; VL, ventrolateral nucleus; VM, ventromedial nucleus; VPM, ventroposteromedial nucleus; ZI, zona incerta. Scale bar, 200 μm.

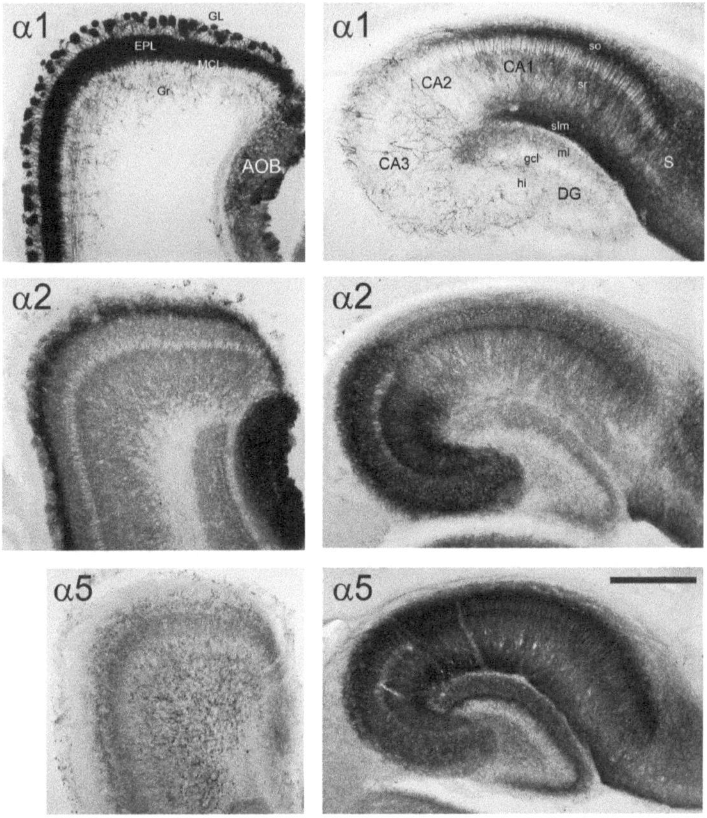

Figure 3 Higher magnification view of the differential distribution of the α1, α2, and α5 subunits in the mouse olfactory bulb (left column) and hippocampal formation (right column) at postnatal day 5 (sagittal sections). In olfactory bulb, each subunit already has its adult-like distribution pattern, with the α1 subunit being highly prominent in mitral (and tufted) cells and their dendrites in the external plexiform layer (EPL) and glomeruli (Gl), as well as short-axon cells in the granule cell layer (Gr); the α2 and α5 subunits are expressed in granule cells, albeit with differential subcellular localization; in addition, the α2 subunit is highly prominent in the accessory olfactory bulb (AOB). In the hippocampal formation, each subunit exhibits striking regional differences (e.g., in CA2), as well as age-dependent features, such as the α1 subunit gradient in CA1, the contrasting α2 subunit staining intensity in CA3 and subiculum (S), and the strongest α5 subunit staining in stratum oriens of CA1. *Abbreviations*: DG, dentate gyrus; gcl, granule cell layer; hi, hilus of the dentate gyrus; MCL, mitral cell layer; ml, molecular layer; so, stratum oriens; sp, stratum pyramidale; sr, stratum radiatum. Scale bar, 200 μm.

To test for a role of thalamocortical afferents in regulating the expression of the α1 and α5 subunits in opposite manners, we first examined at which stage an area-specific expression of the α1 subunit was first apparent in rat neocortex. A distinct parcellation was seen at E20, coinciding with the arrival of thalamocortical afferents in the cortical plate (Paysan et al., 1997). Therefore, we disrupted this projection by electrolytic lesions of the dorsal lateral geniculate nucleus on the day of birth. This lesion was sufficient to prevent both the increase in α1 subunit and the decrease in α5 subunit immunoreactivity, suggesting coordinated regulation of gene expression patterns in layer III–IV of the cortex (Paysan et al., 1997). It is now known that several transcription factors are regulated by thalamocortical afferents in the developing cortical plate (O'Leary, Chou, & Sahara, 2007). However, to our knowledge, it has not been investigated whether these transcription factors regulate GABA$_A$R expression.

We have also monitored the developmental expression of major α subunit variants in primary cultures of hippocampal neurons derived from E17 to E18 rat embryos. Remarkably, the results revealed a close replication of the *in vivo* situation, with a predominance of α2 subunit staining at postsynaptic sites (including the axon-initial segment) and α5 subunit extrasynaptically in pyramidal cells, absence of α3 subunit immunoreactivity, and strong α1 subunit staining in morphologically identified interneurons (Brünig, Scotti, Sidler, & Fritschy, 2002). Therefore, although extrinsic influences fine-tune the expression of GABA$_A$R in a layer-specific manner in the neocortex, most likely by influencing specific transcription factors, cortical neurons likely have an intrinsic program determining the repertoire of GABA$_A$R subunits that they predominantly express.

The early expression of GABA$_A$R in developing CNS neurons and circuits, coupled with experimental evidence that they regulate key steps of neuronal proliferation, differentiation, and survival, suggests that a disruption of the major receptors expressed at fetal stages might have deleterious consequences on brain maturation. While targeted deletion of any α subunit variant leads to the disappearance of the corresponding GABA$_A$R subtype because α subunits are required for GABA$_A$R assembly and cell-surface targeting, none of the α subunit gene-KO mice exhibits clear signs of brain malformation or structural alteration. There are two possible reasons for this unexpected outcome. First, since the majority of neurons express multiple α subunit variants, compensation by the remaining receptors might occur. Along this line, targeted disruption of the α1 subunit is accompanied by significant upregulation of the α2 and α3 subunit, for example, both on the regional level and within single neurons (Kralic et al., 2006; Schneider Gasser, Duveau, Prenosil, & Fritschy, 2007). Second, it is conceivable that a majority of $GABA_AR$ expressed in developing neurons are composed of two different α subunit variants, allowing to preserve receptor assembly (albeit with a modified composition) in the absence of one of these subunits. To our knowledge, there is no published report on mice lacking multiple $GABA_AR$ subunit genes to attempt to discriminate among these possibilities. However, our own studies have shown several examples in which compensation for a missing α subunit variant is limited. For example, Purkinje cells in the adult cerebellum only express α1 subunit-containing $GABA_AR$. Accordingly, in *Gabra1*-null mice, they completely lack GABAergic currents (both phasic and tonic) (Fritschy, Panzanelli, Kralic, Vogt, & Sassoè-Pognetto, 2006). During development, they transiently express the α3 subunit, at the onset of synaptogenesis; this transient expression remains present in *Gabra1*-null mice, allowing the formation of the first GABAergic synapses, but it subsides with essentially the same time course as in wild-type mice (Patrizi et al., 2008). In the ventrobasal complex of the thalamus, thalamocortical relay neurons express both α1-$GABA_AR$ at postsynaptic sites and extrasynaptic α4-$GABA_AR$. In adult *Gabra1*-null mice, these neurons completely lack postsynaptic GABAergic currents (evoked and miniature), but exhibit enhanced tonic inhibition (and increased α4 subunit immunoreactivity) (Kralic et al., 2006; Peden et al., 2008). These observations indicate that α4-$GABA_AR$ are unable to replace α1-$GABA_AR$ at postsynaptic sites, possibly because they cannot interact with postsynaptic scaffolding proteins. As noted above, thalamocortical neurons in early postnatal brain mainly express the α2 subunit, which is gradually replaced by the α1 subunit. In *Gabra1*-KO mice, the transient expression of α2-$GABA_AR$ has the same time course as in wild-type littermates. The decay kinetics of α2-$GABA_AR$ in immature brain is significantly slower than those of α1-$GABA_AR$ in mature brain. The gradual acceleration of decay kinetics during maturation was long considered to reflect the shift from α2- to α1-$GABA_AR$. However, most strikingly, this acceleration of decay kinetics also takes place in *Gabra1*-null mice, indicating that it has another origin (Peden et al., 2008). Taken together, these data underscore the fact that $GABA_ARs$ differing in subunit composition represent distinct functional entities that are not readily interchanged. The lack of developmental phenotype in the absence of a major receptor subtypes points to multiple compensatory responses, which likely include adaptations of intrinsic properties of neurons, upregulation of $GABA_AR$-mediating complementary forms of transmission (tonic versus phasic), and possibly rewiring of neuronal circuits.

1.4. GABA$_A$ receptors and adult neurogenesis

In addition to CNS ontogeny, adult neurogenesis provides another opportunity to investigate the relevance of GABA$_A$R heterogeneity for neuronal development, for two main reasons: first, because GABAergic transmission regulates major steps of adult neurogenesis, quite similarly to its role during ontogeny; therefore, there is evidence for the control of precursor cell proliferation, fate-decision, and migration, and for the regulation of neuronal differentiation and synaptic integration (Ge, Pradhan, Ming, & Song, 2007; Nissant & Pallotto, 2011; Overstreet, Bromberg, Benson, & Westbrook, 2005; Sernagor, Chabrol, Bony, & Cancedda, 2010); second, because adult-born neurons differentiate and establish synaptic connections in a comparatively stable environment, allowing to study the effects of targeted genetic manipulations affecting GABA$_A$R and determine their role in neuronal differentiation and synaptic integration. Furthermore, like in developing brain, evidence is compelling that GABA first mediates tonic actions by activating extrasynaptic receptors on precursor cells, followed by synaptic inhibitory transmission, and that its initial effects are depolarizing, owing to delayed upregulation of KCC2 (Chancey et al., 2013; Ge et al., 2006; Jagasia et al., 2009; Overstreet et al., 2005). Accordingly, distinct GABA$_A$R subtypes act in progenitor cells and in mature adult-born neurons. It should be noted, however, that current knowledge of the GABA$_A$R subunit repertoire expressed by stem and precursor cells, as well as neuroblasts and differentiating neurons, remains fragmentary, making it difficult to assign specific functional roles to distinct receptor subtypes.

In rodents, adult neurogenesis occurs in the subventricular zone of the lateral ventricles and in the subgranular zone of the dentate gyrus. Besides representing an important facet of brain plasticity, it offers, as noted above, an attractive paradigm for investigating the role of GABA$_A$R-mediated regulation of neuronal maturation and functional integration into preexisting synaptic circuits. Thus, analyzing the consequences of GABA$_A$R inactivation in radial glia-like stem cells revealed that local parvalbumin-positive fast-spiking interneurons in the dentate gyrus determine their mode of division (symmetric versus asymmetric) as well as neuronal versus glial fate (Song et al., 2012); furthermore, these receptors are zolpidem insensitive, which is suggestive of the selective expression of α5-GABA$_A$R in stem cells. Supporting the notion that GABA$_A$R subtypes have distinct functional roles, we have provided evidence that extrasynaptic (α4, δ) and postsynaptic (α2) GABA$_A$R subtypes regulate distinct phases of adult neurogenesis in the dentate gyrus (Duveau et al., 2011). Thus, the negative control exerted by GABA on neural precursor cell proliferation (Platel, Lacar, & Bordey, 2007) requires the presence of α4-$GABA_AR$. Targeted deletion of the δ subunit had no effect, as expected from the delayed expression of this subunit during ontogeny (but see Whissell et al., 2013). Targeted deletion of α4-$GABA_ARs$ also impaired dendritic growth and final positioning of adult-born granule cells; in contrast, α2-KO newborn neurons exhibited delayed pruning of dendritic branches, presumably to maintain inhibitory-excitatory balance upon maturation of glutamatergic inputs (Duveau et al., 2011). Newborn granule cells are endowed with functional postsynaptic $GABA_AR$ around 7 days postmitosis, exhibiting synaptic currents activated by Ivy/neuroglia form interneurons (Markwardt, Dieni, Wadiche, & Overstreet-Wadiche, 2011, Overstreet et al., 2005). The timing of formation of basket cell and axo-axonic cell synapses has not been determined. The moderate effects seen on their morphological differentiation upon targeted deletion of the α2 subunit is most likely explained by the expression of additional subunits contributing to postsynaptic receptors, notably the α1 subunit in granule cells.

A more dramatic phenotype was observed upon selective α2 subunit inactivation in adult-born olfactory bulb granule cells (Pallotto et al., 2012), which represent the main interneuron subtype of the olfactory bulb, continuously generated from mitotic precursor cells in the subventricular zone (Carleton, Petreanu, Lansord, Alvarez-Buylla, & Lledo, 2003). As α2-$GABA_AR$ provide most synaptic inhibition to these cells, their inactivation profoundly altered dendritic development, spine formation, and maturation of glutamatergic inputs. Also, modulation of dendritic differentiation by environmental enrichment or deprivation, which is prominent in wild-type adult-born granule cells (Saghatelyan et al., 2005), was abrogated in adult-born α2-KO granule cells (Pallotto et al., 2012). The severity of these effects highlights the central role played by $GABA_AR$ in regulating neuronal differentiation. It is the most remarkable in view of the fact that olfactory bulb granule cells express several subunits contributing to extrasynaptic $GABA_AR$, notably α5 and δ. Once again, these data underscore the fact that postsynaptic and extrasynaptic $GABA_AR$ subtypes are not interchangeable and are expressed independently of each other. An intriguing aspect of these findings is that the deficit in morphological differentiation in adult-born granule cells lacking the α2 subunit becomes evident after a period of about 2 weeks, suggesting that it is related to a failure to form functional GABAergic synapses. However, at the time of their appearance, GABAergic transmission most likely is inhibitory; making it difficult to conceive which intracellular signaling cascades controlling dendritic growth and spine formation is affected in the absence of functional synaptic GABAergic transmission. An alternative, but not mutually exclusive, explanation is that α2-$GABA_AR$ in granule cells are required for specific protein–protein interactions, which in turn are involved in the activation of signaling cascades controlling cell differentiation (Harvey et al., 2004; Tyagarajan & Fritschy, 2014). According to such model, $GABA_AR$ would occupy a high position in the hierarchy of protein interactions required for assembly of functional synapses (Poulopoulos et al., 2009; Saiepour et al., 2010).

1.5. Significance for CNS diseases

Considering the preeminent role of $GABA_AR$-mediated transmission in major steps of brain development and neuronal maturation, one might expect various neurological and psychiatric diseases to be associated with a disturbance of GABAergic transmission during ontogeny. Mutations in several $GABA_AR$ subunits, which typically affect receptor assembly and cell-surface trafficking, have been found in idiopathic epilepsies, notably generalized epilepsy with febrile seizures (Macdonald, Kang, & Gallagher, 2010), but it is unclear to which extent developmental anomalies due to defective GABAergic signaling contribute to the etiology of these diseases. Other forms of epilepsy emerge from cortical dysplasia or heterotopia, due to abnormal neuronal differentiation and migration. Although mutations associated with these disorders do not directly affect the GABAergic system, as seen at the example of mice with an Eml1 mutation (Kielar et al., 2014), and cortical dysplasia can be induced by ablating selective populations of progenitor cells (see, for example, Abbah, Braga, & Juliano, 2014), heterotopic neurons exhibit reduced frequency and amplitude of GABAergic currents (Zhu & Roper, 2000), altered expression of KCC2 (Costa et al., 2001), and tend to express the $GABA_AR$ subunit repertoire corresponding to the layer where they should have migrated, rather than the layer where the heterotopia is located (Hagemann, Fritschy, Witte, & Redecker, 2001). This conclusion is also supported by birth-dating analysis in a genetic model of cortical heterotopia (Croquelois et al., 2009). In dysplastic cortex, there is also evidence for increased GABAergic transmission and altered KCC2 expression, which contribute to slow down neuronal migration (Abbah & Juliano, 2014). In this respect, it might be worth noting that focal temporal lobe epilepsies in human have been associated with abnormal expression of the α3 subunit in layer III of the neocortex human cerebral cortex (Loup et al., 2006). Furthermore, human cortical white matter contains a considerable number of neurons expressing distinct $GABA_AR$ subunits depending on their morphology. In individuals without a history of neurological disease these neurons unlikely represent abnormally curtailed migration; however, their morphology and $GABA_AR$ subunit repertoire are markedly altered in epileptic patients (Loup, Picard, Yonekawa, Wieser, & Fritschy, 2009). Therefore, these observations suggest more complex information processing in human neocortex and cortical white matter than in rodents and the possibility that cortical dysplasia establish abnormal connections with these deeply located neurons in the white matter (Ackman et al., 2009).

There are several lines of evidence implicating disordered GABAergic transmission in schizophrenia (Costa et al., 2001; Lewis, Hashimoto, & Volk, 2005), including defective Reelin signaling from Cajal-Retzius cells, reduced expression of GABA synthesizing enzymes and GABA reuptake transporter type 1, and compensatory upregulation of $\alpha 2$-$GABA_AR$ on the axon-initial segment of cortical pyramidal cells (Volk et al., 2002). These changes are specific to the disease, as they are not observed in patients with major depression, or in nonhuman primates chronically treated with antipsychotic drugs. While being reported typically in dorsolateral prefrontal cortex, it is not fully clear whether they are solely restricted to specific neocortical areas. The contributions of abnormal neurodevelopmental mechanisms to these various alterations of GABAergic transmission are under intense scrutiny and include a large variety of possible mechanisms (Volk & Lewis, 2013). A well-characterized example is provided by experimental manipulation of neuregulin 1—ErbB4 signaling in fast-spiking parvalbumin-positive interneurons. This interaction is essential for both formation of axo-axonic synapses onto neocortical pyramidal cells and glutamatergic synapses onto these interneurons (Del Pino et al., 2013; Fazzari et al., 2010). Thus, these wiring anomalies have long ranging effects on cortical excitability, oscillatory activity, and cognitive functions recapitulating major features of schizophrenia. Strikingly, disrupting postsynaptic clustering of $\alpha 2$-$GABA_AR$ in medial prefrontal cortex by overexpression of a dominant-negative construct likewise reduces γ frequency oscillations and causes cognitive deficits in adult mice (Hines et al., 2013).

Fragile X syndrome and autism-spectrum disorders are other examples of diseases for which experimental studies reveal a developmentally regulated deficit of GABAergic transmission. In FMR1-null mice, the deficit affects receptors mediating extrasynaptic transmission, notably in the amygdala and cerebral cortex (Paluszkiewicz, Martin, & Huntsman, 2011), providing also a rationale for treatment using subunit-selective drugs (Rudolph & Möhler, 2014). In a model of autism-spectrum disorder associated with a mutation in neuroligin-3, evidence points toward increased excitatory GABAergic activity in developing hippocampus, suggesting abnormal circuit formation and imbalance between excitatory and inhibitory transmission in adulthood (Pizzarelli & Cherubini, 2011).

Our own studies in mutant mice lacking α $GABA_AR$ subunit gene strongly expressed during brain development revealed that the system is resilient enough to prevent the appearance of obvious alterations in brain cytoarchitecture. The only $GABA_AR$ mutants suffering from major developmental problems after birth and early postnatal mortality are mice lacking the γ2 or β3 subunits (Ferguson et al., 2007; Gunther et al., 1995). However, even in these animals, brain cytoarchitecture is apparently normal. Yet, at the example of schizophrenia models, subtle alterations in synaptic connectivity involving specific interneurons can be sufficient to induce marked functional and behavioral alterations, reinforcing the notion that abnormal GABAergic signaling during brain development, as well as during the critical periods of plasticity (see Chapter on "GABA Circuit Control of Critical Period Brain Development" by Takao Hensch), might be a widespread cause underlying neurological and psychiatric diseases. As a corollary, therapeutic interventions that enhance $GABA_AR$-mediated transmission might be expected to disturb brain development and neuronal maturation. In particular, there is concern that general anesthetics cause long-lasting alterations in neuronal maturation and circuit formation (De Roo et al., 2009). Likewise, abnormal neurosteroid production of signaling, e.g., in relation to stress or ethanol consumption, is expected to both affect GABAergic transmission and the formation of proper neuronal networks, especially when occurring at critical stages of development (perinatal period, puberty) (Shen et al., 2010).

Another key issue related to therapeutic intervention is the fact that enhancing the depolarizing action of GABA in immature neuronal circuits might be counterproductive. This problem is of particular importance for the treatment of epileptic seizures in neonates (Löscher, Puskarjov, & Kaila, 2013; Ruusuvuori et al., 2013), in which enhancing $GABA_AR$-mediated transmission by allosteric modulation worsens the symptoms due to the excitatory action of GABA. Accordingly, alternative strategies are being sought, in particular to attempt to act on the Cl^- transporters that determine the effects of GABA (Puskarjov, Kahle, Ruusuvuori, & Kaila, 2014).

2. CONCLUSION

This brief overview emphasizes the key role played by $GABA_AR$, and especially by the multiple molecular mechanisms regulating their expression and subcellular trafficking and localization, in developing neurons for the formation and proper function of neuronal circuits. The heterogeneity of $GABA_AR$, and the coexistence of multiple subtypes in various neuronal populations likely holds the key of a functional specialization that we have only begun to unravel, but which might be of fundamental relevance for enabling therapeutic intervention targeted to only specific $GABA_AR$ subtypes. The multidisciplinary approach followed by Hanns Möhler throughout his carrier has been instrumental in highlighting many of the features of $GABA_AR$ that are now common knowledge in textbooks, but which each represent quantum leaps in our understanding of this complex and highly fascinating system.

CONFLICT OF INTEREST

The author has no conflicts of interest to declare.

ACKNOWLEDGMENTS

The author's work described in this review was supported by the Swiss National Science Foundation.

REFERENCES

Abbah, J., Braga, M., & Juliano, S. (2014). Targeted disruption of layer 4 during development increases $GABA_A$ receptor neurotransmission in the neocortex. *Journal of Neurophysiology, 111*, 323–335.

Abbah, J., & Juliano, S. (2014). Altered migratory behavior of interneurons in a model of cortical dysplasia: The influence of elevated $GABA_A$ activity. *Cerebral Cortex, 24*, 2297–2308.

Ackman, J., Aniksztejn, L., Crépel, V., Becq, H., Pellegrino, C., Cardoso, C., et al. (2009). Abnormal network activity in a targeted genetic model of human double cortex. *The Journal of Neuroscience, 29*, 313–327.

Barbin, G., Pollard, H., Gaiarsa, J. L., & Ben-Ari, Y. (1993). Involvement of $GABA_A$ receptors in the outgrowth of cultured hippocampal neurons. *Neuroscience Letters, 152*, 150–154.

Barker, J. L., & Nicoll, R. A. (1972). Gamma-aminobutyric acid: Role in primary afferent depolarization. *Science, 176*, 1043–1045.

Behar, T. N., Schaffner, A. E., Scott, C. A., Greene, C. L., & Barker, J. L. (2000). $GABA_A$ receptor antagonists modulate postmitotic cell migration in slice cultures of embryonic rat cortex. *Cerebral Cortex, 10*, 899–909.

Ben-Ari, Y., Cherubini, E., Corradetti, R., & Gaiarsa, J. (1989). Giant synaptic potentials in immature rat CA3 hippocampal neurones. *Journal of Physiology, 416*, 303–325.

Ben-Ari, Y., Gaiarsa, J., Tyzio, R., & Khazipov, R. (2007). GABA: A pioneer transmitter that excites immature neurons and generates primitive oscillations. *Physiological Reviews, 87*, 1215–1284.

Ben-Ari, Y., Woodin, M., Sernagor, E., Cancedda, L., Vinay, L., Rivera, C., et al. (2012). Refuting the challenges of the developmental shift of polarity of GABA actions: GABA more exciting than ever! *Frontiers in Cellular Neuroscience, 6*, 35.

Benke, D., Fritschy, J. M., Trzeciak, A., Bannwarth, W., & Mohler, H. (1994). Distribution, prevalence and drug-binding profile of GABA$_A$-receptors subtypes differing in β-subunit isoform. *Journal of Biological Chemistry, 269*, 27100–27107.

Benke, D., Mertens, S., Trzeciak, A., Gillessen, D., & Mohler, H. (1991). GABA$_A$ receptors display association of γ2-subunit with α1- and β2/3 subunits. *Journal of Biological Chemistry, 266*, 4478–4483.

Bortone, D., & Polleux, F. (2009). KCC2 expression promotes the termination of cortical interneuron migration in a voltage-sensitive calcium-dependent manner. *Neuron, 62*, 53–71.

Bregestovski, P., & Bernard, C. (2012). Excitatory GABA: How a correct observation may turn out to be an experimental artifact. *Frontiers in Pharmacology, 3*, 65.

Brünig, I., Scotti, E., Sidler, C., & Fritschy, J. M. (2002). Intact sorting, targeting, and clustering of γ-aminobutyric acid A receptor subtypes in hippocampal neurons in vitro. *Journal of Comparative Neurology, 443*, 43–45.

Carleton, A., Petreanu, L. T., Lansord, R., Alvarez-Buylla, A., & Lledo, P. M. (2003). Becoming a new neuron in the adult olfactory bulb. *Nature Neuroscience, 6*, 507–518.

Chancey, J., Adlaf, E., Sapp, M., Pugh, P., Wadiche, J., & Overstreet-Wadiche, L. (2013). GABA depolarization is required for experience-dependent synapse unsilencing in adult-born neurons. *The Journal of Neuroscience, 33*, 6614–6622.

Cherubini, E., Gaiarsa, J. L., & Ben-Ari, Y. (1991). GABA: An excitatory transmitter in early postnatal life. *Trends in Neurosciences, 14*, 515–519.

Costa, E., Davis, J., Grayson, D., Guidotti, A., Pappas, G., & Pesold, C. (2001). Dendritic spine hypoplasticity and downregulation of reelin and GABAergic tone in schizophrenia vulnerability. *Neurobiology of Disease, 8*, 723–742.

Croquelois, A., Giuliani, F., Savary, C., Kielar, M., Amiot, C., Schenk, F., et al. (2009). Characterization of the HeCo mutant mouse: A new model of subcortical band heterotopia associated with seizures and behavioral deficits. *Cerebral Cortex, 19*, 563–575.

Cuzon Carlson, V., & Yeh, H. (2011). GABA$_A$ receptor subunit profiles of tangentially migrating neurons derived from the medial ganglionic eminence. *Cerebral Cortex, 21*, 1792–1802.

Cuzon, V., Yeh, P., Cheng, Q., & Yeh, H. (2006). Ambient GABA promotes cortical entry of tangentially migrating cells derived from the medial ganglionic eminence. *Cerebral Cortex, 16*, 1377–1388.

Czarnecki, A., Le Corronc, H., Rigato, C., Le Bras, B., Couraud, F., Scain, A., et al. (2014). Acetylcholine controls GABA-, glutamate-, and glycine-dependent giant depolarizing potentials that govern spontaneous motoneuron activity at the onset of synaptogenesis in the mouse embryonic spinal cord. *The Journal of Neuroscience, 34*, 6389–6404.

De Blas, A. (1996). Brain GABA$_A$ receptors studied with subunit-specific antibodies. *Molecular Neurobiology, 12*, 55–71.

De Roo, M., Klauser, P., Briner, A., Nikonenko, I., Mendez, P., Dayer, A., et al. (2009). Anesthetics rapidly promote synaptogenesis during a critical period of brain development. *PloS One, 4*, e7043.

Del Pino, I., García-Frigola, C., Dehorter, N., Brotons-Mas, J., Alvarez-Salvado, E., Martínez de Lagrán, M., et al. (2013). Erbb4 deletion from fast-spiking interneurons causes schizophrenia-like phenotypes. *Neuron, 79*, 1152–1168.

Duveau, V., Laustela, S., Barth, L., Gianolini, F., Vogt, K. E., Keist, R., et al. (2011). Spatiotemporal specificity of $GABA_A$ receptor-mediated regulation of adult hippocampal neurogenesis. *European Journal of Neuroscience*, *34*, 362–373.

Ewert, M., Shivers, B. D., Luddens, H., Mohler, H., & Seeburg, P. H. (1990). Subunit selectivity and epitope characterization of mAbs directed against the $GABA_A$/benzodiazepine receptor. *Journal of Cell Biology*, *110*, 2043–2048.

Farrant, M., & Nusser, Z. (2005). Variations on an inhibitory theme: Phasic and tonic activation of $GABA_A$ receptors. *Nature Reviews. Neuroscience*, *6*, 215–229.

Fazzari, P., Paternain, A., Valiente, M., Pla, R., Luján, R., Lloyd, K., et al. (2010). Control of cortical GABA circuitry development by Nrg1 and ErbB4 signalling. *Nature*, *464*, 1376–1380.

Ferguson, C., Hardy, S. L., Werner, D. F., Hileman, S. M., Delorey, T. M., & Homanics, G. E. (2007). New insight into the role of the β3 subunit of the $GABA_A$-R in development, behavior, body weight regulation, and anesthesia revealed by conditional gene knockout. *BMC Neuroscience*, *8*, 85.

Fiszman, M., Borodinsky, L., & Neale, J. (1999). GABA induces proliferation of immature cerebellar granule cells grown in vitro. *Developmental Brain Research*, *115*, 1–8.

Fritschy, J. M., Benke, D., Mertens, S., Oertel, W. H., Bachi, T., & Mohler, H. (1992). Five subtypes of type A γ-aminobutyric acid receptors identified in neurons by double and triple immunofluorescence staining with subunit-specific antibodies. *Proceedings of the National Academy of Sciences of the United States of America*, *89*, 6726–6730.

Fritschy, J. M., & Mohler, H. (1995). $GABA_A$-receptor heterogeneity in the adult rat brain: Differential regional and cellular distribution of seven major subunits. *Journal of Comparative Neurology*, *359*, 154–194.

Fritschy, J. M., Panzanelli, P., Kralic, J. E., Vogt, K. E., & Sassoè-Pognetto, M. (2006). Differential dependence of axo-dendritic and axo-somatic GABAergic synapses on $GABA_A$ receptors containing the α1 subunit in Purkinje cells. *The Journal of Neuroscience*, *26*, 3245–3255.

Fritschy, J. M., Paysan, J., Enna, A., & Mohler, H. (1994). Switch in the expression of rat $GABA_A$-receptor subtypes during postnatal development: An immunohistochemical study. *The Journal of Neuroscience*, *14*, 5302–5324.

Fritschy, J. M., Weinmann, O., Wenzel, A., & Benke, D. (1998). Synapse-specific localization of NMDA- and $GABA_A$-receptor subunits revealed by antigen-retrieval immunohistochemistry. *Journal of Comparative Neurology*, *390*, 194–210.

Furukawa, T., Yamada, J., Akita, T., Matsushima, Y., Yanagawa, Y., & Fukuda, A. (2014). Roles of taurine-mediated tonic $GABA_A$ receptor activation in the radial migration of neurons in the fetal mouse cerebral cortex. *Frontiers in Cellular Neuroscience*, *8*, 88.

Gambarana, C., Beattie, C. E., Rodriguez, Z. R., & Siegel, R. E. (1991). Region-specific expression of messenger RNAs encoding $GABA_A$ receptor subunits in the developing rat brain. *Neuroscience*, *45*, 423–432.

Ge, S., Goh, E. L., Sailor, K. A., Kitabatake, Y., Ming, G. L., & Song, H. (2006). GABA regulates synaptic integration of newly generated neurons in the adult brain. *Nature*, *439*, 589–593.

Ge, S., Pradhan, D. A., Ming, G. L., & Song, H. (2007). GABA sets the tempo for activity-dependent adult neurogenesis. *Trends in Neurosciences*, *30*, 1–8.

Ghosh, A., Carnahan, J., & Greenberg, M. E. (1994). Requirement for BDNF in activity-dependent survival of cortical neurons. *Science*, *263*, 1618–1623.

Gunther, U., Benson, J., Benke, D., Fritschy, J. M., Reyes, G. H., Knoflach, F., et al. (1995). Benzodiazepine-insensitive mice generated by targeted disruption of the γ2-subunit gene of γ-aminobutyric acid type A receptors. *Proceedings of the National Academy of Sciences of the United States of America*, *92*, 7749–7753.

Hagemann, G., Fritschy, J. M., Witte, O. W., & Redecker, C. (2001). Differentiation of GABA$_A$ receptors in subcoritcal heterotopias: Subunit distribution of displaced cortex reflects original commitment. *NeuroReport, 12*, 3413–3417.

Haring, P., Stahli, C., Schoch, P., Takacs, B., Staehelin, T., & Mohler, H. (1985). Monoclonal antibodies reveal structural homogeneity of γ-aminobutyric acid/benzodiazepine receptors in different brain areas. *Proceedings of the National Academy of Sciences of the United States of America, 82*, 4837–4841.

Harvey, K., Duguid, I. C., Alldred, M. J., Beatty, S. E., Ward, H., Keep, N. H., et al. (2004). The GDP-GTP exchange factor collybistin: An essential determinant of neuronal gephyrin clustering. *The Journal of Neuroscience, 24*, 5816–5826.

Heck, N., Kilb, W., Reiprich, P., Kubota, H., Furukawa, T., Fukuda, A., et al. (2007). GABA$_A$ receptors regulate neocortical neuronal migration in vitro and in vivo. *Cerebral Cortex, 17*, 138–148.

Hines, R., Hines, D., Houston, C., Mukherjee, J., Haydon, P., Tretter, V., et al. (2013). Disrupting the clustering of GABA$_A$ receptor α2 subunits in the frontal cortex leads to reduced γ-power and cognitive deficits. *Proceedings of the National Academy of Sciences of the United States of America, 110*, 16628–16633.

Hörtnagl, H., Tasan, R., Wieselthaler, A., Kirchmair, E., Sieghart, W., & Sperk, G. (2013). Patterns of mRNA and protein expression for 12 GABA$_A$ receptor subunits in the mouse brain. *Neuroscience, 236*, 345–372.

Hosokawa, Y., Sciancalepore, M., Stratta, F., Martina, M., & Cherubini, E. (1994). Developmental changes in spontaneus GABA$_A$-mediated synaptic events in rat hippocampal CA3 neurons. *European Journal of Neuroscience, 6*, 805–813.

Inada, H., Watanabe, M., Uchida, T., Ishibashi, H., Wake, H., Nemoto, T., et al. (2011). GABA regulates the multidirectional tangential migration of GABAergic interneurons in living neonatal mice. *PloS One, 6*, e27048.

Jagasia, R., Steib, K., Englberger, E., Herold, S., Faus-Kessler, T., Saxe, M., et al. (2009). GABA-cAMP response element-binding protein signaling regulates maturation and survival of newly generated neurons in the adult hippocampus. *The Journal of Neuroscience, 24*, 7966–7977.

Kasugai, Y., Swinny, J. D., Roberts, J. D., Dalezios, Y., Fukazawa, Y., Sieghart, W., et al. (2010). Quantitative localisation of synaptic and extrasynaptic GABA$_A$ receptor subunits on hippocampal pyramidal cells by freeze-fracture replica immunolabelling. *European Journal of Neuroscience, 32*, 1868–1888.

Kielar, M., Tuy, F., Bizzotto, S., Lebrand, C., de Juan Romero, C., Poirier, K., et al. (2014). Mutations in Eml1 lead to ectopic progenitors and neuronal heterotopia in mouse and human. *Nature Neuroscience, 17*, 923–933.

Kirmse, K., Witte, O., & Holthoff, K. (2010). GABA depolarizes immature neocortical neurons in the presence of the ketone body ß-hydroxybutyrate. *The Journal of Neuroscience, 30*, 16002–16007.

Kralic, J. E., Sidler, C., Parpan, F., Homanics, G., Morrow, A. L., & Fritschy, J. M. (2006). Compensatory alteration of inhibitory synaptic circuits in thalamus and cerebellum of GABA$_A$ receptor α1 subunit knockout mice. *Journal of Comparative Neurology, 495*, 408–421.

Laurie, D. J., Seeburg, P. H., & Wisden, W. (1992). The distribution of 13 GABA$_A$ receptor subunit mRNAs in the rat brain. II. Olfactory bulb and cerebellum. *The Journal of Neuroscience, 12*, 1063–1076.

Lewis, D. A. (2000). GABAergic local circuit neurons and prefrontal cortical dysfunction in schizophrenia. *Brain Research Reviews, 31*, 270–276.

Lewis, D. A., Hashimoto, T., & Volk, D. W. (2005). Cortical inhibitory neurons and schizophrenia. *Nature Reviews. Neuroscience, 6*, 312–324.

Liu, J. P., Morrow, A. L., Devaud, L., Grayson, D. R., & Lauder, J. M. (1997). $GABA_A$ receptors mediate trophic effects of GABA on embryonic brainstem monoamine neurons in vitro. *The Journal of Neuroscience*, *17*, 2420–2428.

Löscher, W., Puskarjov, M., & Kaila, K. (2013). Cation-chloride cotransporters NKCC1 and KCC2 as potential targets for novel antiepileptic and antiepileptogenic treatments. *Neuropharmacology*, *69*, 62–74.

Loturco, J. J., Owens, D. F., Heath, M. J. S., Davis, M. B. E., & Kriegstein, A. R. (1995). GABA and glutamate depolarize cortical progenitor cells and inhibit DNA synthesis. *Neuron*, *15*, 1287–1298.

Loup, F., Picard, F., André, V. M., Kehrli, P., Yonekawa, Y., Wieser, H. G., et al. (2006). Altered expression of α3-containing $GABA_A$ receptors in the neocortex of patients with focal epilepsy. *Brain*, *129*, 3277–3289.

Loup, F., Picard, F., Yonekawa, Y., Wieser, H., & Fritschy, J. (2009). Selective changes in $GABA_A$ receptor subtypes in white matter neurons of patients with focal epilepsy. *Brain & Development*, *132*, 2449–2463.

Luhmann, H., Kirischuk, S., Sinning, A., & Kilb, W. (2014). Early GABAergic circuitry in the cerebral cortex. *Current Opinion in Neurobiology*, *26C*, 72–78.

Ma, W., & Barker, J. L. (1995). Complementary expressions of transcripts encoding GAD67 and $GABA_A$ receptor α4, β1, and γ1 subunits in the proliferative zone of the embryonic rat central nervous system. *The Journal of Neuroscience*, *15*, 2547–2560.

Ma, W., Saunders, P. A., Somogyi, R., Poulter, M. O., & Barker, J. L. (1993). Ontogeny of $GABA_A$ receptor subunit mRNAs in rat spinal cord and dorsal root ganglia. *Journal of Comparative Neurology*, *338*, 337–359.

Macdonald, R., Kang, J., & Gallagher, M. (2010). Mutations in $GABA_A$ receptor subunits associated with genetic epilepsies. *Journal of Physiology*, *588*, 1861–1869.

Maric, D., Maric, I., Wen, X., Fritschy, J. M., Sieghart, W., Barker, J. L., et al. (1999). $GABA_A$ receptor subunit composition and functional properties of Cl^- channels with differential sensitivity to zolpidem in embryonic rat hippocampal cells. *The Journal of Neuroscience*, *19*, 4921–4937.

Markwardt, S. J., Dieni, C. V., Wadiche, J. I., & Overstreet-Wadiche, L. (2011). Ivy/neurogliaform interneurons coordinate activity in the neurogenic niche. *Nature Neuroscience*, *14*, 1407–1409.

Merz, K., Herold, S., & Lie, D. C. (2011). CREB in adult neurogenesis—Master and partner in the development of adult-born neurons? *European Journal of Neuroscience*, *33*, 1078–1086.

Mody, I. (2001). Distinguishing between $GABA_A$ receptors responsible for tonic and phasic conductances. *Neurochemical Research*, *26*, 907.

Mohler, H., Benke, D., Mertens, S., & Fritschy, J. M. (1992). $GABA_A$-receptor subtypes differing in α-subunit composition display unique pharmacological profiles. In G. Biggio, A. Concas, & E. Costa (Eds.), *GABAergic synaptic transmission* (pp. 41–53). New York: Raven Press.

Mohler, H., Fritschy, J. M., & Rudolph, U. (2002). A new benzodiazepine pharmacology. *The Journal of Pharmacology and Experimental Therapeutics*, *300*, 2–8.

Möhler, H., & Okada, T. (1977). Benzodiazepine receptor: Demonstration in the central nervous system. *Science Signaling*, *198*, 849–851.

Mohler, H., Richards, J. G., & Wu, J. Y. (1981). Autoradiographic localization of benzodiazepine receptors in immunocytochemically identified gamma-aminobutyrergic synapses. *Proceedings of the National Academy of Sciences of the United States of America*, *78*, 1935–1938.

Nissant, A., & Pallotto, M. (2011). Integration and maturation of newborn neurons in the adult olfactory bulb–from synapses to function. *European Journal of Neuroscience*, *33*, 1069–1077.

Notter, T., Panzanelli, P., Pfister, S., Mircsof, D., & Fritschy, J. (2013). A protocol for concurrent high-quality immunohistochemical and biochemical analyses in adult mouse central nervous system. *European Journal of Neuroscience, 39*, 165–175.
Nusser, Z., & Somogyi, P. (1994). Compartmentalised distribution of immunoreactivity for GABA$_A$ receptors on the surface of neurons in relation to single or multiple sources of GABAergic synapses. *Journal of Neurochemistry, 63*(Suppl. 1), S39.
Obrietan, K., Gao, X. B., & Van Den Pol, A. N. (2002). Excitatory actions of GABA increase BDNF expression via a MAPK-CREB-dependent mechanism—A positive feedback circuit in developing neurons. *Journal of Neurophysiology, 88*, 1005–1015.
O'Leary, D., Chou, S., & Sahara, S. (2007). Area patterning of the mammalian cortex. *Neuron, 56*, 252–269.
Olsen, R. W., & Sieghart, W. (2008). International union of pharmacology. LXX. Subtypes of γ-aminobutyric acid$_A$ receptors: Classification on the basis of subunit composition, pharmacology, and function. Update. *Pharmacological Reviews, 60*, 243–260.
Overstreet, W. L., Bromberg, D. A., Benson, A. L., & Westbrook, G. L. (2005). GABAergic signaling to newborn neuron in dentate gyrus. *Journal of Neurophysiology, 94*, 4528–4532.
Pallotto, M., Nissant, A., Fritschy, J. M., Rudolph, U., Sassoè-Pognetto, M., Panzanelli, P., et al. (2012). Early formation of GABAergic synapses governs the development of adult-born neurons in the olfactory bulb. *The Journal of Neuroscience, 32*, 9103–9115.
Paluszkiewicz, S., Martin, B., & Huntsman, M. (2011). Fragile X syndrome: The GABAergic system and circuit dysfunction. *Developmental Neuroscience, 33*, 349–364.
Patrizi, A., Scelfo, B., Viltono, L., Briatore, F., Fukaya, M., Watanabe, M., et al. (2008). Synapse formation and clustering of neuroligin-2 in the absence of GABA$_A$ receptors. *Proceedings of the National Academy of Sciences of the United States of America, 105*, 13151–13156.
Paysan, J., Bolz, J., Mohler, H., & Fritschy, J. M. (1994). The GABA$_A$-receptor α1-subunit: An early marker of cortical parcellation. *Journal of Comparative Neurology, 350*, 133–149.
Paysan, J., Kossel, A., Bolz, J., & Fritschy, J. M. (1997). Area-specific regulation of γ-aminobutyric acid A receptor subtypes by thalamic afferents in developing rat neocortex. *Proceedings of the National Academy of Sciences of the United States of America, 94*, 6995–7000.
Peden, D. R., Petitjean, C. M., Herd, M. B., Durakoglugil, M., Rosahl, T. W., Wafford, K., et al. (2008). Developmental maturation of synaptic and extrasynaptic GABA$_A$ receptors in mouse thalamic ventrobasal neurones. *Journal of Physiology, 586*, 965–987.
Pizzarelli, R., & Cherubini, E. (2011). Alterations of GABAergic signaling in autism spectrum disorders. *Neural Plasticity, 2011*, 297153.
Platel, J. C., Lacar, B., & Bordey, A. (2007). GABA and glutamate signaling: Homeostatic control of adult forebrain neurogenesis. *Journal of Molecular Histology, 38*, 602–610.
Poulopoulos, A., Aramuni, G., Meyer, G., Soykan, T., Hoon, M., Papadopoulos, T., et al. (2009). Neuroligin 2 drives postsynaptic assembly at perisomatic inhibitory synapses through gephyrin and collybistin. *Neuron, 63*, 628–642.
Poulter, M. O., Barker, J. L., O'Carroll, A. M., Lolait, S. J., & Mahan, L. C. (1992). Differential and transient expression of GABA$_A$ receptor alpha-subunit mRNAs in the developing rat CNS. *The Journal of Neuroscience, 12*, 2888–2900.
Puskarjov, M., Kahle, K., Ruusuvuori, E., & Kaila, K. (2014). Pharmacotherapeutic targeting of cation-chloride cotransporters in neonatal seizures. *Epilepsia, 55*, 806–818.
Rheims, S., Holmgren, C., Chazal, G., Mulder, J., Harkany, T., Zilberter, T., et al. (2009). GABA action in immature neocortical neurons directly depends on the availability of ketone bodies. *Journal of Neurochemistry, 110*, 1330–1338.
Richards, J. G., Schoch, P., Haring, P., Takacs, B., & Mohler, H. (1987). Resolving GABA$_A$/benzodiazepine receptors: Cellular and subcellular localization in the CNS with monoclonal antibodies. *The Journal of Neuroscience*, 7, 1866–1886.

Rudolph, U., & Möhler, H. (2004). Analysis of $GABA_A$ receptor function and dissection of the pharmacology of benzodiazepines and general anesthetics through mouse genetics. *Annual Review of Pharmacology and Toxicology, 44*, 475–498.

Rudolph, U., & Möhler, H. (2014). $GABA_A$ receptor subtypes: Therapeutic potential in Down syndrome, affective disorders, schizophrenia, and autism. *Annual Review of Pharmacology and Toxicology, 54*, 483–507.

Ruusuvuori, E., Huebner, A., Kirilkin, I., Yukin, A., Blaesse, P., Helmy, M., et al. (2013). Neuronal carbonic anhydrase VII provides GABAergic excitatory drive to exacerbate febrile seizures. *EMBO Journal, 32*, 2275–2286.

Saghatelyan, A., Roux, P., Migliore, M., Rochefort, C., Desmaisons, D., Charneau, P., et al. (2005). Activity-dependent adjustments of the inhibitory network in the olfactory bulb following early postnatal deprivation. *Neuron, 46*, 103–116.

Saiepour, L., Fuchs, C., Patrizi, A., Sassoè-Pognetto, M., Harvey, R. J., & Harvey, K. (2010). Complex role of collybistin and gephyrin in $GABA_A$ receptor clustering. *Journal of Biological Chemistry, 285*, 29623–29631.

Sassoè-Pognetto, M., Kirsch, J., Grünert, U., Greferath, U., Fritschy, J. M., Mohler, H., et al. (1995). Colocalization of gephyrin and $GABA_A$-receptor subunits in the rat retina. *Journal of Comparative Neurology, 357*, 1–14.

Schlumpf, M., Richards, J. G., Lichtensteiger, W., & Mohler, H. (1983). An autoradiographic study of the prenatal development of benzodiazepine-binding sites in rat brain. *The Journal of Neuroscience, 3*, 1478–1487.

Schneider Gasser, E. M., Duveau, V., Prenosil, G. A., & Fritschy, J. M. (2007). Reorganization of GABAergic circuits maintains $GABA_A$ receptor-mediated transmission onto CA1 interneurons in α1-subunit-null mice. *European Journal of Neuroscience, 25*, 3287–3304.

Schoch, P., Richards, J. G., Haring, P., Takacs, B., Stahli, C., Staehelin, T., et al. (1985). Co-localization of $GABA_A$ receptors and benzodiazepine receptors in the brain shown by monoclonal antibodies. *Nature, 314*, 168–171.

Serafini, R., Valeyev, A. Y., Barker, J. L., & Poulter, M. O. (1995). Depolarizing GABA-activated Cl^- channels in embryonic rat spinal and olfactory bulb cells. *Journal of Physiology (London), 488*, 371–386.

Sernagor, E., Chabrol, F., Bony, G., & Cancedda, L. (2010). GABAergic control of neurite outgrowth and remodeling during development and adult neurogenesis: General rules and differences in diverse systems. *Frontiers in Cellular Neuroscience, 4*, 11.

Shen, H., Sabaliauskas, N., Sherpa, A., Fenton, A. A., Stelzer, A., Aoki, C., et al. (2010). A critical role for α4βδ $GABA_A$ receptors in shaping learning deficits at puberty in mice. *Science, 327*, 1515–1518.

Shivers, B. D., Killisch, I., Sprengel, R., Sontheimer, H., Kohler, M., Schofield, P. R., et al. (1989). Two novel $GABA_A$ receptor subunits exist in distinct neuronal subpopulations. *Neuron, 3*, 327–337.

Somogyi, P., Takagi, H., Richards, J. G., & Mohler, H. (1989). Subcellular localization of benzodiazepine/$GABA_A$ receptors in the cerebellum of rat, cat, and monkey using monoclonal antibodies. *The Journal of Neuroscience, 9*, 2197–2209.

Song, J., Zhong, C., Bonaguidi, M. A., Sun, G. J., Hsu, D., Gu, Y., et al. (2012). Neuronal circuitry mechanism regulating adult quiescent neural stem-cell fate decision. *Nature, 489*, 150–154.

Spoerri, P. E. (1988). Neurotrophic effects of GABA in cultures of embryonic chick brain and retina. *Synapse, 2*, 11–22.

Tao, X., Finkbeiner, S., Arnold, D. B., Shaywitz, A. J., & Greenberg, M. E. (1998). Ca2+ influx regulates BDNF transcription by a CREB family transcription factor dependent mechanism. *Neuron, 20*, 709–726.

Tyagarajan, S. K., & Fritschy, J. M. (2014). Gephyrin, a master regulator of neuronal function? *Nature Reviews. Neuroscience*, *15*, 141–156.

Tyzio, R., Represa, A., Jorquera, I., Ben-Ari, Y., Gozlan, H., & Aniksztejn, L. (1999). The establishment of GABAergic and glutamatergic synapses on CA1 pyramidal neurons is sequential and correlated with the development of the apical dendrite. *The Journal of Neuroscience*, *19*, 10372–10382.

Valeeva, G., Valiullina, F., & Khazipov, R. (2013). Excitatory actions of GABA in the intact neonatal rodent hippocampus in vitro. *Frontiers in Cellular Neuroscience*, 7, 20.

van Rijnsoever, C., Sidler, C., & Fritschy, J. M. (2005). Internalized GABA$_A$-receptor subunits are transferred to an intracellular pool associated with the postsynaptic density. *European Journal of Neuroscience*, *21*, 327–338.

Vautrin, J., Maric, D., Sukhareva, M., Schaffner, A. E., & Barker, J. L. (2000). Surface-accessible GABA supports tonic and quantal synaptic transmission. *Synapse*, *37*, 38–55.

Volk, D., & Lewis, D. (2013). Prenatal ontogeny as a susceptibility period for cortical GABA neuron disturbances in schizophrenia. *Neuroscience*, *248C*, 154–164.

Volk, D. W., Pierri, J. N., Fritschy, J. M., Auh, S., Sampson, A. R., & Lewis, D. A. (2002). Reciprocal alterations in pre- and postsynaptic inhibitory markers at chandelier cell inputs to pyramidal neurons in schizophrenia. *Cerebral Cortex*, *12*, 1063–1070.

Wang, J., Johnston, P., Ball, B., & Brinton, R. (2005). The neurosteroid allopregnanolone promotes proliferation of rodent and human neural progenitor cells and regulates cell-cycle gene and protein expression. *The Journal of Neuroscience*, *25*, 4706–4718.

Whissell, P., Rosenzweig, S., Lecker, I., Wang, D., Wojtowicz, J., & Orser, B. (2013). γ-aminobutyric acid type A receptors that contain the δ subunit promote memory and neurogenesis in the dentate gyrus. *Annals of Neurology*, *74*, 611–621.

Wieland, H. A., Luddens, H., & Seeburg, P. H. (1992). A single histidine in GABA$_A$ receptors is essential for benzodiazepine agonist binding. *Journal of Biological Chemistry*, *267*, 1426–1429.

Wlodarczyk, A., Sylantyev, S., Herd, M., Kersanté, F., Lambert, J., Rusakov, D., et al. (2013). GABA-independent GABA$_A$ receptor openings maintain tonic currents. *The Journal of Neuroscience*, *33*, 3905–3914.

Zhu, W. J., & Roper, S. N. (2000). Reduced inhibition in an animal model of cortical dysplasia. *The Journal of Neuroscience*, *20*, 8925–8931.

CHAPTER THREE

Regulation of Cell Surface GABA$_B$ Receptors: Contribution to Synaptic Plasticity in Neurological Diseases

Dietmar Benke[1], Karthik Balakrishnan, Khaled Zemoura

Institute of Pharmacology and Toxicology, University of Zurich, Zurich, Switzerland

[1]Corresponding author: e-mail address: benke@pharma.uzh.ch

Contents

Abstract

γ-Amino butyric acid (GABA$_B$) receptors are heterodimeric G protein-coupled receptors expressed throughout the central nervous system in virtually all neurons. They control the excitability of neurons via activation of different downstream effector systems in pre- and postsynaptic neurons and as such regulate all major brain functions including synaptic plasticity, neuronal network activity, and neuronal development. Accordingly, GABA$_B$ receptors have been implicated in a variety of neurological disorders and thus are regarded as promising drug targets. A key factor determining the extent of GABA$_B$

Advances in Pharmacology, Volume 73
ISSN 1054-3589
http://dx.doi.org/10.1016/bs.apha.2014.11.002

receptor-mediated inhibition is the level of receptors at the cell surface available for signaling. There is increasing evidence that cell surface expression of functional $GABA_B$ receptors is affected in neurological diseases. This diminishes inhibitory control of neuronal excitation and may contribute to the disease state. Here, we discuss recent findings on mechanisms involved in regulating cell surface expression of $GABA_B$ receptors in addiction, neuropathic pain, and brain ischemia.

ABBREVIATIONS

AMPK 5′-adenosine monophosphate-activated protein kinase
CaMKII calcium/calmodulin-dependent kinase II
CNS central nervous system
ER endoplasmic reticulum
ERAD ER-associated protein degradation
ESCRT endosomal sorting complex required for transport
GABA γ-amino butyric acid
KCTD K^+ channel tetramerization domain
PKA cAMP-dependent protein kinase
PKC protein kinase C
PP2A protein phosphatase 2A
sIPSC slow inhibitory postsynaptic currents
VTA ventral tegmental area

1. INTRODUCTION

γ-Amino butyric acid (GABA) mediates slow and prolonged inhibitory neurotransmission via the metabotropic $GABA_B$ receptors. $GABA_B$ receptors were discovered more than three decades ago by their insensitivity to the $GABA_A$ receptor antagonist bicuculline and their sensitivity to the GABA derivative baclofen (β-parachlorophenyl GABA), which in turn does not activate $GABA_A$ receptors (Bowery et al., 1979, 1980).

$GABA_B$ receptors are expressed abundantly throughout the central nervous system (CNS) in a variety of distinct types of neurons (Fritschy et al., 1999, 2004) to control their excitability. Accordingly, they are involved in all main brain functions including synaptic plasticity (for a review, see Pinard, Seddik, & Bettler, 2010), regulation of neuronal network activity (reviewed by Craig & McBain, 2014; Kohl & Paulsen, 2010), and neuronal development (reviewed by Gaiarsa, Kuczewski, & Porcher, 2011; Gaiarsa & Porcher, 2013).

$GABA_B$ receptors are located pre- as well as postsynaptically in both inhibitory and excitatory neurons. They are not directly present in the active zone or postsynaptic density of synapses but rather located at perisynaptic and extrasynaptic sites (Boyes & Bolam, 2003; Chen, Boyes, Yung, & Bolam, 2004; Kulik et al., 2003; Lacey et al., 2005; Lujan, Shigemoto, Kulik, & Juiz, 2004). This localization of $GABA_B$ receptors implies that activation either requires spillover of GABA originating from intense synaptic activity (Scanziani, 2000) or activity-dependent release of GABA from glia cells or neuronal dendrites (Angulo, Le Meur, Kozlov, Charpak, & Audinat, 2008; Velez-Fort, Audinat, & Angulo, 2012; Zilberter, Kaiser, & Sakmann, 1999). Recent measurements at the level of single synapses revealed that even basal neuronal activity can locally generate sufficient ambient GABA concentrations for low-level tonic activation of presynaptic $GABA_B$ receptors at individual synaptic contacts (Laviv et al., 2010).

1.1. Structural organization of $GABA_B$ receptors

$GABA_B$ receptors belong to the class C of G protein-coupled receptors and are obligatory heterodimers assembled from $GABA_{B1}$ and $GABA_{B2}$ subunits (Jones et al., 1998; Kaupmann et al., 1998; White et al., 1998). Structurally, both subunits are composed of a large extracellular domain containing a Venus flytrap-like structure, seven transmembrane domains, and a large intracellular C-terminal domain with a leucine zipper forming a coiled-coil domain (Fig. 1). Two variants of $GABA_{B1}$, $GABA_{B1a}$ and $GABA_{B1b}$, are generated by alternative promoter usage (Steiger, Bandyopadhyay, Farb, & Russek, 2004), which differ solely in their N-terminal domain by the inclusion of two so-called sushi repeats (protein–protein interaction domains) in $GABA_{B1a}$. These domains have been implicated in axonal targeting (Biermann et al., 2010) and increased cell surface stability of receptors containing $GABA_{B1a}$ (Hannan, Wilkins, & Smart, 2012). $GABA_B$ receptor subtypes containing $GABA_{B1a}$ or $GABA_{B1b}$ display distinct localizations in GABAergic and glutamatergic neurons and serve different physiological functions (for a review, see Gassmann & Bettler, 2012).

The binding site for orthosteric ligands is constituted by the Venus flytrap structure of $GABA_{B1}$, while the corresponding structure in $GABA_{B2}$ is inactive (Bernard, Guedin, & Hibert, 2001; Galvez et al., 1999, 2000; Geng, Bush, Mosyak, Wang, & Fan, 2013). Instead, $GABA_{B2}$ contains a binding site for allosteric modulators associated with the transmembrane domain (Binet et al., 2004).

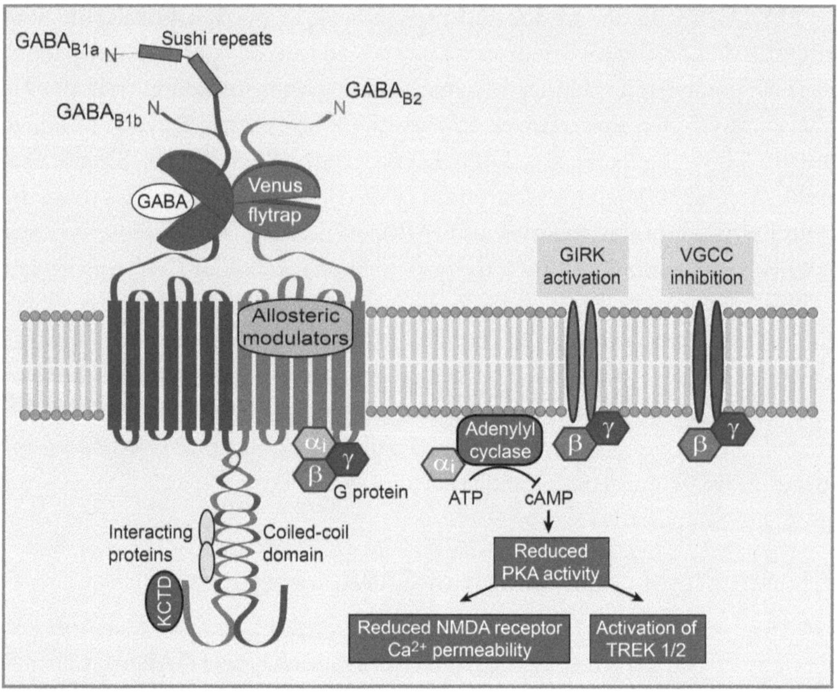

Figure 1 Structural organization of $GABA_B$ receptors and effector systems. Functional $GABA_B$ receptors are composed of the two subunits $GABA_{B1}$ (red (black in the print version)) and $GABA_{B2}$ (blue (dark gray in the print version)). $GABA_{B1}$ exists in two isoforms, $GABA_{B1a}$ and $GABA_{B1b}$, which solely differ in their N-terminal domain ($GABA_{B1a}$ contains two sushi repeats missing in $GABA_{B1}$). The Venus flytrap-like structure in the N-terminal domain of $GABA_{B1}$ constitutes the binding site for orthosteric ligands, while the transmembrane domain of $GABA_{B2}$ harbors a binding site for allosteric ligands. The leucine zipper in the C-terminal domain of $GABA_{B1}$ and $GABA_{B2}$ form a coiled-coil structure to which various $GABA_B$ receptor interacting proteins bind. KCTD proteins constitutively bind to the C terminus of $GABA_{B2}$ and regulate ligand-binding affinity, activation kinetics, desensitization, and cell surface stability of the receptor complex. $GABA_B$ receptors activate $G_{i/o}$ proteins, which in turn activate inwardly rectifying K^+ channels (GIRK) generating sIPSC, inhibit voltage-gated Ca^{2+} channels (VGCC) to reduce transmitter release, and inhibit adenylyl cyclases eventually reducing cAMP-dependent protein kinase (PKA) activity. Inhibition of PKA activates TREK 1/2 channels (two-pore K^+ channels) contributing to membrane hyperpolarization and reduces Ca^{2+} permeability of NMDA receptors.

Heterodimerization of $GABA_{B1}$ and $GABA_{B2}$, involving the interaction between the C-terminal domains of both subunits, inactivates an endoplasmic reticulum (ER) retention motif (RXRR) in the C-terminal domain of $GABA_{B1}$ and is required for ER exit and forward trafficking of the assembled receptor to the cell surface (Calver et al., 2001; Couve et al., 1998; Gassmann et al., 2005; Margeta-Mitrovic, Jan, & Jan, 2000; Pagano et al., 2001). Interestingly, $GABA_B$ receptors leave the ER not only at somatic ER exit sites but also throughout the dendritic compartment, which requires the transport of unassembled $GABA_B$ receptor subunits along the dendritic ER (Ramirez et al., 2009; Valenzuela et al., 2014). A main function of $GABA_{B2}$ within the receptor heterodimer is the interaction with the G protein (Duthey et al., 2002; Galvez et al., 2001; Havlickova et al., 2002; Margeta-Mitrovic, Jan, & Jan, 2001a, 2001b; Robbins et al., 2001).

A number of proteins interact with $GABA_B$ receptors including 14-3-3ζ, ATF4, CHOP, fibulin-2, β-filamin, KCTD8/12/12b/16, marlin-1, MUPP1, *N*-ethylmaleimide-sensitive fusion (NSF), tamalin, and tenascin (for a review and references, see Lujan & Ciruela, 2012). Apart from fibulin-2 and tenascin, which interact with the N-terminal domain of $GABA_{B1}$, all interacting proteins bind to the C-terminal domain of $GABA_{B1}$ and/or $GABA_{B2}$. Although their precise function(s) remain to be characterized, they have been implicated in the regulation of receptors function, receptor trafficking, and anchoring (Lujan & Ciruela, 2012).

Among the $GABA_B$ receptor interacting proteins, the K^+ channel tetramerization domain (KCTD) proteins are of particular importance because they are constitutively bound to the C terminus of $GABA_{B2}$ and can be regarded as an auxiliary part of the functional receptor complex (Schwenk et al., 2010). They accelerate activation kinetics of $GABA_B$ receptor-induced K^+ currents, modulate the affinity of GABA binding to the receptor (Schwenk et al., 2010), and (KCTD12) increase cell surface stability of $GABA_B$ receptors by reducing their rate of constitutive internalization (Ivankova et al., 2013). In addition, the association with KCTD12 as well as KCTD 12b conveys rapid $GABA_B$ receptor desensitization (Schwenk et al., 2010) by uncoupling βγ subunits of the G protein from their effectors (Turecek et al., 2014).

1.2. $GABA_B$ receptor effector systems

$GABA_B$ receptors activate G proteins of the $G_{i/o}$-type (Campbell, Berrow, & Dolphin, 1993; Mannoury la Cour, Herbelles, Pasteau, de Nanteuil, & Millan, 2008; Menon-Johansson, Berrow, & Dolphin, 1993; Morishita, Kato, & Asano, 1990; Nishikawa, Hirouchi, & Kuriyama, 1997; Odagaki & Koyama, 2001). The main effector systems targeted by $GABA_B$ receptors include adenylyl cyclases, K^+ channels, and Ca^{2+} channels (Fig. 1).

1.2.1 Presynaptic effector systems

Activation of $G_{i/o}$ proteins by presynaptic $GABA_B$ receptors inhibits voltage-gated Ca^{2+} channels (P/Q- and N-type Ca^{2+} channels) via the Gβγ dimer by slowing their activation kinetics (Barral, Toro, Galarraga, & Bargas, 2000; Bean, 1989; Bussieres & El Manira, 1999; Chen & van den Pol, 1998; Guyon & Leresche, 1995; Lambert & Wilson, 1996; Mintz & Bean, 1993; Santos, Carvalho, Macedo, & Carvalho, 1995). This is the main mechanism through which $GABA_B$ receptors inhibit neurotransmitter release. In addition, it has been shown that $GABA_B$ receptor activation increase the energy barrier for vesicle fusion in a $G\alpha_{i/o}$-dependent manner (Rost et al., 2011). Furthermore, $GABA_B$ receptors can activate presynaptic inwardly rectifying K^+ channels, which also contribute to the inhibition of transmitter release (Fernandez-Alacid et al., 2009; Ladera et al., 2008; Saint, Thomas, & Gage, 1990).

1.2.2 Postsynaptic effector systems

Postsynaptic $GABA_B$ receptors activate inwardly rectifying K^+ channels (Kir3.1–3.4 also known as GIRK1–4) via Gβγ thereby generating slow and prolonged inhibitory postsynaptic potentials hyperpolarizing the membrane (Andrade, Malenka, & Nicoll, 1986; Luscher, Jan, Stoffel, Malenka, & Nicoll, 1997). In addition, released Gβγ inhibits postsynaptic voltage-gated Ca^{2+} channels in dendrites and spines, which prevents Ca^{2+} spikes and thus further reduces neuronal excitability (Chalifoux & Carter, 2011b; Kavalali, Zhuo, Bito, & Tsien, 1997; Pérez-Garci, Gassmann, Bettler, & Larkum, 2006; Sabatini & Svoboda, 2000).

The activated $G\alpha_{i/o}$ subunit of the G protein inhibits adenylyl cyclases, resulting in reduced cAMP-dependent protein kinase (PKA) activity. This potentially affects a variety of downstream signaling pathway. Two recently described targets are two-pore K^+ channels (TREK 1/2) and NMDA receptors. The $GABA_B$ receptor-mediated reduction of PKA activity activates two-pore K^+ channels (TREK 1/2) via releasing tonic inhibition of the channels by PKA (Deng et al., 2009; Sandoz, Levitz, Kramer, & Isacoff, 2012). These channels have been shown to contribute to $GABA_B$ receptor-mediated inhibition in neurons of the hippocampus and entorhinal cortex. Finally, $GABA_B$ receptor-mediated inhibition of PKA controls Ca^{2+} permeability of NMDA receptors in spines (Chalifoux & Carter, 2010), thereby controlling NMDA receptor-mediated synaptic plasticity.

2. PHOSPHORYLATION OF $GABA_B$ RECEPTORS

Protein phosphorylation is one major posttranslational modification that regulates the physiology of G protein-coupled receptors including desensitization, endocytosis, recycling, and degradation. Four major serine/threonine protein kinases (PKA, PKC, CaMKII, and AMPK) have been shown to phosphorylate $GABA_B$ receptors at distinct sites to mainly affect cell surface stability and desensitization of the receptors (Fig. 2).

2.1. cAMP-dependent protein kinase

PKA was the first protein kinase shown to directly phosphorylate $GABA_B$ receptors. It phosphorylates $GABA_{B2}$ at serine 892, which reduces

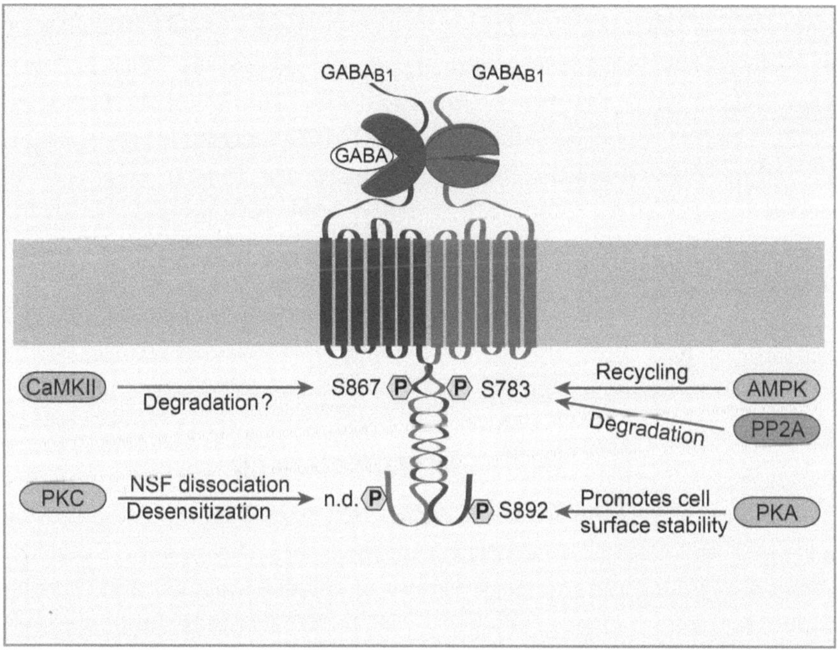

Figure 2 Regulation of $GABA_B$ receptors by phosphorylation. Phosphorylation of serine 867 in the C-terminal domain of $GABA_{B1}$ by CaMKII is involved in the downregulation of the receptors under pathological conditions. PKC-mediated phosphorylation of a so far unidentified site in $GABA_{B1}$ promotes desensitization of the receptors. Phosphorylation of serine 892 in $GABA_{B2}$ reduces desensitization and stabilizes the receptor in the plasma membrane. AMPK phosphorylates $GABA_{B2}$ serine 783, which increases cell surface stability of the receptors most likely by promoting receptor recycling. Dephosphorylation of this site by PP2A directs the receptors to lysosomes for degradation.

desensitization of $GABA_B$ receptor-activated K^+ currents by stabilizing the receptors at the cell surface (Couve et al., 2002). In agreement with these findings, chronic treatment of cultured neurons with baclofen, which reduces the cAMP levels, attenuates PKA-mediated phosphorylation of $GABA_{B2}$ and $GABA_B$ receptor cell surface stability (Couve et al., 2002; Fairfax et al., 2004). PKA phosphorylation appears to be important under normal physiological conditions because $GABA_B$ receptors display a considerable level of basal $GABA_{B2}$ phosphorylation at serine 892 in neurons (Couve et al., 2002). It is however currently not delineated which $G\alpha_s$/$G\alpha_i$-coupled receptors modulate PKA-mediated $GABA_B$ receptor phosphorylation *in vivo*.

2.2. Protein kinase C

Activation of protein kinase C (PKC) decreases $GABA_B$ receptor-mediated inhibition at pre- and postsynaptic sites (Taniyama, Niwa, Kataoka, & Yamashita, 1992; Taniyama, Takeda, Ando, Kuno, & Tanaka, 1991; Taniyama, Takeda, Ando, & Tanaka, 1991; Thompson & Gahwiler, 1992). The PKC-mediated attenuation of $GABA_B$ receptor function is most likely caused by desensitization of the receptors. Pontier et al. (2006) reported that $GABA_B$ receptors interact with the NSF protein, which is a prerequisite for PKC-mediated $GABA_B$ receptor desensitization. Upon chronic activation with baclofen, PKC phosphorylates $GABA_{B1}$ in the C-terminal domain at a so far unidentified site(s), thereby inducing the dissociation of NSF from the receptor complex and desensitization of the receptors (Pontier et al., 2006).

2.3. Calcium/calmodulin-dependent kinase II

Calcium/calmodulin-dependent kinase II (CaMKII) is ubiquitously expressed and highly concentrated in the CNS (Erondu & Kennedy, 1985). It is implicated in a large variety of cellular processes and plays an important role synaptic plasticity, such as learning and memory (Yamauchi, 2005).

CaMKII associates with the C-terminal domain of $GABA_{B1}$ and phosphorylates serine 867 in $GABA_{B1}$ upon sustained activation of NMDA receptors (Guetg et al., 2010). NMDA receptor-induced CaMKII phosphorylation of $GABA_{B1}$ plays a central role in the downregulation of $GABA_B$ receptors upon overexcitation of neurons (Guetg et al., 2010), which most likely occurs in ischemia (see Section 4.3). The precise mechanism as to how CaMKII-mediated phosphorylation of $GABA_{B1}$ triggers downregulation of $GABA_B$ receptors is not entirely clear yet but appears to affect trafficking and sorting of $GABA_B$ receptors.

2.4. 5′-Adenosine monophosphate-activated protein kinase

As CaMKII, 5′-adenosine monophosphate-activated protein kinase (AMPK) is highly expressed in the CNS (Turnley et al., 1999). AMPK is activated under conditions of energy deficiency and metabolic stress to restore the energy balance (Ronnett, Ramamurthy, Kleman, Landree, & Aja, 2009).

AMPK had been identified as $GABA_B$ receptor interacting protein using the *Yeast* two-hybrid assay (Kuramoto et al., 2007). AMPK binds to the C-terminal domain of $GABA_{B1}$ and phosphorylates serine 917 and 923 in $GABA_{B1}$*in vitro* as well as serine 783 in $GABA_{B2}$. While the occurrence and relevance of AMPK phosphorylation of $GABA_{B1}$ in neurons have not been established yet, phosphorylation of serine 783 in $GABA_{B2}$ stabilizes $GABA_B$ receptors at the cell surface and thus baclofen-induced K^+ currents (Kuramoto et al., 2007). The mechanism that stabilizes $GABA_B$ receptors in the plasma membrane is currently unknown but most likely involves trafficking of the receptors. It may involve reducing constitutive endocytosis or increasing the insertion rate of receptors from the trans-Golgi network or from recycling endosomes.

Interestingly, in the middle cerebral artery occlusion *in vivo* model of ischemia, which is characterized by metabolic derailment and activation of AMPK, increased phosphorylation of $GABA_{B2}$ serine 783 was observed and in an *in vitro* model of anoxia phosphorylation of $GABA_{B2}$ serine 783 promoted survival of neurons (Kuramoto et al., 2007). This suggests that stabilization of $GABA_B$ receptors at the cell surface by AMPK phosphorylation is a protective response of the neuron to counteract excitotoxicity.

3. DEGRADATION OF $GABA_B$ RECEPTORS

Because precise regulation of protein degradation is fundamental for most cellular processes, eukaryotic cells have evolved complex machineries for this task (Ciechanover, 2006). Proteins are destructed via two main pathways: lysosomes and proteasomes. Lysosomes are specialized vesicular organelles containing proteases, while proteasomes are large protein complexes with proteolytic activity. Proteins residing in the plasma membrane are predominantly degraded in lysosomes and most intracellularly located membrane proteins as well as soluble proteins are destructed by proteasomes. In either case, ubiquitination is involved in targeting the proteins to the respective degradation pathway (Grabbe, Husnjak, & Dikic, 2011; Weissman, Shabek, & Ciechanover, 2011).

$GABA_B$ receptors endocytosed from the plasma membrane are either recycled or sorted to lysosomes for degradation (Fig. 3; Grampp, Notz, Broll, Fischer, & Benke, 2008; Grampp, Sauter, Markovic, & Benke, 2007; Laffray et al., 2007; Vargas et al., 2008; Wilkins, Li, & Smart, 2008). The recruitment of receptors to the lysosomal degradation pathway presumably involves the ubiquitin-dependent endosomal sorting complex required for transport (ESCRT) machinery as knockdown of the tumor susceptibility gene 101 protein (TSG101), a component of ESCRT, inhibited degradation of the receptors (Kantamneni et al., 2008). Sorting of $GABA_B$ receptors to recycling endosomes and lysosomes is precisely controlled and regulated by the activity of glutamate receptors. Sustained activation of glutamate receptors, a condition that occurs in cerebral ischemia, rapidly

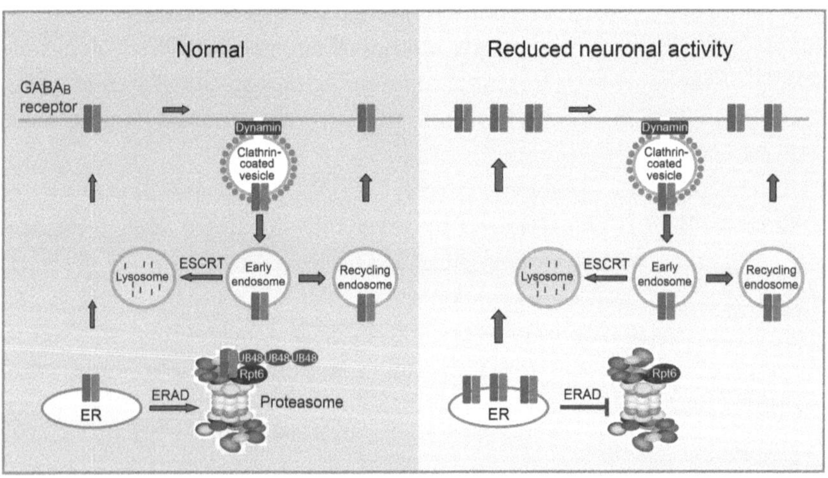

Figure 3 Degradation pathways of $GABA_B$ receptors. $GABA_B$ receptors are constitutively internalized from the plasma membrane via the classical clathrin- and dynamin-dependent pathway. Most of the receptors are recycled to the cell surface, whereas a smaller fraction is sorted via the ESCRT machinery to lysosomes for degradation. Newly synthetized $GABA_B$ receptors in the ER are constitutively degraded by proteasomes via ERAD involving K48-linked ubiquitination (UB48) of lysines 761/771 in $GABA_{B2}$ and interaction of the $GABA_{B2}$ C terminus with the proteasomal AAA-ATPase Rpt6. Chronic reduction of neuronal activity inhibits ERAD-dependent degradation of $GABA_B$ receptors. This upregulates cell surface receptors due to increased levels of receptors in the ER and their export to the plasma membrane.

downregulates $GABA_B$ receptors by shifting the recycling/degradation balance toward lysosomal degradation of the receptors (Guetg et al., 2010; Kantamneni et al., 2014; Maier, Marin, Grampp, Sommer, & Benke, 2010; Terunuma et al., 2010).

In contrast to $GABA_B$ receptors endocytosed from the plasma membrane, the level of receptors in the ER is regulated by proteasomal degradation via the ER-associated protein degradation (ERAD) machinery (Fig. 3; Zemoura & Benke, 2014; Zemoura et al., 2013). Proteasomal degradation of $GABA_B$ receptors requires K48-linked ubiquitination of lysine residues 767/771 in the C-terminal domain of $GABA_{B2}$ and the interaction of the $GABA_{B2}$ C terminus with the proteasomal AAA-ATPase Rpt6. Rpt6 is one of the six AAA-ATPases of the 19S regulatory particle of the proteasome involved in recruiting and unfolding K48-linked ubiquitinated proteins for degradation in the 20S core complex (Rubin et al., 1996). In addition, $GABA_B$ receptors interact with central components of the ERAD machinery: Hrd1 and p97 (Zemoura et al., 2013). Hrd1 is a prototypical ERAD E3 ligase (Smith, Ploegh, & Weissman, 2011) and p97 is believed to mediate retrotranslocation of membrane proteins from the ER membrane to the cytoplasm where proteasomal degradation takes place (Wang, Song, & Li, 2004). Interestingly, the ERAD machinery appears to be directly involved in adjusting the number of cell surface $GABA_B$ receptors according to the activity level of the neurons. Inhibition of the ERAD pathway increased both total and cell surface $GABA_B$ receptor levels, demonstrating that the number of functional receptors is constitutively regulated by ERAD (Zemoura et al., 2013). Moreover, persistent enhancement or depression of neuronal activity decreased and increased, respectively, cell surface $GABA_B$ receptor expression (Zemoura & Benke, 2014). This suggests that neuronal activity controls the number of $GABA_B$ receptors in the ER competent for forward trafficking to the plasma membrane by proteasomal degradation via ERAD. Because this appears to be a relatively slow process, it may play an important role in homeostatic synaptic plasticity.

4. CONTRIBUTION OF ALTERED CELL SURFACE $GABA_B$ RECEPTOR EXPRESSION TO NEUROLOGICAL DISEASES

Excitatory/inhibitory imbalances are frequently associated with neurological disorders. Because $GABA_B$ receptors control the excitability of neurons, it is not surprising that they have been implicated in a variety of neurological diseases. These include addiction, anxiety, autism, brain ischemia, depression, epilepsy, neuropathic pain, and schizophrenia (reviewed in Benarroch, 2012; Bowery, 2006; Cryan & Kaupmann, 2005; Enna & Bowery, 2004; Kumar, Sharma, Kumar, & Deshmukh, 2013). In all these cases, pharmacological manipulation of $GABA_B$ receptor activity exerts beneficial therapeutic effects in animal models of the diseases. This suggests that either $GABA_B$ receptor activity modulates neuronal circuitries in a way that compensates for deficits resulting from the disease state or normal $GABA_B$ receptor function is compromised. Currently, there is only limited information on how $GABA_B$ receptors are affected in neurological diseases. In the following, we discuss potential mechanisms compromising the expression of functional cell surface $GABA_B$ receptors and thus diminish their signaling in drug addiction, neuropathic pain, and cerebral ischemia.

4.1. Drug addiction

There is ample evidence that sustained activation of $GABA_B$ receptors in experimental animal models of addiction as well as in humans reduces the reinforcing properties of drugs of abuse, such as cocaine, heroin, methamphetamine as well as nicotine, and diminishes withdrawal symptoms (for a review and references, see Filip & Frankowska, 2008). The mesocorticolimbic dopaminergic pathway (including the ventral tegmental area (VTA), nucleus accumbens, amygdala, hippocampus, and medial prefrontal cortex) is a main reward circuit involved in the development of addiction (Pierce & Kumaresan, 2006). All drugs of abuse affect the VTA to increase dopamine levels in the mesocorticolimbic system via a variety of distinct mechanisms (reviewed by Creed & Luscher, 2013). For instance, cocaine inhibits dopamine transporters while amphetamines deplete dopamine from synaptic vesicles and induce the release of dopamine by reverse transport.

$GABA_B$ receptors are well expressed in the VTA and control the excitability of dopaminergic neurons (Cruz et al., 2004; Johnson & North, 1992). Systemic administration of baclofen as well as injection of baclofen directly into the VTA diminishes the drug-induced increase in dopamine levels in the nucleus accumbens (Brebner, Ahn, & Phillips, 2005; Fadda, Scherma, Fresu, Collu and Fratta, 2003), confirming $GABA_B$ receptor agonists as promising drugs for the treatment of addiction.

Addictive drugs induce plastic neuronal changes resulting in enhanced excitability of mesolimbic dopamine neurons (Creed & Luscher, 2013). Already a single dose of cocaine or amphetamine results in long-lasting synaptic strengthening by recruitment of AMPA receptors, and repeated cocaine administration diminishes $GABA_A$ receptor-mediated inhibition (Kauer & Malenka, 2007).

There are recent data showing that also pre- (inhibition of GABA release) and postsynaptic $GABA_B$ receptor signaling (activation of GIRK channels) undergoes long-term changes after psychostimulant administration (Padgett et al., 2012). A single injection of methamphetamine or cocaine strongly reduced $GABA_B$ receptor signaling for 7 days. $GABA_B$ receptor-mediated slow inhibitory postsynaptic currents (sIPSC) were selectively impaired in GABAergic neurons, but not in dopaminergic neurons, of the VTA. Quantitative immunoelectron microscopy indicated that methamphetamine treatment induced the internalization of plasma membrane $GABA_B$ receptors together with their effectors, the GIRK channels (Padgett et al., 2012). The mechanism behind the methamphetamine-induced downregulation of cell surface $GABA_B$ receptors involves a reduced phosphorylation of serine 783 in $GABA_{B2}$ (Fig. 4). Phosphorylation of $GABA_{B2}$ by AMPK increases cell surface stability of $GABA_B$ receptors (Kuramoto et al., 2007), most likely by sorting $GABA_B$ receptors to the recycling pathway, whereas its dephosphorylation by protein phosphatase 2A (PP2A) impairs recycling and thus downregulates the receptors from

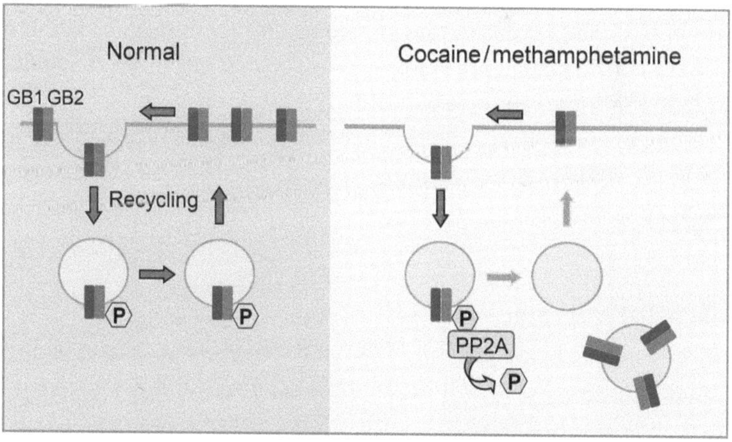

Figure 4 Cocaine/methamphetamine-induced downregulation of $GABA_B$ receptors. Cocaine or methamphetamine administration selectively reduces cell surface receptors in GABAergic neurons of the VTA and layer 5/6 pyramidal neurons of the medial prefrontal cortex. The mechanism involves dephosphorylation of $GABA_{B2}$ (GB2) at serine 783, which appears to prevent recycling of the receptors and leads to their intracellular accumulation.

the plasma membrane (Terunuma et al., 2010). Accordingly, blocking PP2A by injection of okadaic acid in methamphetamine-treated mice restored $GABA_B$ receptor signaling (Padgett et al., 2012). These findings are consistent with a model in which the psychostimulants cocaine and methamphetamine trigger, by a so far unknown mechanism, the dephosphorylation of serine 783 in $GABA_{B2}$. This most likely inhibits recycling of the receptors resulting in the downregulation of cell surface receptors and consequently to diminished $GABA_B$ receptors-mediated inhibition (Fig. 4).

The physiological consequence of the methamphetamine- and cocaine-mediated downregulation of $GABA_B$ receptors in VTA GABAergic neurons is an increased firing rate of these neurons (Padgett et al., 2012). This enhanced activity of GABAergic neurons is expected to increase the inhibition of dopaminergic neurons in the VTA and thus to counteract their psychostimulant-induced hyperactivity. In this regard, this mechanism appears to be an insufficient homeostatic response trying to prevent the hyperactivity of dopaminergic neurons after psychostimulant treatment.

Similar to this mechanism in VTA neurons, repeated cocaine administration downregulates $GABA_B$ receptors in layer 5/6 pyramidal neurons of the medial prefrontal cortex (Hearing et al., 2013). As in the VTA, cocaine administration selectively downregulates cell surface $GABA_B$ receptors in layer 5/6 pyramidal neurons triggered by the dephosphorylation of serine 783 in $GABA_{B2}$. This enhanced the excitability of layer 5/6 pyramidal neurons and glutamatergic output of the prefrontal cortex (Hearing et al., 2013). Because the medial prefrontal cortex projects to the VTA and nucleus accumbens, this increases the excitation of their neurons, which is a prerequisite for the expression of addictive behaviors. Via this mechanism, downregulation of $GABA_B$ receptors appears to contribute to the development of addiction.

4.2. Neuropathic pain

Neuropathic pain originates from injury of the nervous system as well as diseases, such as cancer, diabetes, or multiple sclerosis, and is associated with impaired inhibition of spinal cord dorsal horn neurons (Zeilhofer, Benke, & Yevenes, 2012). $GABA_B$ receptors are abundantly expressed in primary afferent terminals and neurons in the superficial layers of the dorsal horn (Ataka, Kumamoto, Shimoji, & Yoshimura, 2000; Price, Wilkin, Turnbull, & Bowery, 1984; Towers et al., 2000; Yang, Ma, Feng, Dong, & Li, 2002), which is the first relay station for nociceptive signals.

$GABA_B$ receptor-mediated inhibition appears to be involved in controlling pain signals as indicated by the expression of hyperalgesia in mutant mice lacking functional $GABA_B$ receptors (Gassmann et al., 2004; Schuler et al., 2001) and by the analgesic activity of the $GABA_B$ receptor agonist baclofen (Enna & McCarson, 2006). Thus, it is tempting to speculate that downregulation of $GABA_B$ receptor expression might contribute to neuropathic pain. However, there is no clear-cut indication for this hypothesis. The current data range from no change in $GABA_B$ receptor expression in the spinal cord (see, e.g., Engle, Gassman, Sykes, Bettler, & Hammond, 2006) to reduced expression in dorsal root ganglia (Engle, Merrill, Marquez De Prado, & Hammond, 2012) and spinal cord dorsal horn (Wang et al., 2011; Wu, Xu, Pu, Jiang, & Du, 2011). The contradictory findings may result from the different experimental conditions and animal model used. They, however, clearly indicate that global downregulation of $GABA_B$ receptor expression is not a general phenomenon associated with neuropathic pain.

Apart from regulating global $GABA_B$ receptor expression, the number or function of receptors selectively at the cell surface might be affected in neuropathic pain. A recent study favors this scenario by showing that after inducing neuropathic pain in rats by spinal nerve ligation $GABA_B$ receptor heterodimers ($GABA_{B1}/GABA_{B2}$) are selectively disrupted at the cell surface and thereby inactivated (Laffray et al., 2012). Disruption of $GABA_{B1}/GABA_{B2}$ heterodimers is mediated by 14-3-3ζ, which is upregulated in neurons of the superficial layers of the affected dorsal horn (Fig. 5). 14-3-3 Proteins are a family of abundant cytoplasmic proteins regulating numerous cellular processes via protein–protein interactions (Zhao, Meyerkord, Du, Khuri, & Fu, 2011). 14-3-3ζ interacts with $GABA_B$ receptors via a site in the C-terminal domain of $GABA_{B1}$ (Couve et al., 2001). It is currently unknown by which mechanism 14-3-3ζ dissociates an already existing thermodynamically stable $GABA_B$ receptor complex. A possible mechanism may involve posttranslational modification, such as phosphorylation, of the receptor at a site that functions as a switch for dissociation of the heterodimer or, alternatively, the 14-3-3ζ binding site itself is the switch. In any case, the C-terminal domain appears to contain a molecular structure that, if activated, triggers the dissociation of the receptor complex, thereby regulating the number of functional receptors at the cell surface.

Most interestingly, interfering with the 14-3-3ζ/$GABA_B$ receptor interaction using a small synthetic peptide mimicking the $GABA_{B1}$-binding site restored expression of functional $GABA_B$ receptor heterodimers in the

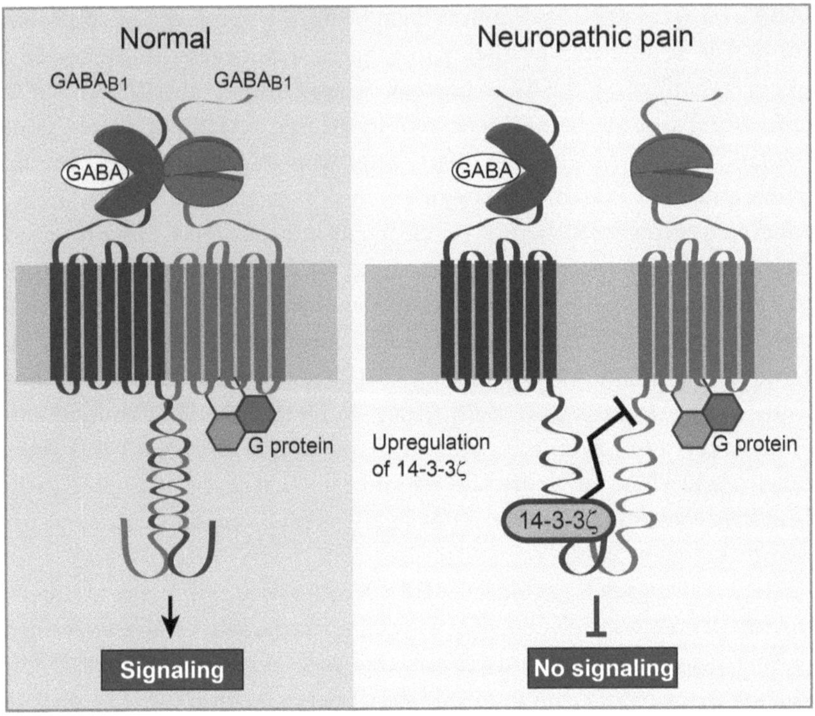

Figure 5 Neuropathic pain-induced disruption of $GABA_B$ receptor heterodimers. Neuropathic pain upregulates 14-3-3ζ in the affected dorsal horn of the spinal cord where it interacts with the C-terminal domain of $GABA_{B1}$ at the plasma membrane. Binding of 14-3-3ζ disrupts the receptor dimer. This functionally inactivates the receptor and leads to diminished $GABA_B$ receptor-mediated inhibition.

plasma membrane of neuropathic rats, enhanced the analgesic activity of baclofen, and was even able to partially relieve the pain in the absence of baclofen (Laffray et al., 2012). This suggests that (1) impaired $GABA_B$ receptor signaling contributes to the expression of neuropathic pain and (2) restoration of normal levels of functional cell surface receptors might be a strategy for the development of a novel, highly selective, therapeutic treatment to relieve neuropathic pain.

4.3. Brain ischemia

In brain ischemia, reduced or blocked blood circulation deprives neurons from oxygen and glucose. This induces numerous devastating cellular reactions eventually leading to neuronal death. One characteristic of ischemia is the excessive release of glutamate in the affected brain area, which overstimulates glutamate receptors and triggers apoptotic death of neurons (Kostandy, 2012; Lipton, 1999). Because $GABA_B$ receptors control presynaptic glutamate release and the activity of postsynaptic glutamate receptors (Chalifoux & Carter, 2011a), one would expect that the increased neuronal excitation during the ischemic insult in turn leads to enhanced activity of $GABA_B$ receptors to counteract the overexcitation. This is not the case and raises the question whether $GABA_B$ receptors *per se* are not able to control the massive excitation during the ischemic insult or whether $GABA_B$ receptor-mediated inhibition is also compromised by the insult. There are solid data from experimental models of cerebral ischemia that sustained activation of $GABA_B$ receptors during and immediately after the ischemic insult reduces neuronal death (Babcock, Everingham, Paden, & Kimura, 2002; Cimarosti, Kantamneni, & Henley, 2009; Dave et al., 2005; Han et al., 2008; Jackson-Friedman, Lyden, Nunez, Jin, & Zweifler, 1997; Kulinskii & Mikhel'son, 2000; Lal, Shuaib, & Ijaz, 1995; Xu, Li, Yin, & Zhang, 2008; Zhang et al., 2007). This documents that $GABA_B$ receptors have the capacity to counteract excitotoxicity at least to a certain extent. However, there are also data indicating that $GABA_B$ receptors are downregulated in ischemia (Cimarosti et al., 2009; Kim, Kim, Yenari, & Chang, 2011; Li et al., 2014; Vollenweider, Bendfeldt, Maetzler, Otten, & Nitsch, 2006). This suggests that impaired $GABA_B$ receptor-mediated inhibition under ischemic conditions is not able to prevent excitotoxicity.

Two recently discovered mechanisms may cause a downregulation of cell surface $GABA_B$ receptors in brain ischemia. In the first mechanism, application of glutamate or NMDA to cultured neurons induced the rapid removal of $GABA_B$ receptors from the cell surface (Fig. 6; Guetg et al., 2010; Kantamneni et al., 2014; Maier et al., 2010; Terunuma et al., 2010). Downregulation of $GABA_B$ receptors was based on the activation of NMDA receptors and Ca^{2+} influx via L-type voltage-gated Ca^{2+} channels. This triggered phosphorylation of $GABA_B$ receptors by CaMKII (Guetg et al., 2010) as well as AMPK and dephosphorylation by PP2A (Terunuma et al., 2010). The phosphorylation and dephosphorylation appear to represent sorting signals for lysosomal degradation of the receptors (Fig. 6). CaMKII phosphorylates $GABA_{B1}$ on Ser867 (Guetg et al., 2010), whereas AMPK initially phosphorylates serine 783 in $GABA_{B2}$, which is then followed by dephosphorylation via PP2A (Terunuma et al., 2010). Both events are essential steps in this mechanism since preventing phosphorylation of $GABA_{B1}$ by

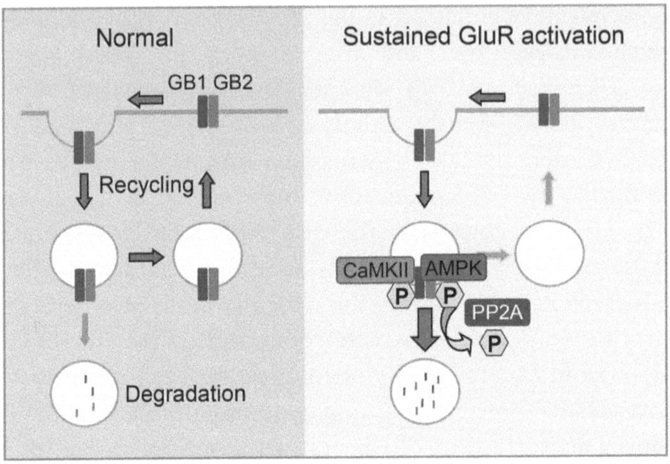

Figure 6 Sustained activation of glutamate receptors downregulates $GABA_B$ receptors. Under normal conditions, $GABA_B$ receptors are constitutively endocytosed and recycled to the plasma membrane with a minor fraction sorted to lysosomes for degradation. After prolonged activation of glutamate receptors, which occurs under ischemic conditions, $GABA_{B1}$ (GB1) is phosphorylated by CaMKII at serine 867 and $GABA_{B2}$ (GB2) by AMPK at serine 783 followed by dephosphorylation via PP2A. These events trigger sorting of internalized to lysosomal degradation instead to the recycling pathway. This leads to downregulation of cell surface receptors and consequently to reduced $GABA_B$ receptor-mediated inhibition.

CaMKII or dephosphorylation of $GABA_{B2}$ by PP2A abolished downregulation of $GABA_B$ receptors. Interestingly, the initial AMPK-mediated phosphorylation of $GABA_{B2}$ at serine 783 appears to be a compensatory but unsuccessful reaction of the neuron to prevent downregulation of the receptors because it stabilizes functional $GABA_B$ receptors in the plasma membrane (Kuramoto et al., 2007), probably by serving as a sorting signal for recycling. The subsequent dephosphorylation of this residue overrides this effect.

A second putative mechanism that downregulates $GABA_B$ receptors under ischemic conditions affects not the recycling/lysosomal degradation pathway but the trafficking of the receptors from the ER to the plasma membrane. One of the important downstream effects of ischemic insults is ER stress and upregulation of the transcription factor CHOP (Kim, Xu, & Reed, 2008). CHOP is only marginally expressed under normal physiological conditions but highly upregulated in response to ER stress and is involved in ischemia-induced apoptosis (Tajiri et al., 2004). In addition to its function as transcription factor, CHOP downregulates cell surface $GABA_B$ receptors by direct interaction with the receptor (Sauter et al., 2005).

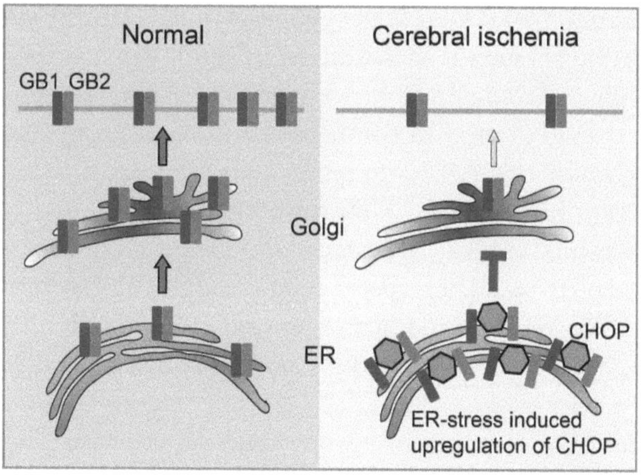

Figure 7 CHOP-induced downregulation of cell surface $GABA_B$ receptors under ischemic conditions. ER stress, which occurs under ischemic conditions, upregulates the transcription factor CHOP, which interacts with $GABA_B$ receptors in the ER. Binding of CHOP to $GABA_B$ receptors interferes with receptor heterodimerization and prevents ER exit and forward trafficking of the receptors to the plasma membrane. This results in accumulation of the receptors in the ER and depletion of receptor from the plasma membrane over time because of ongoing endocytosis and lysosomal degradation of the receptors. GB1: $GABA_{B1}$ and GB2: $GABA_{B2}$.

CHOP binds via its C-terminal leucine zipper specifically to the leucine zipper present in the C-terminal domain of $GABA_{B2}$ and with its N-terminal domain to a so far unidentified site in $GABA_{B1}$. CHOP interacts with $GABA_B$ receptors in the ER, interferes with heterodimerization of the receptors and thereby prevents their ER exit and forward trafficking to the plasma membrane (Fig. 7; Maier et al., 2014). This leads to the intracellular accumulation of the receptors, their reduced expression in the plasma membrane and accordingly to impaired $GABA_B$ receptor-mediated neuronal inhibition. Because this mechanism is operative in the oxygen and glucose deprivation *in vitro* model of cerebral ischemia (Maier et al., 2014), it most likely contributes to excitotoxicity under ischemic conditions *in vivo*.

5. POTENTIAL THERAPEUTIC IMPLICATIONS

Recent work revealed that cell surface availability and function of $GABA_B$ receptors is regulated by various factors including posttranslational modifications (e.g., phosphorylation and ubiquitination) as well as receptor-interacting proteins (e.g., 14-3-3 and CHOP). In particular, downregulation of $GABA_B$ receptors appears to be a contributing factor to neurological diseases such as addiction, neuropathic pain, and cerebral ischemia. Thus, pharmacological enhancement of $GABA_B$ receptor-mediated inhibition is a valid strategy to compensate for this at least partially. In fact, sustained activation of $GABA_B$ receptors with the agonist baclofen has beneficial effects in these diseases. Unfortunately, baclofen—the only drug in the clinic targeting $GABA_B$ receptors—exhibits considerable side effects, which excludes its broad therapeutic application. The recently developed allosteric $GABA_B$ receptor modulators are expected to minimize this problem (Froestl, 2010). However, they are not yet available for therapeutic use in humans.

The general disadvantage of ligands targeting orthosteric and allosteric binding sites of an abundantly expressed receptor involved in a variety of brain functions is that the drugs will address all receptors irrespective of whether they contribute to the disease or are involved in their normal physiological processes in healthy cells. Thus, it is unlikely to avoid unwanted effects with drugs directly targeting ligand-binding sites of such receptors. In an ideal situation, drugs should target only those receptors involved in the disease state and leave those receptors in healthy cells unaffected. In this regard, the mechanisms of $GABA_B$ receptor downregulation in neuropathic pain and cerebral ischemia discussed above provide potential starting points for the development of highly specific therapeutic treatments based on restoring impaired $GABA_B$ receptor expression and function selectively in the diseased neurons by interfering with receptor–protein interactions. Such an approach would target only those receptors involved in the pathological state. In the case of neuropathic pain, interfering with the 14-3-3ζ/$GABA_B$ receptor interaction using a synthetic peptide has already shown to restore normal levels of $GABA_B$ receptors and to partially relieve pain in experimental animals (Laffray et al., 2012). Because under normal physiological conditions $GABA_B$ receptors do not interact with 14-3-3ζ and 14-3-3ζ is selectively upregulated in the affected dorsal spinal neurons only those receptors involved in the disease state are targeted by the interfering peptide.

In case of $GABA_B$ receptor downregulation under ischemic conditions three protein–protein interaction targets can be envisioned: $GABA_{B1}$/CaMKII, $GABA_{B2}$/PP2A, and $GABA_{B2}$/CHOP. However, in these cases it needs to be shown that disruption of these $GABA_B$ receptor interactions indeed restore normal cell surface levels of $GABA_B$ receptors and limits excitotoxicity.

The general feasibility of such an approach as therapeutic intervention has been demonstrated for the NMDA receptor–PSD95 interaction to reduce neuronal cell death in cerebral ischemia using a cell-penetrating synthetic peptide in rats as well as nonhuman primates (Aarts et al., 2002; Cook, Teves, & Tymianski, 2012) and in humans in a phase II clinical trial (Hill et al., 2012).

6. CONCLUSION

$GABA_B$ receptors control the excitability of both inhibitory and excitatory neurons and thus have been implicated in virtually all main brain functions and in various neurological diseases. The maximum possible signaling strength via $GABA_B$ receptors depends on their availability in the plasma membrane, which in turn is largely determined by the balance of trafficking and degradation mechanisms. Currently, we are only at the very beginning to understand the rather complex mechanisms that determine the expression level of $GABA_B$ receptors at the cell surface under physiological and pathological conditions. We expect that unraveling these mechanisms provide the basis for the development of highly specific interventions in diseases where downregulation of $GABA_B$ receptor cell surface expression and the associated loss of inhibitory control is a contributing factor.

Changes in the rates of trafficking and degradation allow the neuron to adjust the number of cell surface $GABA_B$ receptors available for signaling to altered conditions. However, perturbation of this balance in diseases (as discussed for ischemic conditions) can lead to rapid downregulation of $GABA_B$ receptors fostering neuronal overexcitation as a result of diminished inhibitory control of excitatory neurotransmission. In addition, selective upregulation of $GABA_B$ receptor interacting proteins under disease conditions can disrupt the heterodimer at the plasma membrane and thereby inactivate the receptors ($GABA_{B1}$/14-3-3 interaction in neuropathic pain) or prevent/disrupt the heterodimerization of the receptors in the ER, which interfere with forward trafficking of the receptors to the plasma membrane (CHOP/$GABA_B$ receptor interaction under ischemic conditions). The elucidation of the precise mechanisms underlying the downregulation of $GABA_B$ receptors under pathological conditions provides the opportunity to define starting points for the development of therapeutic interventions, which specifically target the receptors in the diseased neurons. Application of small interfering synthetic peptides that disrupt specific diseases-induced protein–protein interactions causing the deregulation of $GABA_B$ receptors should restore normal receptor expression levels and regain normal inhibitory control. This is expected to exert beneficial effects on the course of the diseases.

CONFLICT OF INTEREST

The authors have no conflicts of interest to declare.

REFERENCES

Aarts, M., Liu, Y., Liu, L., Besshoh, S., Arundine, M., Gurd, J. W., et al. (2002). Treatment of ischemic brain damage by perturbing NMDA receptor–PSD-95 protein interactions. *Science*, *298*, 846–850.

Andrade, R., Malenka, R. C., & Nicoll, R. A. (1986). A G protein couples serotonin and $GABA_B$ receptors to the same channels in hippocampus. *Science*, *234*, 1261–1265.

Angulo, M. C., Le Meur, K., Kozlov, A. S., Charpak, S., & Audinat, E. (2008). GABA, a forgotten gliotransmitter. *Progress in Neurobiology*, *86*, 297–303.

Ataka, T., Kumamoto, E., Shimoji, K., & Yoshimura, M. (2000). Baclofen inhibits more effectively C-afferent than Aδ-afferent glutamatergic transmission in substantia gelatinosa neurons of adult rat spinal cord slices. *Pain*, *86*, 273–282.

Babcock, A. M., Everingham, A., Paden, C. M., & Kimura, M. (2002). Baclofen is neuroprotective and prevents loss of calcium/calmodulin-dependent protein kinase II immunoreactivity in the ischemic gerbil hippocampus. *Journal of Neuroscience Research*, *67*, 804–811.

Barral, J., Toro, S., Galarraga, E., & Bargas, J. (2000). GABAergic presynaptic inhibition of rat neostriatal afferents is mediated by Q-type Ca^{2+} channels. *Neuroscience Letters*, *283*, 33–36.

Bean, B. P. (1989). Neurotransmitter inhibition of neuronal calcium currents by changes in channel voltage dependence. *Nature*, *340*, 153–156.

Benarroch, E. E. (2012). $GABA_B$ receptors: Structure, functions, and clinical implications. *Neurology*, *78*, 578–584.

Bernard, P., Guedin, D., & Hibert, M. (2001). Molecular modeling of the GABA/$GABA_B$ receptor complex. *Journal of Medicinal Chemistry*, *44*, 27–35.

Biermann, B., Ivankova-Susankova, K., Bradaia, A., Abdel Aziz, S., Besseyrias, V., Kapfhammer, J. P., et al. (2010). The Sushi domains of $GABA_B$ receptors function as axonal targeting signals. *The Journal of Neuroscience*, *30*, 1385–1394.

Binet, V., Brajon, C., Le Corre, L., Acher, F., Pin, J. P., & Prezeau, L. (2004). The heptahelical domain of $GABA_{B2}$ is activated directly by CGP7930, a positive allosteric modulator of the $GABA_B$ receptor. *The Journal of Biological Chemistry*, *279*, 29085–29091.

Bowery, N. G. (2006). $GABA_B$ receptor: A site of therapeutic benefit. *Current Opinion in Pharmacology*, *6*, 37–43.

Bowery, N. G., Doble, A., Hill, D. R., Hudson, A. L., Shaw, J. S., & Turnbull, M. J. (1979). Baclofen: A selective agonist for a novel type of GABA receptor. *British Journal of Pharmacology*, *67*, 444P–445P.

Bowery, N. G., Hill, D. R., Hudson, A. L., Doble, A., Middlemiss, D. N., Shaw, J., et al. (1980). (−)Baclofen decreases neurotransmitter release in the mammalian CNS by an action at a novel GABA receptor. *Nature*, *283*, 92–94.

Boyes, J., & Bolam, J. P. (2003). The subcellular localization of $GABA_B$ receptor subunits in the rat substantia nigra. *The European Journal of Neuroscience*, *18*, 3279–3293.

Brebner, K., Ahn, S., & Phillips, A. G. (2005). Attenuation of d-amphetamine self-administration by baclofen in the rat: Behavioral and neurochemical correlates. *Psychopharmacology*, *177*, 409–417.

Bussieres, N., & El Manira, A. (1999). $GABA_B$ receptor activation inhibits N- and P/Q-type calcium channels in cultured lamprey sensory neurons. *Brain Research, 847*, 175–185.
Calver, A. R., Robbins, M. J., Cosio, C., Rice, S. Q. J., Babbs, A. J., Hirst, W. D., et al. (2001). The C-terminal domains of the $GABA_B$ receptor subunits mediate intracellular trafficking but are not required for receptor signaling. *The Journal of Neuroscience, 21*, 1203–1210.
Campbell, V., Berrow, N., & Dolphin, A. C. (1993). $GABA_B$ receptor modulation of Ca^{2+} currents in rat sensory neurones by the G protein G_0: Antisense oligonucleotide studies. *The Journal of Physiology, 470*, 1–11.
Chalifoux, J. R., & Carter, A. G. (2010). $GABA_B$ receptors modulate NMDA receptor calcium signals in dendritic spines. *Neuron, 66*, 101–113.
Chalifoux, J. R., & Carter, A. G. (2011a). $GABA_B$ receptor modulation of synaptic function. *Current Opinion in Neurobiology, 21*, 1–6.
Chalifoux, J. R., & Carter, A. G. (2011b). $GABA_B$ receptor modulation of voltage-sensitive calcium channels in spines and dendrites. *The Journal of Neuroscience, 31*, 4221–4232.
Chen, L., Boyes, J., Yung, W. H., & Bolam, J. P. (2004). Subcellular localization of $GABA_B$ receptor subunits in rat globus pallidus. *The Journal of Comparative Neurology, 474*, 340–352.
Chen, G., & van den Pol, A. N. (1998). Presynaptic $GABA_B$ autoreceptor modulation of P/Q-type calcium channels and GABA release in rat suprachiasmatic nucleus neurons. *The Journal of Neuroscience, 18*, 1913–1922.
Ciechanover, A. (2006). Intracellular protein degradation: From a vague idea thru the lysosome and the ubiquitin-proteasome system and onto human diseases and drug targeting. *Hematology/The Education Program of the American Society of Hematology, 2006*, 1–12.
Cimarosti, H., Kantamneni, S., & Henley, J. M. (2009). Ischaemia differentially regulates $GABA_B$ receptor subunits in organotypic hippocampal slice cultures. *Neuropharmacology, 56*, 1088–1096.
Cook, D. J., Teves, L., & Tymianski, M. (2012). Treatment of stroke with a PSD-95 inhibitor in the gyrencephalic primate brain. *Nature, 483*, 213–217.
Couve, A., Filippov, A. K., Connelly, C. N., Bettler, B., Brown, D. A., & Moss, S. J. (1998). Intracellular retention of recombinant $GABA_B$ receptors. *The Journal of Biological Chemistry, 273*, 26361–26367.
Couve, A., Kittler, J. T., Uren, J. M., Calver, A. R., Pangalos, M. N., Walsh, F. S., et al. (2001). Association of $GABA_B$ receptors and members of the 14-3-3 family of signaling proteins. *Molecular and Cellular Neurosciences, 17*, 317–328.
Couve, A., Thomas, P., Calver, A. R., Hirst, W. D., Pangalos, M. N., Walsh, F. S., et al. (2002). Cyclic AMP-dependent protein kinase phosphorylation facilitates $GABA_B$ receptor-effector coupling. *Nature Neuroscience, 5*, 415–424.
Craig, M. T., & McBain, C. J. (2014). The emerging role of $GABA_B$ receptors as regulators of network dynamics: Fast actions from a 'slow' receptor? *Current Opinion in Neurobiology, 26*, 15–21.
Creed, M. C., & Luscher, C. (2013). Drug-evoked synaptic plasticity: Beyond metaplasticity. *Current Opinion in Neurobiology, 23*, 553–558.
Cruz, H. G., Ivanova, T., Lunn, M. L., Stoffel, M., Slesinger, P. A., & Luscher, C. (2004). Bi-directional effects of $GABA_B$ receptor agonists on the mesolimbic dopamine system. *Nature Neuroscience, 7*, 153–159.
Cryan, J. F., & Kaupmann, K. (2005). Don't worry 'B' happy!: A role for $GABA_B$ receptors in anxiety and depression. *Trends in Pharmacological Sciences, 26*, 36–43.
Dave, K. R., Lange-Asschenfeldt, C., Raval, A. P., Prado, R., Busto, R., Saul, I., et al. (2005). Ischemic preconditioning ameliorates excitotoxicity by shifting glutamate/γ-aminobutyric acid release and biosynthesis. *Journal of Neuroscience Research, 82*, 665–673.

Deng, P. Y., Xiao, Z., Yang, C., Rojanathammanee, L., Grisanti, L., Watt, J., et al. (2009). $GABA_B$ receptor activation inhibits neuronal excitability and spatial learning in the entorhinal cortex by activating TREK-2 K^+ channels. *Neuron, 63*, 230–243.

Duthey, B., Caudron, S., Perroy, J., Bettler, B., Fagni, L., Pin, J. P., et al. (2002). A single subunit (GB2) is required for G-protein activation by the heterodimeric $GABA_B$ receptor. *The Journal of Biological Chemistry, 277*, 3236–3241.

Engle, M. P., Gassman, M., Sykes, K. T., Bettler, B., & Hammond, D. L. (2006). Spinal nerve ligation does not alter the expression or function of $GABA_B$ receptors in spinal cord and dorsal root ganglia of the rat. *Neuroscience, 138*, 1277–1287.

Engle, M. P., Merrill, M. A., Marquez De Prado, B., & Hammond, D. L. (2012). Spinal nerve ligation decreases gamma-aminobutyric $acid_B$ receptors on specific populations of immunohistochemically identified neurons in L5 dorsal root ganglion of the rat. *The Journal of Comparative Neurology, 520*, 1663–1677.

Enna, S. J., & Bowery, N. G. (2004). $GABA_B$ receptor alterations as indicators of physiological and pharmacological function. *Biochemical Pharmacology, 68*, 1541–1548.

Enna, S. J., & McCarson, K. E. (2006). The role of GABA in the mediation and perception of pain. *Advances in Pharmacology, 54*, 1–27.

Erondu, N. E., & Kennedy, M. B. (1985). Regional distribution of type II Ca^{2+}/calmodulin-dependent protein kinase in rat brain. *The Journal of Neuroscience, 5*, 3270–3277.

Fadda, P., Scherma, M., Fresu, A., Collu, M., & Fratta, W. (2003). Baclofen antagonizes nicotine-, cocaine-, and morphine-induced dopamine release in the nucleus accumbens of rat. *Synapse, 50*, 1–6.

Fairfax, B. P., Pitcher, J. A., Scott, M. G., Calver, A. R., Pangalos, M. N., Moss, S. J., et al. (2004). Phosphorylation and chronic agonist treatment atypically modulate $GABA_B$ receptor cell surface stability. *The Journal of Biological Chemistry, 279*, 12565–12573.

Fernandez-Alacid, L., Aguado, C., Ciruela, F., Martin, R., Colon, J., Cabanero, M. J., et al. (2009). Subcellular compartment-specific molecular diversity of pre- and post-synaptic GABA-activated GIRK channels in Purkinje cells. *Journal of Neurochemistry, 110*, 1363–1376.

Filip, M., & Frankowska, M. (2008). $GABA_B$ receptors in drug addiction. *Pharmacological Reports, 60*, 755–770.

Fritschy, J. M., Meskenaite, V., Weinmann, O., Honer, M., Benke, D., & Mohler, H. (1999). $GABA_B$ receptor splice variants GB1a and GB1b in rat brain: Developmental regulation, cellular distribution and extrasynaptic localization. *The European Journal of Neuroscience, 11*, 761–768.

Fritschy, J. M., Sidler, C., Parpan, F., Gassmann, M., Kaupmann, K., Bettler, B., et al. (2004). Independent maturation of the $GABA_B$ receptor subunits $GABA_{B1}$ and $GABA_{B2}$ during postnatal development in rodent brain. *The Journal of Comparative Neurology, 477*, 235–252.

Froestl, W. (2010). Novel $GABA_B$ receptor positive modulators: A patent survey. *Expert Opinion on Therapeutic Patents, 20*, 1007–1017.

Gaiarsa, J. L., Kuczewski, N., & Porcher, C. (2011). Contribution of metabotropic $GABA_B$ receptors to neuronal network construction. *Pharmacology & Therapeutics, 132*, 170–179.

Gaiarsa, J. L., & Porcher, C. (2013). Emerging neurotrophic role of $GABA_B$ receptors in neuronal circuit development. *Frontiers in Cellular Neuroscience, 7*, 206.

Galvez, T., Duthey, B., Kniazeff, J., Blahos, J., Rovelli, G., Bettler, B., et al. (2001). Allosteric interactions between GB1 and GB2 subunits are required for optimal $GABA_B$ receptor function. *The EMBO Journal, 20*, 2152–2159.

Galvez, T., Parmentier, M. L., Joly, C., Malitschek, B., Kaupmann, K., Kuhn, R., et al. (1999). Mutagenesis and modeling of the $GABA_B$ receptor extracellular domain support a Venus flytrap mechanism for ligand binding. *The Journal of Biological Chemistry, 274*, 13362–13369.

Galvez, T., Prezeau, L., Milioti, G., Franek, M., Joly, C., Froestl, W., et al. (2000). Mapping the agonist-binding site of $GABA_B$ type 1 subunit sheds light on the activation process of $GABA_B$ receptors. *The Journal of Biological Chemistry*, *275*, 41166–41174.

Gassmann, M., & Bettler, B. (2012). Regulation of neuronal $GABA_B$ receptor functions by subunit composition. *Nature Reviews. Neuroscience*, *13*, 380–394.

Gassmann, M., Haller, C., Stoll, Y., Aziz, S. A., Biermann, B., Mosbacher, J., et al. (2005). The RXR-type endoplasmic reticulum-retention/retrieval signal of $GABA_{B1}$ requires distant spacing from the membrane to function. *Molecular Pharmacology*, *68*, 137–144.

Gassmann, M., Shaban, H., Vigot, R., Sansig, G., Haller, C., Barbieri, S., et al. (2004). Redistribution of $GABA_{B1}$ protein and atypical $GABA_B$ responses in $GABA_{B2}$-deficient mice. *The Journal of Neuroscience*, *24*, 6086–6097.

Geng, Y., Bush, M., Mosyak, L., Wang, F., & Fan, Q. R. (2013). Structural mechanism of ligand activation in human $GABA_B$ receptor. *Nature*, *504*, 254–259.

Grabbe, C., Husnjak, K., & Dikic, I. (2011). The spatial and temporal organization of ubiquitin networks. *Nature Reviews. Molecular Cell Biology*, *12*, 295–307.

Grampp, T., Notz, V., Broll, I., Fischer, N., & Benke, D. (2008). Constitutive, agonist-accelerated, recycling and lysosomal degradation of $GABA_B$ receptors in cortical neurons. *Molecular and Cellular Neurosciences*, *39*, 628–637.

Grampp, T., Sauter, K., Markovic, B., & Benke, D. (2007). γ-Aminobutyric acid type B receptors are constitutively internalized via the clathrin-dependent pathway and targeted to lysosomes for degradation. *The Journal of Biological Chemistry*, *282*, 24157–24165.

Guetg, N., Aziz, S. A., Holbro, N., Turecek, R., Rose, T., Seddik, R., et al. (2010). NMDA receptor-dependent $GABA_B$ receptor internalization via CaMKII phosphorylation of serine 867 in $GABA_{B1}$. *Proceedings of the National Academy of Sciences of the United States of America*, *107*, 13924–13929.

Guyon, A., & Leresche, N. (1995). Modulation by different $GABA_B$ receptor types of voltage-activated calcium currents in rat thalamocortical neurones. *The Journal of Physiology*, *485*, 29–42.

Han, D., Zhang, Q. G., Yong, L., Li, C., Zong, Y. Y., Yu, C. Z., et al. (2008). Co-activation of GABA receptors inhibits the JNK3 apoptotic pathway via the disassembly of the GluR6-PSD95-MLK3 signaling module in cerebral ischemic-reperfusion. *FEBS Letters*, *582*, 1298–1306.

Hannan, S., Wilkins, M. E., & Smart, T. G. (2012). Sushi domains confer distinct trafficking profiles on $GABA_B$ receptors. *Proceedings of the National Academy of Sciences of the United States of America*, *109*, 12171–12176.

Havlickova, M., Prezeau, L., Duthey, B., Bettler, B., Pin, J. P., & Blahos, J. (2002). The intracellular loops of the GB2 subunit are crucial for G-protein coupling of the heteromeric γ-aminobutyrate B receptor. *Molecular Pharmacology*, *62*, 343–350.

Hearing, M., Kotecki, L., Fernandez, Marron, de Velasco, E., Fajardo-Serrano, A., Chung, H. J., et al. (2013). Repeated cocaine weakens GABA-Girk signaling in layer 5/6 pyramidal neurons in the prelimbic cortex. *Neuron*, *80*, 159–170.

Hill, M. D., Martin, R. H., Mikulis, D., Wong, J. H., Silver, F. L., Terbrugge, K. G., et al. (2012). Safety and efficacy of NA-1 in patients with iatrogenic stroke after endovascular aneurysm repair (ENACT): A phase 2, randomised, double-blind, placebo-controlled trial. *Lancet Neurology*, *11*, 942–950.

Ivankova, K., Turecek, R., Fritzius, T., Seddik, R., Prezeau, L., Comps-Agrar, L., et al. (2013). Up-regulation of $GABA_B$ receptor signaling by constitutive assembly with the K^+ channel tetramerization domain-containing protein 12 (KCTD12). *The Journal of Biological Chemistry*, *288*, 24848–24856.

Jackson-Friedman, C., Lyden, P. D., Nunez, S., Jin, A., & Zweifler, R. (1997). High dose baclofen is neuroprotective but also causes intracerebral hemorrhage: A quantal bioassay

study using the intraluminal suture occlusion method. *Experimental Neurology*, *147*, 346–352.
Johnson, S. W., & North, R. A. (1992). Two types of neurone in the rat ventral tegmental area and their synaptic inputs. *The Journal of Physiology*, *450*, 455–468.
Jones, K. A., Borowsky, B., Tamm, J. A., Craig, D. A., Durkin, M. M., Dai, M., et al. (1998). $GABA_B$ receptor function as a heteromeric assembly of the subunits $GABA_{BR1}$ and $GABA_{BR2}$. *Nature*, *396*, 674–679.
Kantamneni, S., Gonzalez-Gonzalez, I. M., Luo, J., Cimarosti, H., Jacobs, S. C., Jaafari, N., et al. (2014). Differential regulation of $GABA_B$ receptor trafficking by different modes of NMDA receptor signalling. *The Journal of Biological Chemistry*, *289*, 6681–6694.
Kantamneni, S., Holman, D., Wilkinson, K. A., Correa, S. A., Feligioni, M., Ogden, S., et al. (2008). GISP binding to TSG101 increases $GABA_B$ receptor stability by down-regulating ESCRT-mediated lysosomal degradation. *Journal of Neurochemistry*, *107*, 86–95.
Kauer, J. A., & Malenka, R. C. (2007). Synaptic plasticity and addiction. *Nature Reviews. Neuroscience*, *8*, 844–858.
Kaupmann, K., Malitschek, B., Schuler, V., Heid, J., Froestl, W., Beck, P., et al. (1998). $GABA_B$ receptor subtypes assemble into functional heteromeric complexes. *Nature*, *396*, 683–687.
Kavalali, E. T., Zhuo, M., Bito, H., & Tsien, R. W. (1997). Dendritic Ca^{2+} channels characterized by recordings from isolated hippocampal dendritic segments. *Neuron*, *18*, 651–663.
Kim, J. Y., Kim, N., Yenari, M. A., & Chang, W. (2011). Mild hypothermia suppresses calcium-sensing receptor (CaSR) induction following forebrain ischemia while increasing $GABA_B$ receptor 1 ($GABA_B$-R1) expression. *Translational Stroke Research*, *2*, 195–201.
Kim, I., Xu, W., & Reed, J. C. (2008). Cell death and endoplasmic reticulum stress: Disease relevance and therapeutic opportunities. *Nature Reviews. Drug Discovery*, *7*, 1013–1030.
Kohl, M. M., & Paulsen, O. (2010). The roles of $GABA_B$ receptors in cortical network activity. *Advances in Pharmacology*, *58*, 205–229.
Kostandy, B. B. (2012). The role of glutamate in neuronal ischemic injury: The role of spark in fire. *Neurological Sciences*, *33*, 223–237.
Kulik, A., Vida, I., Lujan, R., Haas, C. A., Lopez-Bendito, G., Shigemoto, R., et al. (2003). Subcellular localization of metabotropic $GABA_B$ receptor subunits $GABA_{B1a/b}$ and $GABA_{B2}$ in the rat hippocampus. *The Journal of Neuroscience*, *23*, 11026–11035.
Kulinskii, V. I., & Mikhel'son, G. V. (2000). Additivity and independence of neuroprotective effects of $GABA_A$ and $GABA_B$ receptor agonists in complete global cerebral ischemia. *Bulletin of Experimental Biology and Medicine*, *130*, 772–774.
Kumar, K., Sharma, S., Kumar, P., & Deshmukh, R. (2013). Therapeutic potential of $GABA_B$ receptor ligands in drug addiction, anxiety, depression and other CNS disorders. *Pharmacology, Biochemistry, and Behavior*, *110*, 174–184.
Kuramoto, N., Wilkins, M. E., Fairfax, B. P., Revilla-Sanchez, R., Terunuma, M., Tamaki, K., et al. (2007). Phospho-dependent functional modulation of $GABA_B$ receptors by the metabolic sensor AMP-dependent protein kinase. *Neuron*, *53*, 233–247.
Lacey, C. J., Boyes, J., Gerlach, O., Chen, L., Magill, P. J., & Bolam, J. P. (2005). $GABA_B$ receptors at glutamatergic synapses in the rat striatum. *Neuroscience*, *136*, 1083–1095.
Ladera, C., Del Carmen Godino, M., Cabanero, M. J., Torres, M., Watanabe, M., Lujan, R., et al. (2008). Pre-synaptic $GABA_B$ receptors inhibit glutamate release through GIRK channels in rat cerebral cortex. *Journal of Neurochemistry*, *107*, 1506–1517.
Laffray, S., Bouali-Benazzouz, R., Papon, M. A., Favereaux, A., Jiang, Y., Holm, T., et al. (2012). Impairment of $GABA_B$ receptor dimer by endogenous 14-3-3ζ in chronic pain conditions. *The EMBO Journal*, *31*, 3239–3251.

Laffray, S., Tan, K., Dulluc, J., Bouali-Benazzouz, R., Calver, A. R., Nagy, F., et al. (2007). Dissociation and trafficking of rat $GABA_B$ receptor heterodimer upon chronic capsaicin stimulation. *The European Journal of Neuroscience*, *25*, 1402–1416.
Lal, S., Shuaib, A., & Ijaz, S. (1995). Baclofen is cytoprotective to cerebral ischemia in gerbils. *Neurochemical Research*, *20*, 115–119.
Lambert, N. A., & Wilson, W. A. (1996). High-threshold Ca^{2+} currents in rat hippocampal interneurones and their selective inhibition by activation of $GABA_B$ receptors. *The Journal of Physiology*, *492*, 115–127.
Laviv, T., Riven, I., Dolev, I., Vertkin, I., Balana, B., Slesinger, P. A., et al. (2010). Basal GABA regulates $GABA_BR$ conformation and release probability at single hippocampal synapses. *Neuron*, *67*, 253–267.
Li, C. J., Lu, Y., Zhou, M., Zong, X. G., Li, C., Xu, X. L., et al. (2014). Activation of $GABA_B$ receptors ameliorates cognitive impairment via restoring the balance of HCN1/HCN2 surface expression in the hippocampal CA1 area in rats with chronic cerebral hypoperfusion. *Molecular Neurobiology*, *50*, 704–720.
Lipton, P. (1999). Ischemic cell death in brain neurons. *Physiological Reviews*, *79*, 1431–1568.
Lujan, R., & Ciruela, F. (2012). $GABA_B$ receptors-associated proteins: Potential drug targets in neurological disorders? *Current Drug Targets*, *13*, 129–144.
Lujan, R., Shigemoto, R., Kulik, A., & Juiz, J. M. (2004). Localization of the $GABA_B$ receptor 1a/b subunit relative to glutamatergic synapses in the dorsal cochlear nucleus of the rat. *The Journal of Comparative Neurology*, *475*, 36–46.
Luscher, C., Jan, L. Y., Stoffel, M., Malenka, R. C., & Nicoll, R. A. (1997). G protein-coupled inwardly rectifying K^+ channels (GIRKs) mediate postsynaptic but not presynaptic transmitter actions in hippocampal neurons. *Neuron*, *19*, 687–695.
Maier, P. J., Marin, I., Grampp, T., Sommer, A., & Benke, D. (2010). Sustained glutamate receptor activation down-regulates $GABA_B$ receptors by shifting the balance from recycling to lysosomal degradation. *The Journal of Biological Chemistry*, *285*, 35606–35614.
Maier, P. J., Zemoura, K., Acuna, M. A., Yevenes, G. E., Zeilhofer, H. U., & Benke, D. (2014). Ischemia-like oxygen and glucose deprivation mediates down-regulation of cell surface γ-aminobutyric $acid_B$ receptors via the ER stress-induced transcription factor CHOP. *The Journal of Biological Chemistry*, *289*, 12896–12907.
Mannoury la Cour, C., Herbelles, C., Pasteau, V., de Nanteuil, G., & Millan, M. J. (2008). Influence of positive allosteric modulators on $GABA_B$ receptor coupling in rat brain: A scintillation proximity assay characterisation of G protein subtypes. *Journal of Neurochemistry*, *105*, 308–323.
Margeta-Mitrovic, M., Jan, Y. N., & Jan, L. Y. (2000). A trafficking checkpoint controls $GABA_B$ receptor heterodimerization. *Neuron*, *27*, 97–106.
Margeta-Mitrovic, M., Jan, Y. N., & Jan, L. Y. (2001a). Function of GB1 and GB2 subunits in G protein coupling of $GABA_B$ receptors. *Proceedings of the National Academy of Sciences of the United States of America*, *98*, 14649–14654.
Margeta-Mitrovic, M., Jan, Y. N., & Jan, L. Y. (2001b). Ligand-induced signal transduction within heterodimeric $GABA_B$ receptor. *Proceedings of the National Academy of Sciences of the United States of America*, *98*, 14643–14648.
Menon-Johansson, A. S., Berrow, N., & Dolphin, A. C. (1993). Go transduces $GABA_B$-receptor modulation of N-type calcium channels in cultured dorsal root ganglion neurons. *Pflügers Archiv*, *425*, 335–343.
Mintz, I. M., & Bean, B. P. (1993). $GABA_B$ receptor inhibition of P-type Ca^{2+} channels in central neurons. *Neuron*, *10*, 889–898.
Morishita, R., Kato, K., & Asano, T. (1990). $GABA_B$ receptors couple to G proteins Go, Go* and Gi1 but not to Gi2. *FEBS Letters*, *271*(1–2), 231–235.
Nishikawa, M., Hirouchi, M., & Kuriyama, K. (1997). Functional coupling of Gi subtype with $GABA_B$ receptor/adenylyl cyclase system: Analysis using a reconstituted system

with purified GTP-binding protein from bovine cerebral cortex. *Neurochemistry International*, *31*, 21–25.
Odagaki, Y., & Koyama, T. (2001). Identification of Gα subtype(s) involved in γ-aminobutyric acid$_B$ receptor-mediated high-affinity guanosine triphosphatase activity in rat cerebral cortical membranes. *Neuroscience Letters*, *297*, 137–141.
Padgett, C. L., Lalive, A. L., Tan, K. R., Terunuma, M., Munoz, M. B., Pangalos, M. N., et al. (2012). Methamphetamine-evoked depression of GABA$_B$ receptor signaling in GABA neurons of the VTA. *Neuron*, *73*, 978–989.
Pagano, A., Rovelli, G., Mosbacher, J., Lohmann, T., Duthey, B., Stauffer, D., et al. (2001). C-terminal interaction is essential for surface trafficking but not for heteromeric assembly of GABA$_B$ receptors. *The Journal of Neuroscience*, *21*, 1189–1202.
Pérez-Garci, E., Gassmann, M., Bettler, B., & Larkum, M. E. (2006). The GABA$_{B1b}$ isoform mediates long-lasting inhibition of dendritic Ca^{2+} spikes in layer 5 somatosensory pyramidal neurons. *Neuron*, *50*, 603–616.
Pierce, R. C., & Kumaresan, V. (2006). The mesolimbic dopamine system: The final common pathway for the reinforcing effect of drugs of abuse? *Neuroscience and Biobehavioral Reviews*, *30*, 215–238.
Pinard, A., Seddik, R., & Bettler, B. (2010). GABA$_B$ receptors: Physiological functions and mechanisms of diversity. *Advances in Pharmacology*, *58*, 231–255.
Pontier, S. M., Lahaie, N., Ginham, R., St-Gelais, F., Bonin, H., Bell, D. J., et al. (2006). Coordinated action of NSF and PKC regulates GABA$_B$ receptor signaling efficacy. *The EMBO Journal*, *25*, 2698–2709.
Price, G. W., Wilkin, G. P., Turnbull, M. J., & Bowery, N. G. (1984). Are baclofen-sensitive GABA$_B$ receptors present on primary afferent terminals of the spinal cord? *Nature*, *307*, 71–74.
Ramirez, O. A., Vidal, R. L., Tello, J. A., Vargas, K. J., Kindler, S., Härtel, S., et al. (2009). Dendritic assembly of heteromeric γ-aminobutyric acid type B receptor subunits in hippocampal neurons. *The Journal of Biological Chemistry*, *284*, 13077–13085.
Robbins, M. J., Calver, A. R., Filippov, A. K., Hirst, W. D., Russell, R. B., Wood, M. D., et al. (2001). GABA$_{B2}$ is essential for G-protein coupling of the GABA$_B$ receptor heterodimer. *The Journal of Neuroscience*, *21*, 8043–8052.
Ronnett, G. V., Ramamurthy, S., Kleman, A. M., Landree, L. E., & Aja, S. (2009). AMPK in the brain: Its roles in energy balance and neuroprotection. *Journal of Neurochemistry*, *109*(Suppl. 1), 17–23.
Rost, B. R., Nicholson, P., Ahnert-Hilger, G., Rummel, A., Rosenmund, C., Breustedt, J., et al. (2011). Activation of metabotropic GABA receptors increases the energy barrier for vesicle fusion. *Journal of Cell Science*, *124*, 3066–3073.
Rubin, D. M., Coux, O., Wefes, I., Hengartner, C., Young, R. A., Goldberg, A. L., et al. (1996). Identification of the gal4 suppressor Sug1 as a subunit of the yeast 26S proteasome. *Nature*, *379*, 655–657.
Sabatini, B. L., & Svoboda, K. (2000). Analysis of calcium channels in single spines using optical fluctuation analysis. *Nature*, *408*, 589–593.
Saint, D. A., Thomas, T., & Gage, P. W. (1990). GABA$_B$ agonists modulate a transient potassium current in cultured mammalian hippocampal neurons. *Neuroscience Letters*, *118*, 9–13.
Sandoz, G., Levitz, J., Kramer, R. H., & Isacoff, E. Y. (2012). Optical control of endogenous proteins with a photoswitchable conditional subunit reveals a role for TREK1 in GABA$_B$ signaling. *Neuron*, *74*, 1005–1014.
Santos, A. E., Carvalho, C. M., Macedo, T. A., & Carvalho, A. P. (1995). Regulation of intracellular [Ca^{2+}] and GABA release by presynaptic GABA$_B$ receptors in rat cerebrocortical synaptosomes. *Neurochemistry International*, *27*, 397–406.
Sauter, K., Grampp, T., Fritschy, J. M., Kaupmann, K., Bettler, B., Mohler, H., et al. (2005). Subtype-selective interaction with the transcription factor CCAAT/enhancer-binding

protein (C/EBP) homologous protein (CHOP) regulates cell surface expression of $GABA_B$ receptors. *The Journal of Biological Chemistry, 280*, 33566–33572.

Scanziani, M. (2000). GABA spillover activates postsynaptic $GABA_B$ receptors to control rhythmic hippocampal activity. *Neuron, 25*, 673–681.

Schuler, V., Luscher, C., Blanchet, C., Klix, N., Sansig, G., Klebs, K., et al. (2001). Epilepsy, hyperalgesia, impaired memory, and loss of pre- and postsynaptic $GABA_B$ responses in mice lacking $GABA_{B1}$. *Neuron, 31*, 47–58.

Schwenk, J., Metz, M., Zolles, G., Turecek, R., Fritzius, T., Bildl, W., et al. (2010). Native $GABA_B$ receptors are heteromultimers with a family of auxiliary subunits. *Nature, 465*, 231–235.

Smith, M. H., Ploegh, H. L., & Weissman, J. S. (2011). Road to ruin: Targeting proteins for degradation in the endoplasmic reticulum. *Science, 334*, 1086–1090.

Steiger, J. L., Bandyopadhyay, S., Farb, D. H., & Russek, S. J. (2004). cAMP response element-binding protein, activating transcription factor-4, and upstream stimulatory factor differentially control hippocampal $GABA_{BR1a}$ and $GABA_{BR1b}$ subunit gene expression through alternative promoters. *The Journal of Neuroscience, 24*, 6115–6126.

Tajiri, S., Oyadomari, S., Yano, S., Morioka, M., Gotoh, T., Hamada, J.-I., et al. (2004). Ischemia-induced neuronal cell death is mediated by the endoplasmic reticulum stress pathway involving CHOP. *Cell Death and Differentiation, 11*, 403–415.

Taniyama, K., Niwa, M., Kataoka, Y., & Yamashita, K. (1992). Activation of protein kinase C suppresses the γ-aminobutyric $acid_B$ receptor-mediated inhibition of the vesicular release of noradrenaline and acetylcholine. *Journal of Neurochemistry, 58*, 1239–1245.

Taniyama, K., Takeda, K., Ando, H., Kuno, T., & Tanaka, C. (1991). Expression of the $GABA_B$ receptor in Xenopus oocytes and inhibition of the response by activation of protein kinase C. *FEBS Letters, 278*, 222–224.

Taniyama, K., Takeda, K., Ando, H., & Tanaka, C. (1991). Expression of the $GABA_B$ receptor in Xenopus oocytes and desensitization by activation of protein kinase C. *Advances in Experimental Medicine and Biology, 287*, 413–420.

Terunuma, M., Vargas, K. J., Wilkins, M. E., Ramirez, O. A., Jaureguiberry-Bravo, M., Pangalos, M. N., et al. (2010). Prolonged activation of NMDA receptors promotes dephosphorylation and alters postendocytic sorting of $GABA_B$ receptors. *Proceedings of the National Academy of Sciences of the United States of America, 107*, 13918–13923.

Thompson, S. M., & Gahwiler, B. H. (1992). Comparison of the actions of baclofen at pre- and postsynaptic receptors in the rat hippocampus in vitro. *The Journal of Physiology, 451*, 329–345.

Towers, S., Princivalle, A., Billinton, A., Edmunds, M., Bettler, B., Urban, L., et al. (2000). $GABA_B$ receptor protein and mRNA distribution in rat spinal cord and dorsal root ganglia. *The European Journal of Neuroscience, 12*, 3201–3210.

Turecek, R., Schwenk, J., Fritzius, T., Ivankova, K., Zolles, G., Adelfinger, L., et al. (2014). Auxiliary $GABA_B$ receptor subunits uncouple G protein βγ subunits from effector channels to induce desensitization. *Neuron, 82*, 1032–1044.

Turnley, A. M., Stapleton, D., Mann, R. J., Witters, L. A., Kemp, B. E., & Bartlett, P. F. (1999). Cellular distribution and developmental expression of AMP-activated protein kinase isoforms in mouse central nervous system. *Journal of Neurochemistry, 72*, 1707–1716.

Valenzuela, J. I., Jaureguiberry-Bravo, M., Salas, D. A., Ramirez, O. A., Cornejo, V. H., Lu, H. E., et al. (2014). Transport along the dendritic endoplasmic reticulum mediates the trafficking of $GABA_B$ receptors. *Journal of Cell Science, 127*, 3382–3395.

Vargas, K. J., Terunuma, M., Tello, J. A., Pangalos, M. N., Moss, S. J., & Couve, A. (2008). The availability of surface $GABA_B$ receptors is independent of γ-aminobutyric acid but controlled by glutamate in central neurons. *The Journal of Biological Chemistry, 283*, 24641–24648.

Velez-Fort, M., Audinat, E., & Angulo, M. C. (2012). Central role of GABA in neuron–glia interactions. *The Neuroscientist, 18*, 237–250.

Vollenweider, F., Bendfeldt, K., Maetzler, W., Otten, U., & Nitsch, C. (2006). $GABA_B$ receptor expression and cellular localization in gerbil hippocampus after transient global ischemia. *Neuroscience Letters, 395*, 118–123.

Wang, Q., Song, C., & Li, C. C. (2004). Molecular perspectives on p97-VCP: Progress in understanding its structure and diverse biological functions. *Journal of Structural Biology, 146*, 44–57.

Wang, X. L., Zhang, Q., Zhang, Y. Z., Liu, Y. T., Dong, R., Wang, Q. J., et al. (2011). Downregulation of $GABA_B$ receptors in the spinal cord dorsal horn in diabetic neuropathy. *Neuroscience Letters, 490*, 112–115.

Weissman, A. M., Shabek, N., & Ciechanover, A. (2011). The predator becomes the prey: Regulating the ubiquitin system by ubiquitylation and degradation. *Nature Reviews. Molecular Cell Biology, 12*, 605–620.

White, J. H., Wise, A., Main, M. J., Green, A., Fraser, N. J., Disney, G. H., et al. (1998). Heterodimerization is required for the formation of functional $GABA_B$ receptors. *Nature, 396*, 679–682.

Wilkins, M. E., Li, X., & Smart, T. G. (2008). Tracking cell surface $GABA_B$ receptors using an α-bungarotoxin tag. *The Journal of Biological Chemistry, 283*, 34745–34752.

Wu, J., Xu, Y., Pu, S., Jiang, W., & Du, D. (2011). p38/MAPK inhibitor modulates the expression of dorsal horn $GABA_B$ receptors in the spinal nerve ligation model of neuropathic pain. *Neuroimmunomodulation, 18*, 150–155.

Xu, J., Li, C., Yin, X. H., & Zhang, G. Y. (2008). Additive neuroprotection of $GABA_A$ and $GABA_B$ receptor agonists in cerebral ischemic injury via PI-3 K/Akt pathway inhibiting the ASK1-JNK cascade. *Neuropharmacology, 54*, 1029–1040.

Yamauchi, T. (2005). Neuronal Ca^{2+}/calmodulin-dependent protein kinase II-discovery, progress in a quarter of a century, and perspective: Implication for learning and memory. *Biological & Pharmaceutical Bulletin, 28*, 1342–1354.

Yang, K., Ma, W. L., Feng, Y. P., Dong, Y. X., & Li, Y. Q. (2002). Origins of $GABA_B$ receptor-like immunoreactive terminals in the rat spinal dorsal horn. *Brain Research Bulletin, 58*, 499–507.

Zeilhofer, H. U., Benke, D., & Yevenes, G. E. (2012). Chronic pain states: Pharmacological strategies to restore diminished inhibitory spinal pain control. *Annual Review of Pharmacology and Toxicology, 52*, 111–133.

Zemoura, K., & Benke, D. (2014). Proteasomal degradation of γ-aminobutyric $acid_B$ receptors is mediated by the interaction of the $GABA_{B2}$ C terminus with the proteasomal ATPase Rtp6 and regulated by neuronal activity. *The Journal of Biological Chemistry, 289*, 7738–7746.

Zemoura, K., Schenkel, M., Acuna, M. A., Yevenes, G. E., Zeilhofer, H. U., & Benke, D. (2013). Endoplasmic reticulum-associated degradation (ERAD) controls cell surface expression of γ-aminobutyric acid, type B receptors. *The Journal of Biological Chemistry, 288*, 34897–34905.

Zhang, F., Li, C., Wang, R., Han, D., Zhang, Q. G., Zhou, C., et al. (2007). Activation of GABA receptors attenuates neuronal apoptosis through inhibiting the tyrosine phosphorylation of NR2A by Src after cerebral ischemia and reperfusion. *Neuroscience, 150*, 938–949.

Zhao, J., Meyerkord, C. L., Du, Y., Khuri, F. R., & Fu, H. (2011). 14-3-3 proteins as potential therapeutic targets. *Seminars in Cell & Developmental Biology, 22*, 705–712.

Zilberter, Y., Kaiser, K. M., & Sakmann, B. (1999). Dendritic GABA release depresses excitatory transmission between layer 2/3 pyramidal and bitufted neurons in rat neocortex. *Neuron, 24*, 979–988.

CHAPTER FOUR

Restoring the Spinal Pain Gate: $GABA_A$ Receptors as Targets for Novel Analgesics

Hanns Ulrich Zeilhofer[*,†,1], **William T. Ralvenius**[*], **Mario A. Acuña**[*]
[*]Institute of Pharmacology and Toxicology, University of Zurich, Zurich, Switzerland
[†]Institute of Pharmaceutical Sciences, Swiss Federal Institute of Technology (ETH) Zürich, Zurich, Switzerland
[1]Corresponding author: e-mail address: zeilhofer@pharma.uzh.ch

Contents

Abstract

$GABA_A$ receptors ($GABA_A$Rs) and glycine receptors are key elements of the spinal control of nociception and pain. Compromised functioning of these two transmitter systems contributes to chronic pain states. Restoring their proper function through positive allosteric modulators should constitute a rational approach to the treatment of chronic pain syndromes involving diminished inhibitory spinal pain control. Although classical benzodiazepines (i.e., full agonists at the benzodiazepine binding site of $GABA_A$Rs) potentiate synaptic inhibition in spinal pain controlling circuits, they lack clinically relevant

Advances in Pharmacology, Volume 73
ISSN 1054-3589
http://dx.doi.org/10.1016/bs.apha.2014.11.007

analgesic activity in humans. Recent data obtained from experiments in $GABA_AR$ point-mutated mice suggests dose-limiting sedative effects of classical nonspecific benzodiazepines as the underlying cause. Experiments in genetically engineered mice resistant to the sedative effects of classical benzodiazepines and studies with novel less sedating benzodiazepines have indeed shown that profound antihyperalgesia can be obtained at least in preclinical pain models. Present evidence suggests that compounds with high intrinsic activity at α2-$GABA_AR$ and minimal agonistic activity at α1-$GABA_AR$ should possess relevant antihyperalgesic activity without causing unwanted sedation. On-going preclinical studies in genetically engineered mice and clinical trials with more selective benzodiazepine site agonists should soon provide additional insights into this emerging topic.

ABBREVIATIONS

TPA023 7-(1,1-dimethylethyl)-6-(2-ethyl-2H-1,2,4-triazol-3-ylmethoxy)-3-(2-fluorophenyl)-1,2,4-triazolo[4,3-b]pyridazine

TPA023B 6,2′-difluoro-5′-[3-(1-hydroxy-1-methylethyl)imidazo[1,2-b][1,2,4]triazin-7-yl]biphenyl-2-carbonitrile

MRK-409 7-cyclobutyl-6-(2-methyl-2H-1,2,4-triazol-3-ylmethoxy)-3-(2,6-difluorophenyl)-1,2,4-triazolo[4,3-b]pyridazine

MK-0343 7-cyclobutyl-6-(2-methyl-2H-1,2,4-triazol-3-ylmethoxy)-3-(2,6-difluorophenyl)-1,2,4-triazolo[4,3-b]pyridazine

L-838,417 7-*tert*-butyl-3-(2,5-difluorophenyl)-6-(2-methyl-2H-[1,2,4]triazol-3-ylmethoxy)-[1,2,4]triazolo[4,3-b]pyridazine

HZ166 ethyl 8-ethynyl-6-(pyridin-2-yl)-4H-benzo[f]imidazo[1,5-a][1,4]diazepine-3-carboxylate

NS11394 3′-[5-(1-hydroxy-1-methylethyl)-benzoimidazol-1-yl]-biphenyl-2-carbonitrile

CCI chronic constriction injury

SNI spared nerve injury

SNL spinal nerve ligation

TNT tibial nerve transection

CFA complete Freund's adjuvant

$GABA_AR$ γ-aminobutyric acid type A receptor

α_x-$GABA_AR$ $GABA_A$ receptor containing the α_x subunit

1. INTRODUCTION

Chronic pain is a severe medical condition affecting millions of patients worldwide. It is almost generally accepted that neuronal and synaptic plasticity occurring at different levels of the neuraxis are major contributors to chronic pain (Luo, Kuner, & Kuner, 2014; Sandkühler, 2009; Zeilhofer, Witschi, & Johansson, 2009) (for a schematic illustration of the pain pathway, see Fig. 1). Some of these neuroplastic changes occur already in the peripheral terminals of nociceptors, which sense noxious stimuli arriving at the skin or in other peripheral tissues and convey them to the central nervous system. The central terminals of these nociceptors innervate the substantia gelatinosa (lamina II) of the spinal dorsal horn, or the trigeminal nucleus of the brainstem in case of those nociceptors coming from the facial skin or the meninges. From there, signals are propagated through various relay stations in the brainstem, midbrain, and thalamus to several cortical areas which give rise to the conscious sensation of pain. One site that has attracted particular attention in pain-related neuroplasticity is the spinal dorsal horn, which constitutes as the first site of synaptic integration in the pain

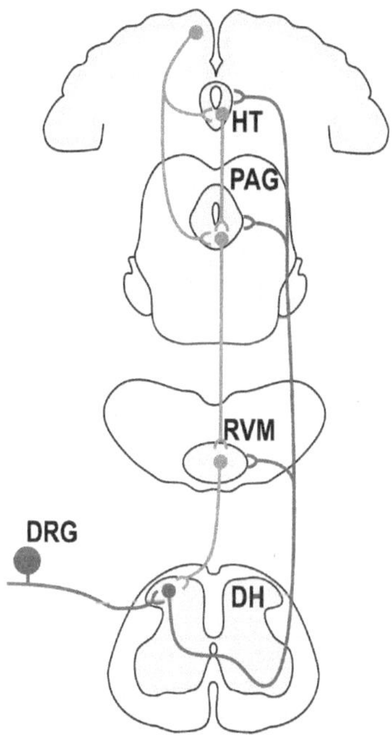

Figure 1 Schematic description of ascending pain pathways and descending antinociceptive fiber tracts. Nociceptive (painful) signals are conveyed by sensory fibers whose cell bodies reside in the dorsal root ganglia (DRGs). They reach the CNS at the level of the spinal dorsal horn (DH), where nociceptor terminals release glutamate to excite postsynaptic second-order neurons. These central neurons transmit the nociceptive information via the brainstem and the midbrain to cortical areas, where the conscious sensation of pain arises. Descending antinociceptive pathways are controlled by cortical areas, which contact the hypothalamus (HT) and the periaqueductal gray (PAG). The PAG in turn controls the rostral ventromedial medulla (RVM), which constitutes the main origin of descending antinociceptive fibers innervating the spinal cord.

pathway. Neurons located in the spinal dorsal horn integrate primary afferent sensory signals of painful and non-painful modalities with input from descending fiber tracts, which can either inhibit or facilitate pain. Inhibitory interneurons have been attributed a critical role in this process already in the gate-control-theory of pain (Fig. 2; Melzack & Wall, 1965). Although some of the synaptic connections proposed in the original scheme do apparently not exist, plenty of evidence indicates that compromising the function of inhibitory dorsal horn neurons induces symptoms reminiscent of chronic pain syndromes in humans. Animals develop an exaggerated sensitivity to painful stimuli (hyperalgesia), they respond with withdrawal responses upon exposure to stimuli, which are normally not felt as painful (allodynia), and they also show signs of spontaneous discomfort. Many lines of evidence indicate that typical causes of chronic pain such as inflammation or neuropathies compromise the function of inhibitory interneurons in the spinal dorsal horn through different mechanisms (for a review, see Zeilhofer, Benke, & Yévenes, 2012). According to this concept, a facilitation of inhibitory neurotransmission should be a rational strategy for the treatment of many chronic pain states. Yet, none of the established analgesics act- through a facilitation of inhibitory neurotransmission. In the following text, we will review mechanisms of pain-related spinal disinhibition and evidence supporting the concept that novel subtype-selective benzodiazepine agonists would be suitable for the treatment of chronic pain syndromes. In the context of this review, we use the term "benzodiazepine" for all agonists at the benzodiazepine binding site of γ-aminobutyric acid type A receptors

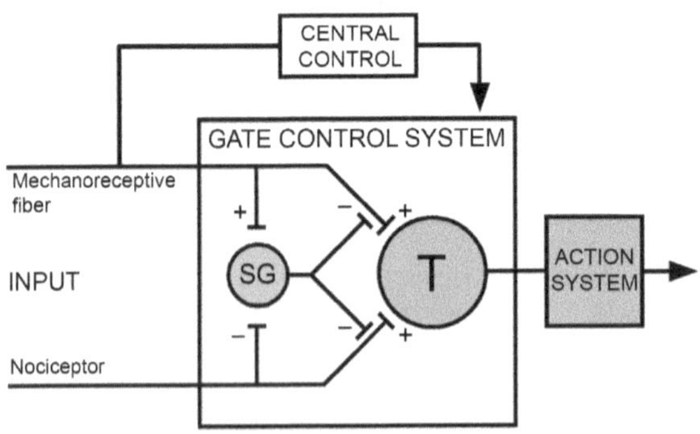

Figure 2 Gate-control theory of pain (Melzack & Wall, 1969). Inhibitory neurons located in the substantia gelatinosa (SG) control the spinal pain gate. According to the original concept these neurons are controlled by non-nociceptive input from mechanosensitive fibers and by nociceptive input in opposite directions. Their activation would in turn control the spinal output system (T).

($GABA_ARs$) independent of their chemical structure. It should also be mentioned here that $GABA_ARs$ exist, which are resistant to modulation by classical benzodiazepines. These receptors contain α4 or α6 subunits instead of α1, α2, α3, or α5, or a γ1 or δ subunit instead of the γ2 subunit. These benzodiazepine-insensitive receptors are quite abundant in several brain regions (e.g., thalamus and cerebellum), but their expression in the spinal cord is sparse.

2. SYNAPTIC DISINHIBITION IN PATHOLOGICAL PAIN

Fast synaptic inhibition in the spinal dorsal horn is mediated by GABA and glycine acting respectively at $GABA_AR$ and strychnine-sensitive GlyRs. Plenty of evidence indicates that blockade of spinal $GABA_ARs$ or GlyRs produces signs of allodynia and spontaneous pain (Beyer, Roberts, & Komisaruk, 1985; Miraucourt, Dallel, & Voisin, 2007; Roberts, Beyer, & Komisaruk, 1986). More recent studies provided insights into the mechanism of this sensitization on the level of dorsal horn neuronal circuits. The most consistent observation in these studies was a strong increase in polysynaptic input onto lamina II neurons after application of the $GABA_AR$ antagonist bicuculline (Baba et al., 2003). A second finding was related to the synaptic input of lamina I projection neurons, which express the neurokinin 1 receptor. These neurons serve an essential role in the relay of pathological pain, as their ablation strongly reduces hyperalgesia induced by inflammation and neuropathy (Nichols et al., 1999). Under normal conditions, these neurons receive sensory input almost exclusively from nociceptors (C and Aδ fibers). Blockade of $GABA_A$ and GlyRs however led to the *de novo* appearance polysynaptic responses from Aβ fibers (Torsney & MacDermott, 2006). These newly appearing polysynaptic connections likely underlie the allodynia seen *in vivo* after spinal application of bicuculline or strychnine. An increase in polysynaptic Aβ fiber input onto substantia gelatinosa (lamina II) may also occur as an endogenous process in chronic pain states (Baba, Doubell, & Woolf, 1999).

Several groups have identified signaling pathways that reduce inhibitory synaptic transmission in inflammatory or neuropathic pain states (Fig. 3). A prostaglandin E2-mediated phosphorylation of superficial dorsal horn GlyRs renders these receptors less responsive to glycine (Ahmadi, Lippross, Neuhuber, & Zeilhofer, 2002; Harvey et al., 2004; Reinold et al., 2005). Peripheral nerve damage leads to a downregulation of the GABA synthesizing enzyme GAD65 in the spinal cord (Moore et al.,

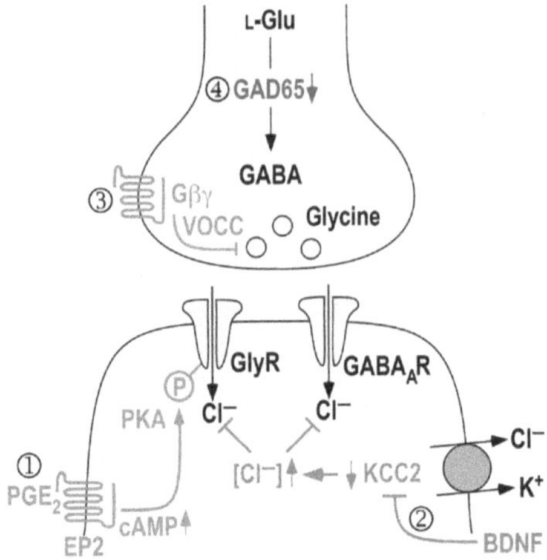

Figure 3 Four signaling pathways leading to spinal disinhibition in pathological pain states. (1) Prostaglandin E2 (PGE_2) produced in the spinal cord in response to peripheral inflammation increases cAMP production after activation EP2 receptors (EP2). The subsequent activation of protein kinase A (PKA) phosphorylates and inhibits GlyR of the superficial dorsal horn. (2) Peripheral nerve damage activates spinal microglia which releases brain-derived neurotrophic factor (BDNF). BDNF downregulates the expression of the potassium/chloride exporter KCC2 leading to an increase in intracellular chloride ($[Cl^-]_i$). As a consequence GABAergic and glycinergic input becomes less inhibitory (or even excitatory). (3) Several neuromodulators including endocannabinoids reduce presynaptic GABA and glycine release rendering dorsal horn neurons more excitable. (4) Peripheral nerve damage leads to the downregulation of the GABA synthesizing enzyme GAD65 and possible to reduced GABA content in inhibitory dorsal horn neurons.

2002), and both inflammation and nerve injury cause an epigenetic downregulation of the same enzyme in the brainstem (Zhang, Cai, Zou, Bie, & Pan, 2011). A large number of neuromodulators interfere with the release of GABA and glycine from inhibitory dorsal horn neurons via activation of G protein-coupled receptors and inhibition of Ca^{2+} channels (Zeilhofer, Wildner, & Yévenes, 2012). An endocannabinoid and CB1 receptor-mediated inhibition of glycine and/or GABA release contributes to spinal sensitization evoked by extensive nociceptive input to the dorsal horn (Pernia-Andrade et al., 2009). Microglia activated in the dorsal horn in response to peripheral nerve damage downregulates the expression of the potassium and chloride co-exporter KCC2 in superficial dorsal horn neurons, thereby shifting the reversal potential of GABA and glycine evoked chloride currents to more depolarized values. This shift renders glycinergic and GABAergic input less inhibitory (Coull et al., 2003, 2005; Keller, Beggs,

Salter, & De Koninck, 2007), or, if the shift is sufficiently large, glycinergic and GABAergic input may even become excitatory and trigger action potentials in postsynaptic neurons (Coull et al., 2003).

Pharmacological enhancement of GABAergic synaptic transmission in the dorsal horn should be able to reverse pathological pain states that result from reduced presynaptic GABA release or from reduced responsiveness of postsynaptic $GABA_A$Rs. Some of the disinhibitory processes discussed above do specifically reduce glycinergic inhibition prompting the question whether a potentiation of GABAergic responses would be able to restore proper inhibition in these cases. Many inhibitory dorsal horn neurons, co-release GABA and glycine from the same terminals and even from the same vesicles (Bohlhalter, Möhler, & Fritschy, 1994; Colin, Rostaing, Augustin, & Triller, 1998; Feng et al., 2005; Todd & Sullivan, 1990; Todd, Watt, Spike, & Sieghart, 1996). In most dorsal horn neurons, inhibitory postsynaptic responses are mediated by $GABA_A$R and GlyRs (Baccei & Fitzgerald, 2004; Yoshimura & Nishi, 1995) and even in cells, in which no GABAergic component is visible under normal conditions, a GABAergic IPSC component can be revealed with benzodiazepines and neurosteroids (Keller, Breton, Schlichter, & Poisbeau, 2004; Keller, Coull, Chery, Poisbeau, & De Koninck, 2001). It is thus conceivable that pharmacological enhancement of GABAergic neurotransmission would also compensate for reduced glycinergic transmission.

The situation is more complex in those cases where disinhibition results from changes in the transmembrane chloride gradient. As long as the activation of $GABA_A$R or GlyRs remains below the threshold of action potential activation, potentiation of $GABA_A$R or GlyR may still remain inhibitory. However, as soon as the chloride equilibrium potential reaches the action potential threshold, potentiation of $GABA_A$R or GlyR would increase the risk of paradoxical GABAergic and glycinergic excitation (Prescott, Sejnowski, & De Koninck, 2006). We discuss this issue below in the context of preclinical studies on subtype-selective benzodiazepines.

3. SPINAL $GABA_A$R SUBTYPES MEDIATING ANTIHYPERALGESIA: EVIDENCE FROM GENETICALLY ENGINEERED MICE

Analgesic or antihyperalgesic actions of benzodiazepines occur after local spinal injection suggesting that these effects are mediated by $GABA_A$Rs expressed in the spinal cord. To identify the $GABA_A$R subtypes responsible for these antihyperalgesic effects, "knock-in" mice were investigated, in which the α1, α2, α3, or α5-GABA$_A$R subunits had been rendered diazepam-insensitive through the introduction of a histamine to arginine (H/R) point mutation (Knabl et al., 2008; for information on the different point-mutated mouse strains, see Rudolph & Möhler, 2004). In wild-type mice, intrathecal diazepam strongly reduced hyperalgesia in models of inflammatory or neuropathic pain, but had no effects on acute nociceptive pain. This "antihyperalgesic" activity was unchanged in mice, which carried the H/R point mutation in the α1 subunit, but strongly reduced in mice, which carried the point mutation in the α2 subunit. Mice with point-mutated α3 or α5 subunits showed reduced antihyperalgesic activity in some but not in all tests. The different subtypes of benzodiazepine-sensitive GABA$_A$Rs contribute to spinal antihyperalgesia with the rank order $\alpha2 > \alpha3 \geq \alpha5 \gg \alpha1$. The GABA$_A$R subtype-dependence of spinal antihyperalgesia hence matched well with the expression of the different α subunits in the superficial dorsal horn (Bohlhalter, Weinmann, Möhler, & Fritschy, 1996; Lorenzo et al., 2014; Paul, Zeilhofer, & Fritschy, 2012).

The lack of a contribution from α1-GABA$_A$Rs, which mediate the sedative effects of diazepam (Rudolph et al., 1999), demonstrates that antihyperalgesia by benzodiazepines can be studied in the absence of confounding sedation as long as α1-GABA$_A$Rs are not activated. Such experiments were performed in mice carrying H/R point mutations in the α2, α3, or α5 subunits in addition to α1 (Knabl, Zeilhofer, Crestani, Rudolph, & Zeilhofer, 2009). These experiments showed that antihyperalgesia could also be obtained after systemic diazepam (and in the absence of sedation) and that α2 and α3-GABA$_A$R subtypes were the most relevant subtypes also for antihyperalgesia following systemic administration.

So far, GABA$_A$R point-mutated mice have been used to assess the antihyperalgesic properties of benzodiazepines in three pain models, i.e., against zymosan A-induced inflammatory hyperalgesia, against neuropathic hyperalgesia induced by chronic constriction injury (CCI) of the sciatic nerve and in the formalin test. GABA$_A$R subtypes mediating antihyperalgesia are thus clearly different from those mediating sedation (Rudolph et al., 1999), amnesia (Rudolph et al., 1999), and the rewarding properties of classical benzodiazepines (Tan et al., 2010). In the case of the α2- and α3-GABA$_A$Rs, there is a clear overlap with the receptors mediating anxiolysis (Löw et al., 2000) and muscle relaxation (Crestani et al., 2001), and, in case of α5-GABA$_A$R, possibly also with those responsible for benzodiazepine-induced cognitive impairment (Dawson et al., 2006). For a comparison of the contribution of the different $GABA_AR$ subtypes to desired hyperalgesia and other effects, see Fig. 4E.

The antihyperalgesic efficacy of diazepam after systemic administration prompts two questions. First, how important are spinal versus supraspinal CNS areas for antihyperalgesia by systemic benzodiazepines? Second, do central effects such as a reversal of anxiety-induced hyperalgesia (Andre et al., 2005; Vidal & Jacob, 1982) indirectly contribute to the antihyperalgesia by systemic benzodiazepines? The latter question appears relevant in particular because $\alpha2$-$GABA_ARs$ mediate not only antihyperalgesia but also anxiolysis (Löw et al., 2000). Both questions were addressed with conditional $GABA_AR$ deficient mice (*hoxb8*-$\alpha2^{-/-}$ mice), which lack the $GABA_AR$ $\alpha2$ subunit specifically from the spinal cord and dorsal root ganglia (DRGs) (up to about segment C4). In these experiments, a recently developed benzodiazepine site agonists (HZ166; Rivas et al., 2009) was employed which exerts antihyperalgesic actions similar to systemic diazepam but with reduced sedative and muscle relaxant properties (Di Lio et al., 2011). Antihyperalgesia was assessed as the change in heat and pin-prick induced withdrawal responses. These withdrawal responses are under strong control from descending pain-modulating fiber tracts from various CNS areas (Carrasquillo & Gereau, 2007; Harris & Westbrook, 1995; Jasmin, Rabkin, Granato, Boudah, & Ohara, 2003; Tatsuo, Salgado, Yokoro, Duarte, & Francischi, 1999). Analysis of such withdrawal responses should

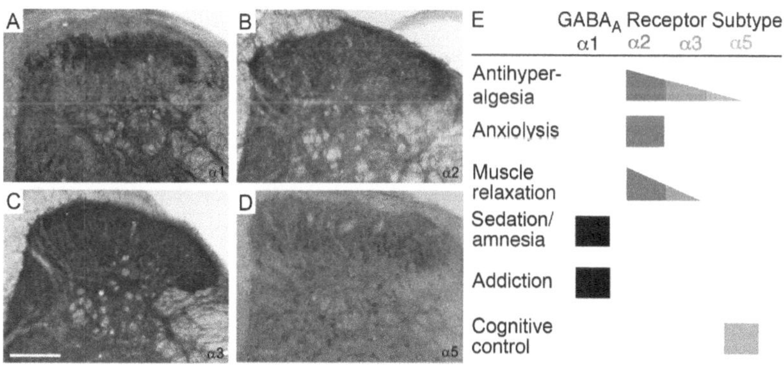

Figure 4 Distribution of $\alpha1$, $\alpha2$, $\alpha3$, and $\alpha5$-$GABA_AR$ subunits in the lumbar spinal cord and contribution of the four subtypes of $GABA_ARs$ to antihyperalgesia. (A–D) Immunocytochemical analysis of the expression of $GABA_AR$ α subunits in the spinal dorsal horn of mice. Scale bar, 100 μm. (E) Contribution of the different $GABA_AR$ subtype to spinal antihyperalgesia and comparison with other behavioral effects of benzodiazepines. *Panels (A–D): reproduced from Paul et al. (2012); Panel E: modified from Zeilhofer, Möhler, and Di Lio (2009).*

hence reveal possible effects of descending (facilitating or inhibitory) pain modulation. Mice lacking the $GABA_AR$ α2 subunits specifically from the spinal cord showed virtually the same reduction in benzodiazepine-induced antihyperalgesia as global α2 (H/R) point-mutated mice confirming that the spinal cord was the most relevant site for the antihyperalgesic action of benzodiazepines also after systemic administration (Paul et al., 2014). This finding also largely ruled out that antihyperalgesia occurred secondary to other effects such as a reversal of anxiety-induced hyperalgesia. In this context, it should also be added that the spinal cords of *hoxb8*-$\alpha 2^{-/-}$ mice completely lacked α2-$GABA_ARs$ indicating that the spinal terminals of fibers descending from supraspinal CNS areas to the spinal cord do not express α2-$GABA_ARs$.

4. MECHANISMS OF SPINAL BENZODIAZEPINE-MEDIATED ANTIHYPERALGESIA

Immunohistochemistry studies have identified specific spinal distribution patterns of $GABA_AR$ subunits (Bohlhalter et al., 1996; Paul et al., 2012; Fig. 4A–D). These receptors are expressed on intrinsic dorsal horn neurons and on the central terminals of primary sensory nociceptors. Spinal antihyperalgesia may therefore originate either from classical postsynaptic inhibition mediated by $GABA_ARs$ on intrinsic dorsal horn neurons or from $GABA_ARs$ on nociceptor terminals which mediate presynaptic inhibition through so-called primary afferent depolarization. Both processes are illustrated in Fig. 5. The availability of a "floxed" α2-$GABA_AR$ allele for conditional gene deletion allowed experiments distinguishing between these two possibilities.

4.1. Contribution of presynaptic inhibition and primary afferent depolarization

To assess the contribution of presynaptic inhibition, the α2-$GABA_AR$ subunit was ablated specifically from nociceptor terminals using an sns::cre BAC transgenic mouse which expresses the cre recombinase under the transcriptional control of the *scn10a* (Nav1.8) gene (Agarwal, Offermanns, & Kuner, 2004). In the case of an inflammatory pain model, the degree of antihyperalgesia by spinally applied diazepam in the nociceptor-specific α2-$GABA_AR$ subunit-deficient (sns-$\alpha 2^{-/-}$) mice fell between those measured in wild-type mice and in global α2-$GABA_AR$ point-mutated mice. In the inflammatory model, the partial loss of diazepam-induced antihyperalgesia

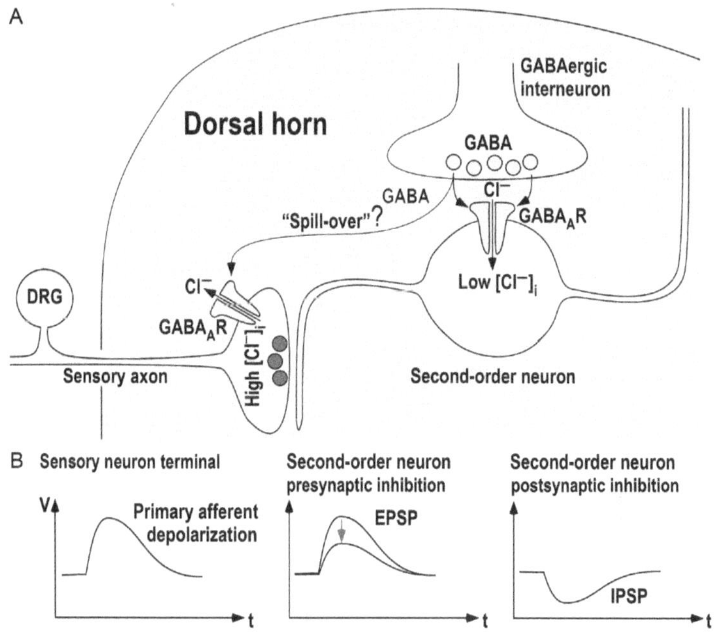

Figure 5 Mechanisms of the GABAergic control of spinal pain transmission. (A) Anatomical arrangement of pre- and postsynaptic inhibition. (B) Activation of $GABA_A$Rs on nociceptor terminals causes chloride efflux and primary afferent depolarization (left panel). This primary afferent depolarization causes presynaptic inhibition of glutamate release from nociceptor terminals (middle panel). Activation of $GABA_A$Rs on second-order neurons (intrinsic) dorsal horn neurons activates chloride influx and causes classical postsynaptic inhibition through hyperpolarization and dendritic shunting (right panel).

in sns-$\alpha2^{-/-}$ mice clearly indicated a contribution of presynaptic inhibition/primary afferent depolarization to antihyperalgesia by intrathecal diazepam. This was different in a neuropathy model in which all three genotypes responded with virtually identical antihyperalgesia (Witschi et al., 2011). This unaltered efficacy either indicates that antihyperalgesia in the neuropathy model was entirely due to postsynaptic inhibition of intrinsic dorsal horn neurons, or that antihyperalgesia occurred through inhibition of cre-negative (non-nociceptive) fibers. It has indeed been shown that inflammatory and neuropathic hyperalgesia depend on different classes of sensory fibers with Nav1.8 (sns) expressing sensory neurons being particularly important for inflammatory pain (Abrahamsen et al., 2008).

The results obtained in nociceptor-specific $\alpha2^{-/-}$ mice show that at least part of the antihyperalgesia originates from enhanced presynaptic inhibition.

The remaining α2-GABA$_A$R-mediated component may result either from inhibition of non-nociceptive fibers or from postsynaptic inhibition of intrinsic dorsal horn neurons. In both cases, the complete loss of α2-GABA$_A$R-mediated antihyperalgesia in *hoxb8*-α2$^{-/-}$ mice unequivocally demonstrates that the major component of benzodiazepine-evoked antihyperalgesia is of spinal origin (Paul et al., 2014). This has not yet been formally proven for the α3- and α5-GABA$_A$R-mediated components. However, both subunits are also enriched in the dorsal horn and their antihyperalgesic actions may thus also come from the spinal cord.

4.1.1 Mechanisms of presynaptic inhibition

It is well established that the spinal terminals of primary sensory neurons carry functional benzodiazepine-sensitive GABA$_A$Rs. Activation of these presynaptic GABA$_A$Rs causes depolarization of sensory neurons rather than hyperpolarization because primary afferent sensory neurons lack an efficient chloride export mechanism (Kanaka et al., 2001; Price, Hargreaves, & Cervero, 2006). As a consequence, the intracellular chloride concentration in these neurons renders the chloride equilibrium potential more positive than the resting membrane potential. This depolarization is however still inhibitory probably because it leads to a voltage-dependent inactivation of Na^+ and Ca^{2+} channels in the axon and the axon terminal, respectively, and subsequently reduces transmitter release (Kullmann et al., 2005). This presynaptic inhibition can occur through axo-axonic synapses. Their existence is firmly established for non-nociceptive primary sensory fibers (Aβ and low threshold Aδ fibers; Ribeiro-da-Silva, 1995). Axo-axonic contacts have also been found in nociceptor terminals, but less frequently than in terminals of non-nociceptive fibers (Alvarez, Kavookjian, & Light, 1993; Ribeiro-Da-Silva, Castro-Lopes, & Coimbra, 1986; Ribeiro-da-Silva, Tagari, & Cuello, 1989). Two recent studies disagree on the presence of gephyrin clusters on nociceptor terminals (Lorenzo et al., 2014; Paul et al., 2012). Because gephyrin is required for postsynaptic clustering of inhibitory neurotransmitter receptors in central neurons, the presence or absence of gephyrin clusters from sensory fiber terminals may be taken as an argument in favor or against the presence of axo-axonic synapses between GABAergic interneurons and nociceptor terminals. Physiological studies have established that primary afferent depolarization and presynaptic inhibition exist also in nociceptors (Lin, Wu, & Willis, 1999; Lin, Zou, & Willis, 2000; Witschi et al., 2011). Nociceptor terminals lacking GABAergic axo-axonic synapses may be subject to presynaptic inhibition through GABA$_A$Rs via "spill-over" of GABA from neighboring synapses and so-called volume transmission (Fig. 5; Rudomin & Schmidt, 1999, for a more recent review see also Zeilhofer, Wildner, et al., 2012).

5. ANTIHYPERALGESIC ACTION OF BENZODIAZEPINES WITH IMPROVED SUBTYPE SPECIFICITY: PRECLINICAL STUDIES

A number of benzodiazepines with reduced activity at α1-$GABA_A$Rs have been developed in the last two decades mainly in the quest for nonsedating anxiolytics (for a comprehensive list, see Rudolph & Knoflach, 2011). Because benzodiazepine-mediated anxiolysis and antihyperalgesia share a similar dependence on $GABA_A$R subtypes, some of these compounds were also tested in pain studies (Table 1). It should be mentioned here that α1-sparing compounds are sometimes referred to as "α2/3 selective" (e.g., Hofmann et al., 2012), this is however incorrect as all of them are also agonists at α5-$GABA_A$Rs. This additional activity at α5-$GABA_A$-Rs should not be forgotten when undesired effects of these compounds are discussed.

NS11394, which has very low activity at α1-$GABA_A$Rs (<10% relative to diazepam) and good partial agonistic activity at α2-, α3-, and α5-$GABA_A$Rs (26–78%, relative to diazepam) (Mirza et al., 2008), exhibited antinociceptive activity at nonsedating doses in several rodent pain models (Hofmann et al., 2012; Munro et al., 2008). The analgesic effect in the formalin test was blocked by the benzodiazepine site antagonist flumazenil (Munro et al., 2008) confirming that it occurred through $GABA_A$Rs. L-838,417, which completely lacks intrinsic activity at α1-$GABA_A$Rs and possesses partial agonistic activity at α2-, α3-, and α5-$GABA_A$Rs (15–32%, relative to diazepam; McKernan et al., 2000), was tested in several rodent pain models where it was not sedative but active against inflammatory and neuropathic hyperalgesia (Knabl et al., 2008; Nickolls et al., 2011) and formalin-induced nociception (Hofmann et al., 2012). In addition, L-838,417 was active against hyperalgesia evoked by skin incision, a model of postoperative hyperalgesia (Reichl et al., 2012), and against capsaicin-induced central pain sensitization (Hansen et al., 2012). Other benzodiazepines with low sedative propensities include HZ166 (Rivas et al., 2009) and TPA023 (Atack et al., 2006). HZ166, which exerts rather high intrinsic activity, was antihyperalgesic in inflammatory and neuropathic mouse models (Di Lio et al., 2011), while TPA023, which is a low

Table 1 Subtype-selective benzodiazepine site agonists in rodent pain models

Compound	Selectivity ratio α2/α1intrinsic activity at α2	Effects in pain models	Reference
Good selectivity and high intrinsic activity at α2			
HZ166[a]	3.1213%	Antihyperalgesic in mouse zymosan A and CCI[b]	Di Lio et al. (2011)
NS11394	6.273%	Antinociceptive in rat formalin and capsaicin test Antihyperalgesic in CFA[c] Inflammation, CCI, and SNI[d] reduced capsaicin-induced secondary hyperalgesia	Munro et al. (2008) and Hansen, Erichsen, Brown, Mirza, and Munro (2012)
Good selectivity and low intrinsic activity at α2			
L-838,417	28 42.7%	Antihyperalgesic in rat zymosan A and CCI Antiallodynic in rat SNL[e] but not TNT[f]Antihyperalgesic but not antiallodynic in rat CFAAntihyperalgesic against incisional pain	Knabl et al. (2008), Nickolls et al. (2011), and Reichl, Augustin, Zahn, and Pogatzki-Zahn (2012)
TPA023	1212%	Antiallodynic in rat SNL; no antihyperalgesia in rat CFALittle effect in rat formalin; hyperalgesic in rat carrageenan and CCI	Nickolls et al. (2011) and Munro, Erichsen, Rae, and Mirza (2011)
No selectivity toward α2			
Zolpidem	0.9210%	Antinociceptive in rat formalin and capsaicin, but only at sedative doses	Munro et al. (2008)
Bretazenil	0.560%	No antihyperalgesia in rat CCI and SNI at nonsedative doses	Munro et al. (2008)

[a]Compound 2 in Rivas et al. (2009).
[b]Chronic constriction injury.
[c]Complete Freund's adjuvant.
[d]Spared nerve injury.
[e]Spinal nerve ligation.
[f]Tibial nerve transection.
Modified from Zeilhofer, Benke, and Yévenes (2012).

intrinsic activity partial agonist, showed comparatively weak anti-allodynic or antihyperalgesic effects (Munro et al., 2011; Nickolls et al., 2011). Other compounds, which have higher intrinsic activities at α1-GABA$_A$Rs than at α2-, α3-, and α5-GABA$_A$Rs, did not exhibit antihyperalgesic activity at non-sedative doses. These results are fully consistent with the results obtained from H/R point-mutated mice.

Novel benzodiazepines for treatment of chronic pain would not only have to have good analgesic or antihyperalgesic efficacy and low, or at least weak, sedative properties, but would also have to avoid other typical side effects such as addiction and tolerance development (i.e., a loss of activity during prolonged treatment).

5.1. Addiction

Reinforcing (or addictive) properties were absent or reduced in the case of partial agonists which lack efficacy at α1-GABA$_A$Rs, including L-838,417 (Rowlett, Platt, Lelas, Atack, & Dawson, 2005) and TPA023 (Abe et al., 2009). Other partial agonists with higher efficacy at α1-GABA$_A$Rs such as TPA123 caused a withdrawal syndrome following upon cessation of self-administration, suggesting that the lack of agonistic activity at α1-GABA$_A$Rs was the relevant parameter rather than a general reduction in agonistic activity (Ator, Atack, Hargreaves, Burns, & Dawson, 2010). These results are consistent with studies in α1-GABA$_A$R point-mutated mice, in which no reinforcement was seen with the full agonist midazolam (Tan et al., 2010).

5.2. Tolerance development against antihyperalgesia

Analgesia by opioids shows a high liability to tolerance development and the same applies to many of the actions of classical benzodiazepines including antihyperalgesia in rats (Witschi & Zeilhofer, unpublished) and mice (Ralvenius, Benke, Rudolph, & Zeilhofer, 2013). Whether tolerance against antihyperalgesia occurs also with more selective or nonsedative benzodiazepines has been studied for three compounds. L-838,417 was devoid of tolerance developments against antihyperalgesic/analgesic actions during chronic 9-day treatment, while an equally effective dose of morphine completely lost its antihyperalgesic activity during the same time period (Knabl et al., 2008). Tolerance development was also not observed against antihyperalgesia by HZ166 (Di Lio et al., 2011), whereas NS11394 lost at least part of its analgesic activity against formalin-induced pain during chronic 8-day treatment (Hofmann et al., 2012). It is at present unknown whether reduced tolerance liability of nonsedative or more selective benzodiazepines is due to their reduced intrinsic activity (partial agonism), to improved subtype-selectivity, or to pharmacokinetic differences. Mice carrying different combinations of point-mutated $GABA_AR$ α subunits should offer a tool to address these possibilities.

6. CLINICAL STUDIES ON ANTIHYPERALGESIA BY BENZODIAZEPINES

The preclinical studies discussed above performed in mice resistant to the sedative effects of benzodiazepines demonstrate that classical benzodiazepines do in principle exert profound antihyperalgesic actions but only at doses, which normally induce strong sedation. Less-sedating benzodiazepines exhibited antihyperalgesic efficacy at nonsedating doses also in wild-type mice.

In human patients, classical nonselective benzodiazepines do not exert relevant analgesic (or antihyperalgesic) actions at clinically used doses. The few publications that found positive evidence for efficacy in pain patients were pilot studies or open trials (e.g., Fishbain, Cutler, Rosomoff, & Rosomoff, 2000; Harkins, Linford, Cohen, Kramer, & Cueva, 1991), which provide only limited evidence. Possible explanations for this lack of analgesic or antihyperalgesic efficacy in humans include (1) species differences (rodents vs. humans), (2) biological differences between preclinical models of pain and pain in human patients, and (3) dose-limiting sedation after systemic administration of classical benzodiazepines. The authors of this review favor the latter possibility for several reasons.

A recent study investigated possible analgesic or antihyperalgesic effects of two classical benzodiazepines (clobazam and clonazepam) in a battery of pain tests in human volunteers (Vuilleumier, Besson, Desmeules, Arendt-Nielsen, & Curatolo, 2013). This study found of a (small) antihyperalgesic effect in several parameters including in the size of the hyperalgesic area induced by intracutaneous capsaicin injection and in several muscle pain-related read-outs. Non-selective full agonists such as diazepam, clonazepam, midazolam typically induce strong sedation or sleep already at receptor occupancies between 15% and 30% (Fujita et al., 1999; Malizia et al., 1995; Pauli, Farde, Halldin, & Sedvall, 1991; Shinotoh et al., 1989). Compounds with a better $\alpha 2/\alpha 1$-selectivity ratio should permit higher levels of receptor occupancy and hence higher α2-$GABA_AR$ activation before reaching dose-limiting sedation.

The data available from the few benzodiazepines with improved selectivity profile that went into clinical trials provide some insight as to what degree of selectivity would be needed to avoid sedation in humans. One such study investigated MK-0343 (also known as MRK-409), which has very weak agonistic activity at α1-$GABA_AR$s (18% relative to the full agonist chlordiazepoxide; Atack et al., 2006). This compound was sedative in humans although previous preclinical tests in rodents had not shown any evidence for sedative properties (Atack et al., 2006; de Haas et al., 2008). A related compound (TPA023B) fully devoid of agonistic activity at α1-$GABA_AR$s did not produce sedation in man (Atack, Hallett, et al., 2011; Atack, Wafford, et al., 2011). These results suggest that the human brain is more susceptible to sedative actions of benzodiazepines than that of mice and rats (or, alternatively one might argue that our tests to assess sedation on rodents are less sensitive than those in man). In both cases, the available data indicate that sedation in humans can be avoided with compounds fully devoid of intrinsic activity at α1-$GABA_AR$s. Concert Pharmaceuticals is currently performing a phase 1 clinical trial on a deuterated version of L-838,417 (now called CTP-354), which has more favorable pharmacokinetics in humans than L-838,417 (http://www.concertpharma.com/CTP354Phase1Initiation.htm). It would be very informative to see whether CTP-354 or TPA023B, which has an intrinsic activity at α2-$GABA_AR$s even higher than that of L-838,417 and no agonistic activity at α1-$GABA_AR$s, possess antihyperalgesic or analgesic activity in human volunteers or pain patients.

7. OPEN QUESTIONS

7.1. Which $GABA_AR$ subtypes should be targeted for optimal analgesia with minimal side-effects?

Work in the $GABA_AR$ H/R point-mutated mice suggests that α2-, α3-, and α5-$GABA_AR$s contribute to spinal benzodiazepine antihyperalgesia. It is however not clear whether only one subtype (i.e., α2-$GABA_AR$s) should be targeted for optimal antihyperalgesia or whether simultaneous activity at more than one subunit would be advantageous. In the absence of fully selective subtype-specific drugs, investigations on mice carrying more than one point-mutated $GABA_AR$ subtype should be informative. When taking undesired effects into consideration, one would probably try to avoid positive allosteric modulation of α5-GABA$_A$Rs as this receptor subtype might confer cognitive impairment. Another reason to avoid activity at α5-GABA$_A$Rs comes from a recent report suggesting that inverse agonistic activity at α5-GABA$_A$Rs could be analgesic (Munro et al., 2011). So far, one would not expect undesired effects from activity at α3-GABA$_A$Rs, however, it is at present unknown if development of tolerance against the α2-mediated antihyperalgesic activity can be avoided by sparing activity at other GABA$_A$R subtypes. This is again a question, which is potentially accessible to studies examining mutant mice carrying combinations of H/R point-mutated GABA$_A$R subunits.

The genetic approaches described in this review are very well suited for experiments addressing the function of precisely defined GABA$_A$R subtypes. It is however not clear whether structural differences in the benzodiazepine binding site of different GABA$_A$R subtypes are large enough to permit development of subtype-selective benzodiazepines for each subtype or subtype combination. In particular differences between α2 and α3 subunits may be too small to permit full subtype specificity. Searching for modulatory sites at GABA$_A$Rs different from the classical benzodiazepine binding site might offer an alternative very intriguing, yet so far largely unexplored, opportunity.

7.2. Mixed GABA$_A$Rs with more than one type of α subunit

One potential limitation of the H/R point mutation approach is the behavior of GABA$_A$Rs with more than one type of α subunit. More than 25% of α1 containing GABA$_A$Rs harbor a second α subunit different from α1 (mainly α2 or α3), and the majority of α2 and α3 GABA$_A$Rs are mixed receptors (Benke et al., 2004). Because only one of the two α subunits can interact with the γ subunit to form the high-affinity benzodiazepine binding site, the responses of these mixed receptors to fully subtype-selective compounds will be determined by the type of α subunit which associates with the γ subunit (Minier & Sigel, 2004). This association may occur randomly but biochemical data suggest a certain rank order of the α subunits for interaction with the γ2 subunit (Balic, Rudolph, Fritschy, Möhler, & Benke, 2009). Moreover, biochemical experiments suggest that the H/R point mutation in the α subunit not only abolishes modulation by diazepam but also impairs the interaction with the γ subunit. This may eventually lead to false negative results in experiments with H/R point-mutated mice and may contribute to some of the discrepancies between predictions made from the H/R point-mutated mice and results obtained with subtype-selective agents (Sigel & Steinmann, 2012; Skolnick, 2012). Experiments comparing the phenotypes of single point-mutated mice and of triple point-mutated mice in which only a single subtype remains benzodiazepine-sensitive may provide additional insights also here.

8. CONCLUSION

There is compelling evidence from preclinical studies in rodents to support that non-sedative benzodiazepines with improved subtype specificity exert antihyperalgesic effects. Available clinical data are consistent with this view. Current knowledge suggests that robust antihyperalgesic activity with low sedative properties requires a high intrinsic activity at α2-$GABA_A$Rs (or possibly also at α3-/α5-$GABA_A$Rs) and very low activity at α1-$GABA_A$Rs. The optimal profile of such a drug in terms of $GABA_A$R subtype specificity is however still not known. On-going preclinical studies and clinical trials will hopefully soon provide additional insights.

CONFLICT OF INTEREST STATEMENT

The authors declare that they have no conflict of interest.

ACKNOWLEDGMENT

The research of H. U. Z. has been supported by grants from the Swiss National Science Foundation and by an Advanced Investigator Grant from the European Research Council (DHISP; 250128).

REFERENCES

Abe, K., Kato, G., Katafuchi, T., Tamae, A., Furue, H., & Yoshimura, M. (2009). Responses to 5-HT in morphologically identified neurons in the rat substantia gelatinosa in vitro. *Neuroscience*, *159*(1), 316–324. http://dx.doi.org/10.1016/j.neuroscience.2008.12.021, S0306-4522(08)01798-3 [pii].

Abrahamsen, B., Zhao, J., Asante, C. O., Cendan, C. M., Marsh, S., Martinez-Barbera, J. P., et al. (2008). The cell and molecular basis of mechanical, cold, and inflammatory pain. *Science*, *321*(5889), 702–705. http://dx.doi.org/10.1126/science.1156916, 321/5889/702 [pii].

Agarwal, N., Offermanns, S., & Kuner, R. (2004). Conditional gene deletion in primary nociceptive neurons of trigeminal ganglia and dorsal root ganglia. *Genesis*, *38*(3), 122–129. http://dx.doi.org/10.1002/gene.20010.

Ahmadi, S., Lippross, S., Neuhuber, W. L., & Zeilhofer, H. U. (2002). PGE_2 selectively blocks inhibitory glycinergic neurotransmission onto rat superficial dorsal horn neurons. *Nature Neuroscience*, *5*(1), 34–40.

Alvarez, F. J., Kavookjian, A. M., & Light, A. R. (1993). Ultrastructural morphology, synaptic relationships, and CGRP immunoreactivity of physiologically identified C-fiber terminals in the monkey spinal cord. *The Journal of Comparative Neurology, 329*(4), 472–490. http://dx.doi.org/10.1002/cne.903290405.

Andre, J., Zeau, B., Pohl, M., Cesselin, F., Benoliel, J. J., & Becker, C. (2005). Involvement of cholecystokininergic systems in anxiety-induced hyperalgesia in male rats: Behavioral and biochemical studies. *The Journal of Neuroscience, 25*(35), 7896–7904. http://dx.doi.org/10.1523/JNEUROSCI.0743-05.2005, 25/35/7896 [pii].

Atack, J. R., Hallett, D. J., Tye, S., Wafford, K. A., Ryan, C., Sanabria-Bohorquez, S. M., et al. (2011). Preclinical and clinical pharmacology of TPA023B, a $GABA_A$ receptor α2/α3 subtype-selective partial agonist. *Journal of Psychopharmacology, 25*(3), 329–344. http://dx.doi.org/10.1177/0269881109354928, 0269881109354928 [pii].

Atack, J. R., Wafford, K. A., Street, L. J., Dawson, G. R., Tye, S., Van Laere, K., et al. (2011). MRK-409 (MK-0343), a $GABA_A$ receptor subtype-selective partial agonist, is a non-sedating anxiolytic in preclinical species but causes sedation in humans. *Journal of Psychopharmacology, 25*(3), 314–328. http://dx.doi.org/10.1177/0269881109354927, 0269881109354927 [pii].

Atack, J. R., Wafford, K. A., Tye, S. J., Cook, S. M., Sohal, B., Pike, A., et al. (2006). TPA023 [7-(1,1-Dimethylethyl)-6-(2-ethyl-2H-1,2,4-triazol-3-ylmethoxy)-3-(2-fluorophenyl)-1,2,4-triazolo[4,3-b]pyridazine], an agonist selective for α2- and α3-containing $GABA_A$ receptors, is a nonsedating anxiolytic in rodents and primates. *The Journal of Pharmacology and Experimental Therapeutics, 316*(1), 410–422.

Ator, N. A., Atack, J. R., Hargreaves, R. J., Burns, H. D., & Dawson, G. R. (2010). Reducing abuse liability of $GABA_A$/benzodiazepine ligands via selective partial agonist efficacy at α1 and α2/3 subtypes. *The Journal of Pharmacology and Experimental Therapeutics, 332*(1), 4–16. http://dx.doi.org/10.1124/jpet.109.158303, jpet.109.158303 [pii].

Baba, H., Doubell, T. P., & Woolf, C. J. (1999). Peripheral inflammation facilitates Aβ fiber-mediated synaptic input to the substantia gelatinosa of the adult rat spinal cord. *The Journal of Neuroscience, 19*(2), 859–867.

Baba, H., Ji, R. R., Kohno, T., Moore, K. A., Ataka, T., Wakai, A., et al. (2003). Removal of GABAergic inhibition facilitates polysynaptic A fiber-mediated excitatory transmission to the superficial spinal dorsal horn. *Molecular and Cellular Neurosciences, 24*(3), 818–830.

Baccei, M. L., & Fitzgerald, M. (2004). Development of GABAergic and glycinergic transmission in the neonatal rat dorsal horn. *The Journal of Neuroscience, 24*(20), 4749–4757.

Balic, E., Rudolph, U., Fritschy, J. M., Möhler, H., & Benke, D. (2009). The α5(H105R) mutation impairs α5 selective binding properties by altered positioning of the α5 subunit in $GABA_A$ receptors containing two distinct types of α subunits. *Journal of Neurochemistry, 110*(1), 244–254. http://dx.doi.org/10.1111/j.1471-4159.2009.06119.x, JNC6119 [pii].

Benke, D., Fakitsas, P., Roggenmoser, C., Michel, C., Rudolph, U., & Möhler, H. (2004). Analysis of the presence and abundance of $GABA_A$ receptors containing two different types of α-subunits in murine brain using point-mutated α-subunits. *The Journal of Biological Chemistry, 279*(42), 43654–43660. http://dx.doi.org/10.1074/jbc.M407154200, M407154200 [pii].

Beyer, C., Roberts, L. A., & Komisaruk, B. R. (1985). Hyperalgesia induced by altered glycinergic activity at the spinal cord. *Life Sciences, 37*(9), 875–882.

Bohlhalter, S., Möhler, H., & Fritschy, J. M. (1994). Inhibitory neurotransmission in rat spinal cord: Co-localization of glycine- and $GABA_A$-receptors at GABAergic synaptic contacts demonstrated by triple immunofluorescence staining. *Brain Research, 642*(1–2), 59–69.

Bohlhalter, S., Weinmann, O., Möhler, H., & Fritschy, J. M. (1996). Laminar compartmentalization of $GABA_A$-receptor subtypes in the spinal cord: An immunohistochemical study. *The Journal of Neuroscience, 16*(1), 283–297.

Carrasquillo, Y., & Gereau, R. W. (2007). Activation of the extracellular signal-regulated kinase in the amygdala modulates pain perception. *The Journal of Neuroscience*, *27*(7), 1543–1551. http://dx.doi.org/10.1523/JNEUROSCI.3536-06.2007, 27/7/1543 [pii].

Colin, I., Rostaing, P., Augustin, A., & Triller, A. (1998). Localization of components of glycinergic synapses during rat spinal cord development. *The Journal of Comparative Neurology*, *398*(3), 359–372.

Coull, J. A., Beggs, S., Boudreau, D., Boivin, D., Tsuda, M., Inoue, K., et al. (2005). BDNF from microglia causes the shift in neuronal anion gradient underlying neuropathic pain. *Nature*, *438*(7070), 1017–1021.

Coull, J. A., Boudreau, D., Bachand, K., Prescott, S. A., Nault, F., Sik, A., et al. (2003). Trans-synaptic shift in anion gradient in spinal lamina I neurons as a mechanism of neuropathic pain. *Nature*, *424*(6951), 938–942.

Crestani, F., Löw, K., Keist, R., Mandelli, M., Möhler, H., & Rudolph, U. (2001). Molecular targets for the myorelaxant action of diazepam. *Molecular Pharmacology*, *59*(3), 442–445.

Dawson, G. R., Maubach, K. A., Collinson, N., Cobain, M., Everitt, B. J., MacLeod, A. M., et al. (2006). An inverse agonist selective for α5 subunit-containing $GABA_A$ receptors enhances cognition. *The Journal of Pharmacology and Experimental Therapeutics*, *316*(3), 1335–1345. http://dx.doi.org/10.1124/jpet.105.092320, jpet.105.092320 [pii].

de Haas, S. L., de Visser, S. J., van der Post, J. P., Schoemaker, R. C., van Dyck, K., Murphy, M. G., et al. (2008). Pharmacodynamic and pharmacokinetic effects of MK-0343, a $GABA_A$ α2,3 subtype selective agonist, compared to lorazepam and placebo in healthy male volunteers. *Journal of Psychopharmacology*, *22*(1), 24–32. http://dx.doi.org/10.1177/0269881107082108, 22/1/24 [pii].

Di Lio, A., Benke, D., Besson, M., Desmeules, J., Daali, Y., Wang, Z. J., et al. (2011). HZ166, a novel $GABA_A$ receptor subtype-selective benzodiazepine site ligand, is antihyperalgesic in mouse models of inflammatory and neuropathic pain. *Neuropharmacology*, *60*(4), 626–632. http://dx.doi.org/10.1016/j.neuropharm.2010.11.026, S0028-3908(10)00325-4 [pii].

Feng, Y. P., Li, Y. Q., Wang, W., Wu, S. X., Chen, T., Shigemoto, R., et al. (2005). Morphological evidence for GABA/glycine-cocontaining terminals in synaptic contact with neurokinin-1 receptor-expressing neurons in the sacral dorsal commissural nucleus of the rat. *Neuroscience Letters*, *388*(3), 144–148. http://dx.doi.org/10.1016/j.neulet.2005.06.068, S0304-3940(05)00754-8 [pii].

Fishbain, D. A., Cutler, R. B., Rosomoff, H. L., & Rosomoff, R. S. (2000). Clonazepam open clinical treatment trial for myofascial syndrome associated chronic pain. *Pain Medicine*, *1*(4), 332–339.

Fujita, M., Woods, S. W., Verhoeff, N. P., Abi-Dargham, A., Baldwin, R. M., Zoghbi, S. S., et al. (1999). Changes of benzodiazepine receptors during chronic benzodiazepine administration in humans. *European Journal of Pharmacology*, *368*(2–3), 161–172.

Hansen, R. R., Erichsen, H. K., Brown, D. T., Mirza, N. R., & Munro, G. (2012). Positive allosteric modulation of $GABA_A$ receptors reduces capsaicin-induced primary and secondary hypersensitivity in rats. *Neuropharmacology*, *63*(8), 1360–1367. http://dx.doi.org/10.1016/j.neuropharm.2012.08.002, S0028-3908(12)00399-1 [pii].

Harkins, S., Linford, J., Cohen, J., Kramer, T., & Cueva, L. (1991). Administration of clonazepam in the treatment of TMD and associated myofascial pain: A double-blind pilot study. *Journal of Craniomandibular Disorders*, *5*(3), 179–186.

Harris, J. A., & Westbrook, R. F. (1995). Effects of benzodiazepine microinjection into the amygdala or periaqueductal gray on the expression of conditioned fear and hypoalgesia in rats. *Behavioral Neuroscience*, *109*(2), 295–304.

Harvey, R. J., Depner, U. B., Wässle, H., Ahmadi, S., Heindl, C., Reinold, H., et al. (2004). GlyR α3: An essential target for spinal PGE2-mediated inflammatory pain sensitization. *Science*, *304*(5672), 884–887.

Hofmann, M., Kordas, K. S., Gravius, A., Bolcskei, K., Parsons, C. G., Dekundy, A., et al. (2012). Assessment of the effects of NS11394 and L-838417, α2/3 subunit-selective $GABA_A$ [corrected] receptor-positive allosteric modulators, in tests for pain, anxiety, memory and motor function. *Behavioural Pharmacology*, *23*(8), 790–801. http://dx.doi.org/10.1097/FBP.0b013e32835a7c7e.

Jasmin, L., Rabkin, S. D., Granato, A., Boudah, A., & Ohara, P. T. (2003). Analgesia and hyperalgesia from GABA-mediated modulation of the cerebral cortex. *Nature*, *424*(6946), 316–320.

Kanaka, C., Ohno, K., Okabe, A., Kuriyama, K., Itoh, T., Fukuda, A., et al. (2001). The differential expression patterns of messenger RNAs encoding K-Cl cotransporters (KCC1,2) and Na-K-2Cl cotransporter (NKCC1) in the rat nervous system. *Neuroscience*, *104*(4), 933–946.

Keller, A. F., Beggs, S., Salter, M. W., & De Koninck, Y. (2007). Transformation of the output of spinal lamina I neurons after nerve injury and microglia stimulation underlying neuropathic pain. *Molecular Pain*, *3*, 27.

Keller, A. F., Breton, J. D., Schlichter, R., & Poisbeau, P. (2004). Production of 5α-reduced neurosteroids is developmentally regulated and shapes $GABA_A$ miniature IPSCs in lamina II of the spinal cord. *The Journal of Neuroscience*, *24*(4), 907–915.

Keller, A. F., Coull, J. A., Chery, N., Poisbeau, P., & De Koninck, Y. (2001). Region-specific developmental specialization of GABA-glycine cosynapses in laminas I-II of the rat spinal dorsal horn. *The Journal of Neuroscience*, *21*(20), 7871–7880.

Knabl, J., Witschi, R., Hösl, K., Reinold, H., Zeilhofer, U. B., Ahmadi, S., et al. (2008). Reversal of pathological pain through specific spinal $GABA_A$ receptor subtypes. *Nature*, *451*, 330–334.

Knabl, J., Zeilhofer, U. B., Crestani, F., Rudolph, U., & Zeilhofer, H. U. (2009). Genuine antihyperalgesia by systemic diazepam revealed by experiments in $GABA_A$ receptor point-mutated mice. *Pain*, *141*(3), 233–238. http://dx.doi.org/10.1016/j.pain.2008.10.015, S0304-3959(08)00656-8 [pii].

Kullmann, D. M., Ruiz, A., Rusakov, D. M., Scott, R., Semyanov, A., & Walker, M. C. (2005). Presynaptic, extrasynaptic and axonal $GABA_A$ receptors in the CNS: Where and why? *Progress in Biophysics & Molecular Biology*, *87*(1), 33–46. http://dx.doi.org/10.1016/j.pbiomolbio.2004.06.003, S0079-6107(04)00060-4 [pii].

Lin, Q., Wu, J., & Willis, W. D. (1999). Dorsal root reflexes and cutaneous neurogenic inflammation after intradermal injection of capsaicin in rats. *Journal of Neurophysiology*, *82*(5), 2602–2611.

Lin, Q., Zou, X., & Willis, W. D. (2000). Aδ and C primary afferents convey dorsal root reflexes after intradermal injection of capsaicin in rats. *Journal of Neurophysiology*, *84*(5), 2695–2698.

Lorenzo, L., Godin, A., Wang, F., St-Louis, M., Carbonetto, S., Wiseman, P., et al. (2014). Gephyrin clusters are absent from small diameter primary afferent terminals despite the presence of $GABA_A$ receptors. *The Journal of Neuroscience*, *34*(24), 8300–8317. http://dx.doi.org/10.1523/JNEUROSCI.0159-14.2014.

Löw, K., Crestani, F., Keist, R., Benke, D., Brünig, I., Benson, J. A., et al. (2000). Molecular and neuronal substrate for the selective attenuation of anxiety. *Science*, *290*(5489), 131–134.

Luo, C., Kuner, T., & Kuner, R. (2014). Synaptic plasticity in pathological pain. *Trends in Neurosciences*, *37*(6), 343–355. http://dx.doi.org/10.1016/j.tins.2014.04.002, S0166-2236(14)00061-7 [pii].

Malizia, A., Forse, G., Haida, A., Gunn, R., Melichar, J., Poole, K., et al. (1995). A new human (psycho)pharmacology tool: The multiple organs coincidences counter (MOCC). *Journal of Psychopharmacology*, *9*(4), 294–306. http://dx.doi.org/10.1177/026988119500900402, 9/4/294 [pii].

McKernan, R. M., Rosahl, T. W., Reynolds, D. S., Sur, C., Wafford, K. A., Atack, J. R., et al. (2000). Sedative but not anxiolytic properties of benzodiazepines are mediated by the $GABA_A$ receptor α1 subtype. *Nature Neuroscience*, *3*(6), 587–592.

Melzack, R., & Wall, P. D. (1965). Pain mechanisms: A new theory. *Science*, *150*(699), 971–979.

Minier, F., & Sigel, E. (2004). Positioning of the α-subunit isoforms confers a functional signature to γ-aminobutyric acid type A receptors. *Proceedings of the National Academy of Sciences of the United States of America*, *101*(20), 7769–7774. http://dx.doi.org/10.1073/pnas.0400220101, 0400220101 [pii].

Miraucourt, L. S., Dallel, R., & Voisin, D. L. (2007). Glycine inhibitory dysfunction turns touch into pain through PKCγ interneurons. *PLoS One*, *2*(11), e1116.

Mirza, N. R., Larsen, J. S., Mathiasen, C., Jacobsen, T. A., Munro, G., Erichsen, H. K., et al. (2008). NS11394 [3'-[5-(1-hydroxy-1-methyl-ethyl)-benzoimidazol-1-yl]-biphenyl-2-carbonitr ile], a unique subtype-selective $GABA_A$ receptor positive allosteric modulator: In vitro actions, pharmacokinetic properties and in vivo anxiolytic efficacy. *The Journal of Pharmacology and Experimental Therapeutics*, *327*(3), 954–968. http://dx.doi.org/10.1124/jpet.108.138859, jpet.108.138859 [pii].

Moore, K. A., Kohno, T., Karchewski, L. A., Scholz, J., Baba, H., & Woolf, C. J. (2002). Partial peripheral nerve injury promotes a selective loss of GABAergic inhibition in the superficial dorsal horn of the spinal cord. *The Journal of Neuroscience*, *22*(15), 6724–6731.

Munro, G., Erichsen, H. K., Rae, M. G., & Mirza, N. R. (2011). A question of balance–positive versus negative allosteric modulation of $GABA_A$ receptor subtypes as a driver of analgesic efficacy in rat models of inflammatory and neuropathic pain. *Neuropharmacology*, *61*(1–2), 121–132. http://dx.doi.org/10.1016/j.neuropharm.2011.03.017, S0028-3908(11)00126-2 [pii].

Munro, G., Lopez-Garcia, J. A., Rivera-Arconada, I., Erichsen, H. K., Nielsen, E. O., Larsen, J. S., et al. (2008). Comparison of the novel subtype-selective $GABA_A$ receptor-positive allosteric modulator NS11394 [3'-[5-(1-hydroxy-1-methyl-ethyl)-benzoimidazol-1-yl]-biphenyl-2-carbonitr ile] with diazepam, zolpidem, bretazenil, and gaboxadol in rat models of inflammatory and neuropathic pain. *The Journal of Pharmacology and Experimental Therapeutics*, *327*(3), 969–981. http://dx.doi.org/10.1124/jpet.108.144568, jpet.108.144568 [pii].

Nichols, M. L., Allen, B. J., Rogers, S. D., Ghilardi, J. R., Honore, P., Luger, N. M., et al. (1999). Transmission of chronic nociception by spinal neurons expressing the substance P receptor. *Science*, *286*(5444), 1558–1561, 7999 [pii].

Nickolls, S., Mace, H., Fish, R., Edye, M., Gurrell, R., Ivarsson, M., et al. (2011). A comparison of the α2/3/5 selective positive allosteric modulators L-838,417 and TPA023 in preclinical models of inflammatory and neuropathic pain. *Advances in Pharmaceutical Sciences*, *2011*, 608912. http://dx.doi.org/10.1155/2011/608912.

Paul, J., Yévenes, G. E., Benke, D., Di Lio, A., Ralvenius, W. T., Witschi, R., et al. (2014). Antihyperalgesia by α2-$GABA_A$ receptors occurs via a genuine spinal action and does not involve supraspinal sites. *Neuropsychopharmacology*, *39*(2), 477–487. http://dx.doi.org/10.1038/npp.2013.221, npp2013221 [pii].

Paul, J., Zeilhofer, H. U., & Fritschy, J. M. (2012). Selective distribution of $GABA_A$ receptor subtypes in mouse spinal dorsal horn neurons and primary afferents. *The Journal of Comparative Neurology*, *520*(17), 3895–3911. http://dx.doi.org/10.1002/cne.23129.

Pauli, S., Farde, L., Halldin, C., & Sedvall, G. (1991). Occupancy of the central benzodiazepine receptors during benzodiazepine treatment determined by PET. *European Neuropsychopharmacology*, *1*, 229–231.

Pernia-Andrade, A. J., Kato, A., Witschi, R., Witschi, R., Nyilas, R., Katona, I., et al. (2009). Spinal endocannabinoids and CB1 receptors mediate C-fiber-induced heterosynaptic pain sensitization. *Science*, *325*(5941), 760–764. http://dx.doi.org/10.1126/science.1171870, 325/5941/760 [pii].

Prescott, S. A., Sejnowski, T. J., & De Koninck, Y. (2006). Reduction of anion reversal potential subverts the inhibitory control of firing rate in spinal lamina I neurons: Towards a biophysical basis for neuropathic pain. *Molecular Pain*, *2*, 32.

Price, T. J., Hargreaves, K. M., & Cervero, F. (2006). Protein expression and mRNA cellular distribution of the NKCC1 cotransporter in the dorsal root and trigeminal ganglia of the rat. *Brain Research*, *1112*(1), 146–158.

Ralvenius, W. T., Benke, D., Rudolph, U., & Zeilhofer, H. U. (2013). Antihyperalgesic effects of systemic diazepam in $GABA_A$ receptor point-mutated mice carrying only one diazepam-sensitive $GABA_A$ receptor subtype *Neuroscience 2013 Abstracts*, 447.416/II443.

Reichl, S., Augustin, M., Zahn, P. K., & Pogatzki-Zahn, E. M. (2012). Peripheral and spinal GABAergic regulation of incisional pain in rats. *Pain*, *153*(1), 129–141. http://dx.doi.org/10.1016/j.pain.2011.09.028, S0304-3959(11)00585-9 [pii].

Reinold, H., Ahmadi, S., Depner, U. B., Layh, B., Heindl, C., Hamza, M., et al. (2005). Spinal inflammatory hyperalgesia is mediated by prostaglandin E receptors of the EP2 subtype. *The Journal of Clinical Investigation*, *115*(3), 673–679.

Ribeiro-da-Silva, A. (1995). Substantia gelatinosa of the spinal cord. In G. Paxinos (Ed.), *The rat nervous system*. San Diego: Academic Press.

Ribeiro-Da-Silva, A., Castro-Lopes, J. M., & Coimbra, A. (1986). Distribution of glomeruli with fluoride-resistant acid phosphatase (FRAP)-containing terminals in the substantia gelatinosa of the rat. *Brain Research*, *377*(2), 323–329, 0006-8993(86)90875-9 [pii].

Ribeiro-da-Silva, A., Tagari, P., & Cuello, A. C. (1989). Morphological characterization of substance P-like immunoreactive glomeruli in the superficial dorsal horn of the rat spinal cord and trigeminal subnucleus caudalis: A quantitative study. *The Journal of Comparative Neurology*, *281*(4), 497–515. http://dx.doi.org/10.1002/cne.902810402.

Rivas, F. M., Stables, J. P., Murphree, L., Edwankar, R. V., Edwankar, C. R., Huang, S., et al. (2009). Antiseizure activity of novel γ-aminobutyric $acid_A$ receptor subtype-selective benzodiazepine analogues in mice and rat models. *Journal of Medicinal Chemistry*, *52*(7), 1795–1798. http://dx.doi.org/10.1021/jm801652d.

Roberts, L. A., Beyer, C., & Komisaruk, B. R. (1986). Nociceptive responses to altered GABAergic activity at the spinal cord. *Life Sciences*, *39*(18), 1667–1674.

Rowlett, J. K., Platt, D. M., Lelas, S., Atack, J. R., & Dawson, G. R. (2005). Different $GABA_A$ receptor subtypes mediate the anxiolytic, abuse-related, and motor effects of benzodiazepine-like drugs in primates. *Proceedings of the National Academy of Sciences of the United States of America*, *102*(3), 915–920. http://dx.doi.org/10.1073/pnas.0405621102, 0405621102 [pii].

Rudolph, U., Crestani, F., Benke, D., Brünig, I., Benson, J. A., Fritschy, J. M., et al. (1999). Benzodiazepine actions mediated by specific γ-aminobutyric $acid_A$ receptor subtypes. *Nature*, *401*(6755), 796–800.

Rudolph, U., & Knoflach, F. (2011). Beyond classical benzodiazepines: Novel therapeutic potential of $GABA_A$ receptor subtypes. *Nature Reviews Drug Discovery*, *10*(9), 685–697. http://dx.doi.org/10.1038/nrd3502, nrd3502 [pii].

Rudolph, U., & Möhler, H. (2004). Analysis of $GABA_A$ receptor function and dissection of the pharmacology of benzodiazepines and general anesthetics through mouse genetics. *Annual Review of Pharmacology and Toxicology*, *44*, 475–498.

Rudomin, P., & Schmidt, R. F. (1999). Presynaptic inhibition in the vertebrate spinal cord revisited. *Experimental Brain Research*, *129*(1), 1–37.

Sandkühler, J. (2009). Models and mechanisms of hyperalgesia and allodynia. *Physiological Reviews*, *89*(2), 707–758. http://dx.doi.org/10.1152/physrev.00025.2008, 89/2/707 [pii].

Shinotoh, H., Iyo, M., Yamada, T., Inoue, O., Suzuki, K., Itoh, T., et al. (1989). Detection of benzodiazepine receptor occupancy in the human brain by positron emission tomography. *Psychopharmacology*, *99*(2), 202–207.

Sigel, E., & Steinmann, M. E. (2012). Structure, function, and modulation of $GABA_A$ receptors. *The Journal of Biological Chemistry*, *287*(48), 40224–40231. http://dx.doi.org/10.1074/jbc.R112.386664, R112.386664 [pii].

Skolnick, P. (2012). Anxioselective anxiolytics: On a quest for the holy grail. *Trends in Pharmacological Sciences*, *33*(11), 611–620. http://dx.doi.org/10.1016/j.tips.2012.08.003, S0165-6147(12)00144-7 [pii].

Tan, K. R., Brown, M., Labouebe, G., Yvon, C., Creton, C., Fritschy, J. M., et al. (2010). Neural bases for addictive properties of benzodiazepines. *Nature*, *463*(7282), 769–774. http://dx.doi.org/10.1038/nature08758, nature08758 [pii].

Tatsuo, M. A., Salgado, J. V., Yokoro, C. M., Duarte, I. D., & Francischi, J. N. (1999). Midazolam-induced hyperalgesia in rats: Modulation via $GABA_A$ receptors at supraspinal level. *European Journal of Pharmacology*, *370*(1), 9–15.

Todd, A. J., & Sullivan, A. C. (1990). Light microscope study of the coexistence of GABA-like and glycine-like immunoreactivities in the spinal cord of the rat. *The Journal of Comparative Neurology*, *296*(3), 496–505.

Todd, A. J., Watt, C., Spike, R. C., & Sieghart, W. (1996). Colocalization of GABA, glycine, and their receptors at synapses in the rat spinal cord. *The Journal of Neuroscience*, *16*(3), 974–982.

Torsney, C., & MacDermott, A. B. (2006). Disinhibition opens the gate to pathological pain signaling in superficial neurokinin 1 receptor-expressing neurons in rat spinal cord. *The Journal of Neuroscience*, *26*(6), 1833–1843.

Vidal, C., & Jacob, J. J. (1982). Stress hyperalgesia in rats: An experimental animal model of anxiogenic hyperalgesia in human. *Life Sciences*, *31*(12–13), 1241–1244.

Vuilleumier, P. H., Besson, M., Desmeules, J., Arendt-Nielsen, L., & Curatolo, M. (2013). Evaluation of anti-hyperalgesic and analgesic effects of two benzodiazepines in human experimental pain: A randomized placebo-controlled study. *PLoS One*, *8*(3), e43896, 10.1371/journal.pone.0043896 PONE-D-12-21279 [pii].

Witschi, R., Punnakkal, P., Paul, J., Walczak, J.-S., Cervero, F., Fritschy, J.-M., et al. (2011). Presynaptic α2-$GABA_A$ receptors in primary afferent depolarization in spinal pain control. *The Journal of Neuroscience*, *31*(22), 8134–8142.

Yoshimura, M., & Nishi, S. (1995). Primary afferent-evoked glycine- and GABA-mediated IPSPs in substantia gelatinosa neurones in the rat spinal cord in vitro. *The Journal of Physiology*, *482*(Pt 1), 29–38.

Zeilhofer, H. U., Benke, D., & Yévenes, G. E. (2012). Chronic pain states: Pharmacological strategies to restore diminished inhibitory spinal pain control. *Annual Review of Pharmacology and Toxicology*, *52*, 111–133. http://dx.doi.org/10.1146/annurev-pharmtox-010611-134636.

Zeilhofer, H. U., Möhler, H., & Di Lio, A. (2009). GABAergic analgesia—New insights from mutant mice and subtype-selective agonists. *Trends in Pharmacological Sciences*, *30*(8), 397–402. http://dx.doi.org/10.1016/j.tips.2009.05.007.

Zeilhofer, H. U., Wildner, H., & Yévenes, G. E. (2012). Fast synaptic inhibition in spinal sensory processing and pain control. *Physiological Reviews*, *92*(1), 193–235. http://dx.doi.org/10.1152/physrev.00043.2010, 92/1/193 [pii].

Zeilhofer, H. U., Witschi, R., & Johansson, T. (2009). Fast inhibitory transmission of pain in the spinal cord. In M. Malcangio (Ed.), *Synaptic plasticity in pain*. Heidelberg: Springer.

Zhang, Z., Cai, Y. Q., Zou, F., Bie, B., & Pan, Z. Z. (2011). Epigenetic suppression of GAD65 expression mediates persistent pain. *Nature Medicine*, *17*(11), 1448–1455. http://dx.doi.org/10.1038/nm.2442, nm.2442 [pii].

CHAPTER FIVE

GABAergic Control of Depression-Related Brain States

Bernhard Luscher[*,†,‡,1], **Thomas Fuchs**[*,†,‡]
[*]Department of Biology, The Pennsylvania State University, University Park, Pennsylvania, USA
[†]Department of Biochemistry and Molecular Biology, The Pennsylvania State University, University Park, Pennsylvania, USA
[‡]Center for Molecular Investigation of Neurological Disorders, The Huck Institutes of the Life Sciences, The Pennsylvania State University, University Park, Pennsylvania, USA
[1]Corresponding author: e-mail address: bxl25@psu.edu

Contents

Abstract

The GABAergic deficit hypothesis of major depressive disorders (MDDs) posits that reduced γ-aminobutyric acid (GABA) concentration in brain, impaired function of GABAergic interneurons, altered expression and function of $GABA_A$ receptors, and changes in GABAergic transmission dictated by altered chloride homeostasis can contribute to the etiology of MDD. Conversely, the hypothesis posits that the efficacy of currently used antidepressants is determined by their ability to enhance GABAergic neurotransmission. We here provide an update for corresponding evidence from studies of patients and preclinical animal models of depression. In addition, we propose an explanation for the continued lack of genetic evidence that explains the considerable heritability of MDD. Lastly, we discuss how alterations in GABAergic transmission are integral to other hypotheses of MDD that emphasize (i) the role of monoaminergic deficits, (ii) stress-based etiologies, (iii) neurotrophic deficits, and (iv) the neurotoxic and neural circuit-impairing consequences of chronic excesses of glutamate. We propose that

Advances in Pharmacology, Volume 73
ISSN 1054-3589
http://dx.doi.org/10.1016/bs.apha.2014.11.003

altered GABAergic transmission serves as a common denominator of MDD that can account for all these other hypotheses and that plays a causal and common role in diverse mechanistic etiologies of depressive brain states and in the mechanism of action of current antidepressant drug therapies.

ABBREVIATIONS

5-HT 5-hydroxytryptamine
AMPA α-amino-3-hydroxy-5-methyl-4-isoxazolepropionic acid
BDNF brain-derived neurotrophic factor
CA1 cornus ammonis 1
CA3 cornus ammonis 3
CREB cAMP-regulated element-binding protein
CRH corticotropin-releasing hormone
DLPFC dorsolateral prefrontal cortex
E_{Cl} chloride equilibrium potential
ERK extracellular signal-regulated kinase
GABA γ-aminobutyric acid
GABA$_A$R GABA type A receptor
GAD glutamic acid decarboxylase
GSK3β glycogen synthase kinase 3β
GWAS genome-wide association studies
HPA hypothalamus–pituitary–adrenal
KCC2 potassium-chloride cotransporter 2
MDD major depressive disorder
NMDA *N*-methyl-D-aspartic acid
NPY neuropeptide Y
PET positron emission tomography
PV parvalbumin
PVN paraventricular nucleus of the hypothalamus
RGL radial glia-like astrocyte/stem cell
sgACC subgenual anterior cingulate cortex
SSRI selective serotonin reuptake inhibitor
SST somatostatin
TCA tricyclic antidepressant
TrkB tropomyosin-related kinase B
VPN ventral posterior nucleus of the hypothalamus

1. INTRODUCTION

Major depressive disorder (MDD) is a common and highly heterogeneous psychiatric syndrome and a leading cause of total disability (Murray & Lopez, 1996; World Health Organization, 2008). The lifetime prevalence of MDD in the US population has been estimated to be between 13% and 17%

(Hasin, Goodwin, Stinson, & Grant, 2005; Kessler et al., 2003). Depressive disorders are highly comorbid with pathological anxiety, especially generalized anxiety disorder and posttraumatic stress disorder. About 85% of patients with MDD exhibit significant anxiety symptoms, and 58% of patients with a history of depression also suffer from an anxiety disorder (Baldwin, Evans, Hirschfeld, & Kasper, 2002; Gamez, Watson, & Doebbeling, 2007; Gorman, 1996). Currently used antidepressant drug therapies act with a delay of several weeks. Moreover, they are ineffective in that only one-third of patients respond to the first agent prescribed (Keller et al., 2000), and the observed therapeutic effect is superior to placebo in approximately 50% of clinical trials only (Khan, Khan, Walens, Kolts, & Giller, 2003). Among patients who respond to drug treatment only a fraction shows remission, and recurrence is the rule rather than an exception. Thus, there is an enormous unmet need for better antidepressant therapies.

According to DSMIV, the diagnostic criteria for MDD are loosely defined as a cluster of at least five symptoms of which at least one of two core symptoms, depressed mood or loss of interest or pleasure in life activities, is observed in combination with three or four other symptoms during the same 2-week period. These additional symptoms include unintentional weight gain or loss, insomnia or hypersomnia, psychomotor agitation or retardation, fatigue or loss of energy, feelings of worthlessness or guilt, diminished ability to think, concentrate or indecisiveness, and recurrent thoughts of death (American Psychiatric Association, 2000). Most of these symptoms exist on a continuum from normal to pathological, which defies categorical quantification (Angst & Merikangas, 2001). Consequently, two patients can suffer from MDD without sharing any of their diagnostic symptoms. The broad range of symptoms implies that the goal of mapping MDD onto a unique set of abnormal molecules, cells, or neural circuits is impossible to attain. Nevertheless, based on a wide range of approaches, different subregions of the prefrontal cortex, the subgenual anterior cingulate cortex (sgACC), and the hippocampus have emerged as primary sites of pathology. The amygdala and subcortical reward circuits are also implicated (Drevets, 2001; Russo & Nestler, 2013). Particularly informative on the brain substrate of MDD were insights from brain imaging (Murray, Wise, & Drevets, 2011), regional brain volume loss (i.e., Bell-McGinty et al., 2002; Hickie et al., 2005; Koolschijn, van Haren, Lensvelt-Mulders, Hulshoff Pol, & Kahn, 2009; Sacher et al., 2012; Steffens et al., 2000), circuits underlying cognitive impairments (Fujii, Saito, Yanaka, Kosaka,

& Okazawa, 2014), and brain regions and their functional connections that respond to therapeutically effective deep brain stimulation (Holtzheimer & Mayberg, 2011).

Based on family and twin studies, MDD shows significant heritability. The nature of this heritability, however, is unexplained as evidence for specific genes that confer risk for MDD continues to be lacking. Therefore, and in contrast to schizophrenia or autism spectrum disorders, current hypotheses on the etiopathology of MDD must rely on empirical information other than human genetic vulnerabilities. The monoamine hypothesis of MDD dates back more than 50 years and posits that depressive disorders are caused by imbalances in serotonergic, noradrenergic, and possibly dopaminergic transmitter systems. It is derived from the notion that monoamine oxidase inhibitors, tricyclic antidepressants (TCAs), and selective serotonin reuptake inhibitors (SSRIs) have in common that they increase the extracellular concentrations of these transmitters (Bunney & Davis, 1965; Coppen, 1967; Matussek, 1969; Schildkraut, 1995). However, it has become widely accepted that antidepressant mechanisms do not simply reflect increased monoamine transmitter function and that alterations in these transmitter systems are insufficient to explain the complex nature of affective disorders and antidepressant drug mechanisms. More recently, a number of alternative hypotheses have emerged that focus on the underlying biology rather than pharmacology. We here will provide an update on the GABAergic deficit hypothesis of MDD (Luscher, Shen, & Sahir, 2011). We also discuss the continued lack of concrete evidence for genetic alterations that could explain the heritability of MDD. In addition, we review and emphasize the role of GABAergic deficits in the context of the monoamine deficiency hypothesis, stress-based hypothesis (Holsboer, 2001; Kendler, Karkowski, & Prescott, 1999; Pittenger & Duman, 2008), neurotrophic deficit hypothesis (Duman & Monteggia, 2006; Krishnan & Nestler, 2008), and glutamatergic hypothesis of MDD (Paul & Skolnick, 2003; Tokita, Yamaji, & Hashimoto, 2012).

2. THE GABAergic DEFICIT HYPOTHESIS OF MDD

There is an abundance of evidence that MDD is associated with diverse defects in GABAergic transmission. Given the heterogeneity of MDD, it seems likely that none of these individual deficits are representative of MDD. Instead, the currently available data suggest that environmental conditions interact with heritable abnormalities to affect different aspects of GABAergic transmission and converge on an impaired balance of neural excitation and inhibition as a common feature that causally contributes to the psychopathology of MDD. In particular, a first line of clinical evidence suggesting reduced GABAergic transmission in depressed patients is based on positron emission tomography (PET) imaging of [^{11}C]-flumazenil binding (Klumpers et al., 2010). These studies revealed reduced expression of $GABA_A$Rs (GABA type A receptors) in the limbic parahippocampal temporal gyrus and right lateral superior temporal gyrus, which is consistent with evidence for glucose hypermetabolism in the same brain region of patients (Aihara et al., 2007). There is also evidence for reduced $GABA_A$R expression based on downregulation of subunit transcripts. Comparing the frontopolar cortex of depressed suicide victims with that of nondepressed control subjects that had died from other causes, Merali et al. found evidence for reduced expression of α1,3,4 and δ subunit mRNAs (Merali et al., 2004). However, when brains of depressed and nondepressed suicide victims were compared directly, several $GABA_A$R subunit mRNA levels were increased rather than reduced (Choudary et al., 2005; Klempan et al., 2009; Sequeira et al., 2007, 2009). Further experimentation is needed to determine whether differences between transcript studies and PET imaging reflect brain region-specific phenotypes of MDD or whether transcript changes are compensatory for reductions in protein expression. However, the functional expression of $GABA_A$Rs at the cell surface and at synapses is known to be subject to a plethora of posttranslational regulatory mechanisms that ultimately determine receptor function, as illustrated throughout in this chapter and summarized comprehensively in previous reviews (Brickley & Mody, 2012; Jacob, Moss, & Jurd, 2008; Luscher, Fuchs, & Kilpatrick, 2011; Tyagarajan & Fritschy, 2014). In particular, $GABA_A$Rs and associated proteins are targets of diverse signaling cascades that, mostly through changes in phosphorylation state, control their trafficking, stability in the plasma membrane, and diffusional dynamics at synapses, independently of changes in transcript levels. Changes in $GABA_A$R transcript levels therefore must be interpreted with caution.

A second salient feature pointing to GABAergic dysfunction in MDD is based on drastically reduced brain concentrations of γ-aminobutyric acid (GABA) in both occipital cortex and anterior cingulate cortex (Gabbay et al., 2012; Hasler et al., 2007; Sanacora et al., 1999, 2004). This finding is supported by studies of postmortem prefrontal cortex and amygdala showing reduced protein and mRNA encoding glutamic acid decarboxylase (GAD)67, a key enzyme for the synthesis of GABA (Guilloux et al.,

2012; Karolewicz et al., 2010). Reduced GAD67 and GABA levels are further consistent with reduced function or densities of specific subtypes of GABAergic interneurons. Specifically, immunostaining of postmortem brain revealed a drastically reduced density of calbindin-positive GABAergic interneurons mainly in the dorsolateral prefrontal cortex (DLPFC) of MDD patients, along with a significant reduction in size of cell bodies (Rajkowska, O'Dwyer, Teleki, Stockmeier, & Miguel-Hidalgo, 2007). Interestingly, calbindin-positive cells were reduced also in occipital cortex (Maciag et al., 2010), again consistent with GABA reductions in this area. In addition, transcript analyses of the postmortem DLPFC and sgACC revealed association of MDD with reduced expression of somatostatin (SST), a neuropeptide marker representative of about 30% of cortical GABAergic interneurons including dendrite-targeting Martinotti cells (Sibille, Morris, Kota, & Lewis, 2011; Tripp, Kota, Lewis, & Sibille, 2011). SST-positive interneurons show variable colabeling with calbindin (Rossignol, 2011), and it remains to be determined to what extent changes in these two markers are representative of the same neurons. Importantly, SST transcripts and protein were also reduced in the lateral/basolateral/basomedian amygdala of a separate cohort of patients (Guilloux et al., 2012). Furthermore, functional deficits of SST neurons are evidenced by reduced expression of neuropeptide Y (NPY), tachykinin, and cortistatin transcripts, which are coexpressed with SST in the same subpopulation of interneurons (Guilloux et al., 2012). This same study also confirmed earlier evidence that MDD is associated with significantly reduced expression of brain-derived neurotrophic factor (BDNF), a key protein in the etiology of MDD and antidepressant drug action (Sections 4 and 6). By analyzing two different strains of mice with constitutive or activity-dependent decreases in BDNF, Guilloux et al. demonstrated a causal relationship between BDNF functional deficits and reduced expression of markers of SST/NPY-positive interneurons (Guilloux et al., 2012). Abundant evidence from mouse genetics, discussed further below, indicates that BDNF promotes the functional maturation of GABAergic circuits, which is a prerequisite for structural maturation also of dendrites and dendritic spines of glutamatergic neurons.

Importantly, the changes in expression of markers of GABAergic transmission described above are functionally relevant for MDD as indicated by marked reductions in cortical GABAergic inhibition in depressed patients observed by transcranial magnetic stimulation (TMS; Levinson et al., 2010). The importance of GABAergic deficits for MDD is further suggested by analyses of patients subjected to treatment with SSRIs (Bhagwagar et al.,

2004; Kucukibrahimoglu et al., 2009; Sanacora, Mason, Rothman, & Krystal, 2002) or electroconvulsive therapy (Sanacora et al., 2003), which showed that these treatments normalize GABA concentrations in brain and plasma of MDD subjects. Besides GABA, MDD has been shown to involve reduced cerebrospinal fluid concentrations of the neurosteroid 3alpha-hydroxy-5alpha-pregnan-20-one (allopregnanolone; Uzunova et al., 1998). Allopregnanolone and other neurosteroids are endogenously produced high-affinity ligands of $GABA_A$Rs that dynamically potentiate the function of GABA similar to benzodiazepines. Significantly, chronic treatment with fluoxetine (but not imipramine) increased the allopregnanolone concentrations in rats and normalized its concentration in MDD patients (Uzunov, Cooper, Costa, & Guidotti, 1996; Uzunova et al., 1998). Collectively, it is becoming increasingly evident that antidepressant therapies, including drug therapies designed to normalize monoaminergic transmission, ultimately act to enhance GABAergic transmission.

Alterations in biological markers of GABAergic transmission for the most part are not uniquely found in patients with MDD but also broadly implicated in other psychiatric disorders. While there is a consensus that GABA is reduced in MDD (Gabbay et al., 2012; Hasler et al., 2007; Sanacora et al., 1999, 2004), GABA concentrations appear mostly unchanged or increased in schizophrenia (Kaufman et al., 2009; Ongur, Prescot, McCarthy, Cohen, & Renshaw, 2010; Tayoshi et al., 2010). However, one study found GABA levels trending higher in young patients and lower in older patients with schizophrenia compared to healthy controls (Rowland et al., 2013). By contrast, reduced expression of GAD67 is a well-replicated finding in both MDD and schizophrenia (Curley et al., 2013). However, in schizophrenia, GAD67 reductions are selectively found in parvalbumin (PV)-positive interneurons (Hashimoto et al., 2003), a subset of reciprocally interconnected soma-targeting GABAergic neurons whose dysfunction in the prefrontal cortex is thought to underlie abnormal gamma-band oscillations and cognitive deficits of schizophrenia (Carlen et al., 2012). PV-neuron-specific dysfunction in schizophrenia is also evident based on reduced expression of PV mRNA (Hashimoto et al., 2003). By contrast, PV transcript levels are normal in patients with MDD (Guilloux et al., 2012; Rajkowska et al., 2007; Sibille et al., 2011). However, reduced expression of SST is not exclusively found in MDD but has also been reported for schizophrenia (Guillozet-Bongaarts et al., 2014; Morris, Hashimoto, & Lewis, 2008). Collectively, the data suggest that at the cellular and molecular level, GABAergic deficits associated with schizophrenia and

MDD include both disease-specific and common characteristics. This is consistent with depressive symptoms being a common clinical feature of schizophrenia.

Multiple lines of evidence from preclinical studies confirm that impairments of GABAergic transmission observed in MDD patients can be causal for depression-related phenotypes (Luscher, Shen, et al., 2011; Mohler, 2012; Smith & Rudolph, 2012). In particular, mice that were rendered heterozygous for the γ2 subunit ($\gamma2^{+/-}$ mice) exhibit behavioral, cognitive, neuroendocrine, cellular, and pharmacologic alterations expected of an animal model of anxious depression (Crestani et al., 1999; Earnheart et al., 2007; Shen et al., 2010). The phenotypes of these mice include anxiety-like behavior in a number of test paradigms, reduced escape behavior in the forced swim and tail suspension tests that may indicate behavioral despair, and reduced sucrose consumption thought to indicate anhedonia. The $\gamma2^{+/-}$ model further includes constitutively elevated serum corticosterone levels, which point to disinhibition of the hypothalamus–pituitary–adrenal (HPA) axis (Shen et al., 2010) and are reminiscent of corresponding neuroendocrine alterations in MDD (Guerry & Hastings, 2011; Kathol, Jaeckle, Lopez, & Meller, 1989). Other than the anhedonia phenotype (which was not examined), all these phenotypes of the $\gamma2^{+/-}$ model were reproduced in mice in which the same genetic lesion was delimited to glutamatergic neurons of the telencephalon (Earnheart et al., 2007). Thus, the phenotypes of these mice are caused by reduced GABAergic input to glutamatergic neurons, most likely resulting in modest hyperexcitability of these neurons. Moreover, results from the telencephalon-specific γ2-deficient mice indicated that the HPA axis hyperactivity was caused by extrahypothalamic $GABA_AR$ deficits, most likely due to alterations of neural circuits in the hippocampus and neocortex, which normally delimit HPA axis activity (Section 5) and where reductions in γ2-containing $GABA_ARs$ are most pronounced (Crestani et al., 1999; Shen et al., 2010). Importantly, the anxiety- and depressive-like behavioral and neuroendocrine abnormalities of $\gamma2^{+/-}$ mice were reversed by chronic but not acute treatment with the norepinephrine transporter-selective reuptake inhibitor desipramine, thereby confirming that this drug acts over time to overcome deficits in GABAergic inhibition (Shen et al., 2010). Notably, in contrast to desipramine, the SSRI fluoxetine was effective in the novelty suppressed feeding test (a conflict test assessing anxiety) only and failed to normalize despair- and anhedonia-like phenotypes as well as the elevated corticosterone serum concentrations of $\gamma2^{+/-}$ mice. The qualitatively lesser responses of $\gamma2^{+/-}$ mice to fluoxetine than to desipramine are reminiscent of severe subtypes of anxious depressive disorders including melancholic depression, which tend to show greater responsiveness to TCAs than fluoxetine (Bauer et al., 2002; Clerc, Ruimy, & Verdeau-Palles, 1994; Perry, 1996; Roose, Glassman, Attia, & Woodring, 1994; Swartz & Guadagno, 1998; Young et al., 2004) and are generally less responsive to treatment than depressive disorders without anxiety (Fava et al., 2008). Thus, $\gamma2^{+/-}$ mice represent a model for partially drug-resistant melancholic/anxious depression.

The phenotype of $\gamma2^{+/-}$ mice further includes alterations in hippocampus-dependent cognitive tasks such as enhanced hippocampus-dependent trace fear conditioning and impaired ambiguous cue fear conditioning (Crestani et al., 1999). The latter of these two cognitive phenotypes has recently become recognized as a defect in pattern separation, a task that involves forming distinct memory traces for similar experiences or situations. Moreover, in mice, this task is critically dependent on adult hippocampal neurogenesis (reviewed in Aimone, Deng, & Gage, 2011; Clelland et al., 2009). Importantly, defects in pattern separation of emotionally relevant stimuli have emerged as a cognitive phenotype of MDD (Fujii et al., 2014; Leal, Tighe, Jones, & Yassa, 2014). In the context of anxiety disorders, defects in pattern separation have been proposed to underlie the commonly observed phenomenon of stimulus generalization (Kheirbek, Klemenhagen, Sahay, & Hen, 2012). In rodents, efficient pattern separation is controlled by the continued production and proper functional integration of adult-born granule cell neurons in the hippocampus (Clelland et al., 2009; Sahay et al., 2011). These same neurons are also recognized as a cellular substrate for the behavioral action of antidepressant drugs (Samuels & Hen, 2011). Consistent with a role of impaired pattern separation and hippocampal neurogenesis in the pathology of MDD, $\gamma2^{+/-}$ mice exhibit marked deficits in the maturation and survival of adult-born hippocampal granule cells (Earnheart et al., 2007; Ren et al., 2014). Anxiety- and depression-related behavior and maturational deficits of adult-born hippocampal neurons similar to those of $\gamma2^{+/-}$ mice have also been reported for mice lacking the $\alpha2$ subunit (Duveau et al., 2011; Vollenweider, Smith, Keist, & Rudolph, 2011). This subunit invariably coassembles with $\gamma2$ and β subunits to form postsynaptic $GABA_ARs$. These $\alpha2\beta\gamma2$ receptors are most prominently expressed in limbic neural circuits including neocortex, hippocampus, and striatum, and they are known to be responsible for the anxiolytic responses of benzodiazepines (Fritschy & Mohler, 1995; Low et al., 2000).

3. GABAergic TRANSMISSION AND HERITABILITY OF MDD

Based on meta-analyses of high-quality family and twin studies, the heritability of MDD has been determined to be 37–38%, with significantly higher values in women than men (Kendler, Gatz, Gardner, & Pedersen, 2006). However, to date no genes have been confirmed to carry significant risk specifically for MDD (Flint & Kendler, 2014; Hek et al., 2013; Major Depressive Disorder Working Group of the Psychiatric GWAS Consortium et al., 2013). Thus, neither the GABAergic deficit hypothesis nor any other proposed etiology of MDD is currently supported by human genetic evidence. The lack of a genetic foundation for MDD is contrasted by corresponding evidence in bipolar disorder and schizophrenia, where genome-wide association studies (GWAS), twin and adoption studies, and evidence for a role of copy number variance have identified an increasing number of risk genes. Although the heritability of bipolar disorder and schizophrenia is much higher than that of MDD (Lichtenstein et al., 2009; Sullivan, Kendler, & Neale, 2003), there is significant overlap in genetic risk between all three disorders. This is evidenced by familial risk genes for schizophrenia such as DISC1 (Disrupted in Schizophrenia) that confer similar risk also for MDD in the same families (Blackwood et al., 2007), by general familial coclustering of MDD with bipolar disorder and schizophrenia (Aukes et al., 2012) and by significant correlation of single nucleotide polymorphisms across the genome of subjects with MDD, schizophrenia, or bipolar disorder (Cross-Disorder Group of the Psychiatric Genomics Consortium et al., 2013).

In addition to phenotypic heterogeneity and considerable environmental contributions, the failure to identify genetic risk factors that are specific for MDD may be due to a very large number of genes and allelic variations that confer very small effects of vulnerability and in only a subset of combinations (Flint & Kendler, 2014). Alternatively, the difficulty of finding risk genes might be due to a large number of rare mutations that have large effects but function independently. These interpretations imply that information on the genetic basis of MDD, even if successfully attained eventually, might be clinically nonactionable. Indeed, therapeutic approaches for highly heterogeneous polygenic syndromes such as MDD would appear most effective if they were targeting cellular mechanisms disrupted in the majority of patients, irrespective of the function of hitherto hypothetical risk genes. Importantly, there is a third and increasingly plausible explanation for the elusiveness of risk genes for MDD. It is based on recent evidence from mice showing that heritability of emotion-related behavioral traits involves experience-dependent epigenetic modifications of DNA. Such changes can be passed on through the male and female germ line over multiple generations (Dias & Ressler, 2014). However, given that they do not involve alterations in the DNA sequence they are not detectable by GWAS. Environmental risk factors implicated in affective disorders that have been shown to program heritable behavioral traits in mice include fearful experiences in adulthood (Dias & Ressler, 2014) as well as early-life stress (Bohacek et al., 2014; Franklin et al., 2010). Moreover, the detrimental effects of both chronic and early-life stress are likely to involve dysfunction of GABAergic transmission (Section 5; Gunn et al., 2013; Holm et al., 2011; Hu, Zhang, Czeh, Flugge, & Zhang, 2010; Jacobson-Pick, Audet, McQuaid, Kalvapalle, & Anisman, 2012; Martisova et al., 2012; Radley, Gosselink, & Sawchenko, 2009). Genes that encode proteins important for GABAergic transmission that are subject to epigenetic regulation and dysregulated in MDD include BDNF (D'Addario et al., 2013; Karpova, 2014), GAD67 (Dong et al., 2005; Zhang et al., 2010), SST, and corticostatin (Jackson et al., 2011; Rubio et al., 2012), as well as genes encoding $GABA_AR$ subunits (Poulter et al., 2008; Zhao et al., 2012). Epigenetic mechanisms are also increasingly implicated in the regulated production and survival of hippocampal granule cells, which are increasingly recognized as a cellular substrate for emotion regulation and certain forms of cognition (Section 2). Furthermore, the production and maturation of these neurons are under direct GABAergic control as discussed in the following sections.

4. GABAergic TRANSMISSION IN RELATION TO THE MONOAMINE DEFICIENCY HYPOTHESIS OF MDD

Currently used antidepressant drug therapies are universally designed to modulate monoaminergic transmission, by blocking the reuptake or breakdown of serotonin and/or norepinephrine or, less commonly, dopamine. The acute action of these drugs therefore is predicted to increase the occupancy of corresponding receptors and to affect the basal set point of G protein-coupled receptor-signaling pathways. Conversely, the monoaminergic deficit hypothesis of MDD posits that defects in these mechanisms underlie depressive states (Bunney & Davis, 1965; Coppen, 1967; Matussek, 1969; Schildkraut, 1965). However, a common denominator of all approved antidepressant medications is that their therapeutic effects develop with a delay of several weeks, indicating that the mechanisms underlying therapeutic effects do not merely reflect altered monoaminergic transmission. Indeed, elevated serotonin concentrations commonly result in internalization of serotonin receptors (Bhattacharyya, Puri, Miledi, & Panicker, 2002; Guthrie, Murray, Franklin, & Hamblin, 2005; Riad et al., 2004; Schlag, Lou, Fennell, & Dunlop, 2004). Moreover, some antidepressants are known to act not only as uptake inhibitors but also as antagonists of monoamine transmitter receptors, indicating that the chronic effects of elevated extracellular monoamine transmitters in many cases involve reduced rather than elevated function of cognate G protein-coupled receptors (de Boer, 1996; Schreiber & Avissar, 2007). Lastly, pharmacological depletion of serotonin or norepinephrine in healthy humans is insufficient to induce depressive episodes (Ruhe, Mason, & Schene, 2007). Collectively, these studies indicate that antidepressants act slowly and indirectly through remote targets to affect the balance of neurotransmitter systems other than serotonin and norepinephrine. This is consistent with clinical and preclinical evidence that antidepressants can normalize the behavioral and neuroendocrine sequelae of defects in GABAergic transmission (Kucukibrahimoglu et al., 2009; Sanacora et al., 2003, 2002; Shen et al., 2010; Uzunova et al., 1998). Moreover, fluoxetine can also normalize anxiety and depression-related behavioral abnormalities of mice induced by chronic pharmacological increases in cholinergic transmission (Mineur et al., 2013), which is in keeping with the well-established anticholinergic activity of fluoxetine (Chew et al., 2008).

The detailed mechanisms by which antidepressants affect GABAergic transmission remain poorly understood. However, consistent with evidence from patients mentioned earlier, chronic treatment with fluoxetine increases extracellular GABA levels in brain of rats (Goren, Kucukibrahimoglu, Berkman, & Terzioglu, 2007). Moreover, fluoxetine and other SSRIs but not imipramine stimulate the release of neurosteroids (Pinna, Costa, & Guidotti, 2006, 2009). Neurosteroids act to reduce neural excitability by selective potentiation of neuronal tonic inhibition via δ subunit-containing $GABA_A$Rs (Mihalek et al., 1999; Stell, Brickley, Tang, Farrant, & Mody, 2003; Vicini, Losi, & Homanics, 2002). A third potential GABA-related mechanism of antidepressants might involve strengthening of GABAergic inhibition through a hyperpolarizing shift of the chloride equilibrium potential, E_{Cl}. For example, in the context of corticotropin-releasing hormone (CRH) neurons of the ventral posterior nucleus of the hypothalamus (VPN), stress is known to impact on GABAergic inhibition by down-regulation of potassium-chloride cotransporter 2 (KCC2; Section 5), most likely by an NMDAR (*N*-methyl-D-aspartic acid receptor)-mediated mechanism (Lee, Deeb, Walker, Davies, & Moss, 2011; Section 7). Conversely, Bos et al. (2013) recently showed that activation of 5-HT2A receptors facilitates translocation of KCC2 to the plasma membrane, which results in a hyperpolarizing shift of E_{Cl} and thereby renders inhibitory GABAergic inputs through $GABA_A$Rs more efficient. Although these latter experiments were conducted with spinal cord neurons, it is conceivable that SSRI-mediated augmentation of serotonin has similar effects on GABAergic transmission impaired in MDD. However, ultimately the neural network effects of 5-HT2C ligands will depend also on the relative distribution of these receptors between GABAergic and glutamatergic cells. Indeed, antidepressant-like behavioral effects in rodents have been described for both 5-HT2C-selective agonists (Rosenzweig-Lipson et al., 2007) and antagonists (Dekeyne et al., 2008).

The monoaminergic deficit hypothesis of MDD predicts that genetic ablation of serotonin receptors or transporters that mediate antidepressant drug responses results in depression-related phenotypes in mice. However, evidence to that effect is inconclusive. Knockout of 5-HT1A receptors, which are essential for antidepressant effects of SSRIs in mice (Gross et al., 2002), results in a heightened anxious phenotype combined with an antidepressant phenotype (Heisler et al., 1998; Parks, Robinson, Sibille, Shenk, & Toth, 1998; reviewed in Toth, 2003). Moreover, while adult neurogenesis is essential for many of the behavioral effects of antidepressants (David et al., 2009; Santarelli et al., 2003; Section 6), neurogenesis is not affected by knockout of the 5-HT1A receptor gene (Santarelli et al., 2003). Similarly, pharmacologic or genetic depletion of serotonin in mice does not affect basal proliferation of granule cell progenitors, although serotonin is required for potentiation of neurogenesis by exercise (Diaz et al., 2013; Klempin et al., 2013). Moreover, serotonin depletion paradoxically results in dramatically increased survival of adult-born neurons, indicating that it normally promotes apoptosis of granule cell progenitors (Diaz et al., 2013). Immunofluorescent staining for 5-HT1A receptors in the dentate gyrus suggests that these receptors are expressed selectively by radial glia-like astrocytes (RGLs) and hilar GABAergic interneurons but are absent on immature granule cells and barely detectable in mature granule cells

(Klempin et al., 2010). Therefore, the 5-HT1A receptor-dependent effects of serotonin and SSRIs on immature neurons must be indirect through 5-HT1A receptors on GABAergic interneurons (Luscher & Fuchs, 2013). Behavioral state-dependent GABA release by these cells then controls the mitotic activation of RGLs and the differentiation and maturation of granule cell precursors through $\alpha4\beta\gamma2$ and $\alpha2\beta\gamma2$ $GABA_ARs$, respectively (Duveau et al., 2011; Luscher & Fuchs, 2013; Ren et al., 2014).

Upregulation of BDNF is a key feature of all currently used antidepressant drug therapies and correlated with antidepressant efficacy also among antipsychotics (Duman & Li, 2012; Nibuya, Morinobu, & Duman, 1995; reviewed in Schmidt & Duman, 2007). Preclinical evidence from mice suggests that BDNF exerts its effects on neural circuits primarily by affecting the efficacy of GABAergic transmission, with secondary effects on glutamatergic transmission. In particular, the principal loss-of-function phenotype of mice lacking the BDNF receptor TrkB (tropomyosin-related kinase B) consists of marked and selective defects in GABAergic synapse formation (Chen et al., 2011; Rico, Xu, & Reichardt, 2002). BDNF is also particularly important for normal interneuron maturation (Hong, McCord, & Greenberg, 2008; Huang et al., 1999; Kohara et al., 2003; Sakata et al., 2009; Waterhouse et al., 2012). Lastly, BDNF and GABAergic transmission are mechanistically intertwined in their support of adult hippocampal neurogenesis, which serves as a cellular substrate for the behavioral effects of antidepressants (David et al., 2009). These interactions are discussed in further detail in Section 6.

BDNF-TrkB signaling promotes the functional expression of $GABA_ARs$ at the cell surface of both mature and immature neurons (Mizoguchi, Kanematsu, Hirata, & Nabekura, 2003; Porcher et al., 2011). Specifically, BDNF/TrkB signaling controls the phosphorylation state of a pair of Tyr residues in the cytoplasmic loop region of the $GABA_AR$ $\gamma2$ subunit (Vithlani et al., 2013), most likely by Fyn kinase (Jurd, Tretter, Walker, Brandon, & Moss, 2010). Phosphorylation of these residues interferes with clathrin-mediated endocytosis of $GABA_ARs$, thereby strengthening GABAergic synaptic inhibition (Kittler et al., 2008). Increased cell surface expression of $GABA_ARs$ and enhancement of GABAergic synaptic currents is similarly seen upon treatment of frontal cortex brain slices with BDNF (Vithlani et al., 2013). Predictably, mice carrying phospho-tyrosine-mimicking amino acid substitutions of the $\gamma2$ subunit show constitutively elevated cell surface expression of $GABA_ARs$. Intriguingly, these effects are cell type-specific and most notable in the prefrontal cortex and CA3 region of the hippocampus but absent in the CA1 region (Tretter et al., 2009; Vithlani et al., 2013). Increased cell surface expression of $GABA_A$Rs in the same animals was correlated with increased hippocampal neurogenesis and constitutive antidepressant-like behavior, as well as occluded behavioral responsiveness to BDNF (Vithlani et al., 2013). These phenotypes are consistent with and inverse to those of $\gamma 2^{+/-}$ mice characterized by defects in the survival of adult-born hippocampal neurons, depressive-like behavior, and increased behavioral sensitivity to antidepressant drugs (Earnheart et al., 2007; Ren et al., 2014; Shen et al., 2010). Given that BDNF signaling is universally required as a mediator of antidepressant drug responses (Saarelainen et al., 2003; Sairanen, Lucas, Ernfors, Castren, & Castren, 2005), these data suggest that BDNF-mediated enhancement of GABAergic inhibition via γ2-containing $GABA_A$Rs serves as a key mechanism for antidepressant drug treatments.

The accumulation of $GABA_A$Rs at inhibitory synapses is regulated not only by posttranslational modifications of receptor subunits but also by gephyrin, the principal subsynaptic scaffold protein that exerts effective control over the strength of GABAergic synapses (Essrich, Lorez, Benson, Fritschy, & Luscher, 1998; Kneussel et al., 1999; reviewed by Tyagarajan & Fritschy, 2014). Gephyrin accumulation at GABAergic synapses is subject to dynamic regulation by phosphorylation, acetylation (Tyagarajan et al., 2013; Tyagarajan, Ghosh, Yevenes, et al., 2011), S-palmitoylation (Dejanovic et al., 2014), and S-nitrosylation (Dejanovic & Schwarz, 2014). Of particular interest is a pair of interdependent Ser phosphorylation sites of gephyrin that function as targets for glycogen synthase kinase (GSK)3β and extracellular signal-regulated kinase (ERK/MAPK)-mediated phosphorylation. The phosphorylation states of these sites control calpain-mediated proteolytic degradation of gephyrin and hence the strength of GABAergic synaptic inhibition (Tyagarajan, Ghosh, Harvey, & Fritschy, 2011; Tyagarajan et al., 2013; Tyagarajan, Ghosh, Yevenes, et al., 2011). Postmortem studies of suicide victims and patients who suffered from depression suggest that the activities of ERK/MAPK and Wnt/GSK3β (glycogen synthase kinase 3β) pathways are altered in MDD (Duric et al., 2010; Dwivedi et al., 2001, 2007, 2010). Moreover, changes in the activity of these two signaling pathways are independently implicated in bidirectional regulation of depression-related behavior of rodents by stress and antidepressant drug treatment (Duman, Schlesinger, Kodama, Russell, & Duman, 2007; Duric et al., 2010; Gourley et al., 2008; Liu et al., 2013; Okamoto et al., 2010; reviewed in

Voleti & Duman, 2012). Importantly, pharmacological inhibition of GSK3β by the mood-stabilizing agent lithium *in vivo* suppresses calpain-mediated degradation of gephyrin, which potentiates GABAergic synaptic transmission (Tyagarajan, Ghosh, Harvey, et al., 2011). The effect of the phosphorylation state of the GSK3β site on calpain-mediated degradation of gephyrin appears to depend on the phosphorylation state of the adjacent ERK site of gephyrin, thereby leading to functional integration of two otherwise independent signaling pathways (Tyagarajan et al., 2013). Collectively, the data suggest that gephyrin and the $GABA_AR$ γ2 subunit serve as key targets that convey the mood-stabilizing actions of lithium and antidepressants via enhancement of GABAergic transmission.

5. GABAergic TRANSMISSION IN RELATION TO STRESS-BASED ETIOLOGIES OF MDD

Dysregulation of the HPA axis figures among the most consistently observed neuroendocrine abnormalities of MDD, with excessive stress and early-life stress as likely causal factors (Bremne & Vermetten, 2001; Guerry & Hastings, 2011; Stander, Thomsen, & Highfill-McRoy, 2014). In rodents, chronic intermittent exposures to various stressful conditions or chronic administration of corticosterone result in heightened emotion-related behavior and cellular changes expected of animal models of depression. These effects of stress can be prevented or reversed by antidepressant drug treatment (reviewed by Samuels et al., 2011). HPA axis activity is subject to GABAergic inhibitory control by the frontal cortex (Akana, Chu, Soriano, & Dallman, 2001; Diorio, Viau, & Meaney, 1993; Radley et al., 2009) and ventral hippocampus (Cullinan, Herman, & Watson, 1993) at the level of CRH-positive neurons of the paraventricular nucleus (PVN) of the hypothalamus (Diorio et al., 1993; Herman, Cullinan, Morano, Akil, & Watson, 1995; Sapolsky, Krey, & McEwen, 1984). In behaviorally naïve animals, these neurons are effectively inhibited by phasic- and tonic-hyperpolarizing currents mediated by γ2- and δ-subunit-containing $GABA_ARs$, respectively (Lee, Sarkar, & Maguire, 2014; Verkuyl, Hemby, & Joels, 2004). Acute stress leads to the release of neurosteroids that potentiate the function of δ-containing $GABA_ARs$, which limits acute stress-induced HPA axis activation (Sarkar, Wakefield, MacKenzie, Moss, & Maguire, 2011). However, repeated stress leads to down-regulation of the chloride transporter KCC2 and hence to a lasting depolarizing shift in the chloride reversal potential of CRH neurons that renders GABAergic inputs ineffective or even depolarizing (Hewitt, Wamsteeker, Kurz, & Bains, 2009). In addition, stress-induced NE release results in β-adrenergic receptor-mediated upregulation of metabotropic glutamate receptors, which in turn results in a lasting potentiation of GABAergic synaptic inputs to CRH neurons and further enhances GABAergic excitatory input to the PVN (Inoue et al., 2013). Chronic stress-induced presynaptic changes that reduce GABA release onto CRH neurons have also been described (Verkuyl et al., 2004). Collectively, chronic stress-induced pre- and postsynaptic changes render GABAergic inhibition of CRH neurons ineffective or even excitatory, which explains excessive and constitutive activity of the HPA axis and provides an explanation for the most common neuroendocrine abnormality of MDD.

Early-life stress-induced dysfunction leads to pathological changes of the HPA axis and confers considerable risk for MDD and other neuropsychiatric disorders in later life (Bale et al., 2010). Moreover, compared to adulthood the developing brain is far more vulnerable to stress. Recent evidence indicates that stress-induced early-life reprogramming of the HPA axis involves stress-induced release of allopregnanolone and corresponding potentiation of tonic GABAergic inhibition mediated by δ subunit-containing $GABA_A$Rs on CRH neurons of the PVN (Gunn et al., 2013). Increased GABAergic inhibition appears to result in a compensatory developmental increase in glutamatergic innervation of these neurons that persists to adulthood. Excessive glutamatergic activation of CRH neurons in adulthood renders these neurons insensitive to stress-induced increases in GABAergic inhibition, which explains the permanent changes in HPA axis activity observed in animals and possibly people exposed to early-life stress (Gunn et al., 2013). Importantly, the action of neurosteroids on these receptors expressed by hypothalamic CRH neurons is required also for normal physiological responses of mice to stress (Sarkar et al., 2011). Thus, stress-induced allosteric potentiation of tonically active $GABA_A$Rs serves as a key mechanism underlying both normal and pathological developmental programming of HPA axis function.

The hippocampus and frontal cortex are activated along with the PVN in response to acute emotional stress (Li & Sawchenko, 1998) and they represent additional sites of vulnerability to stress (Arnsten, 2009; Calfa, Bussolino, & Molina, 2007; Duman, Malberg, & Nakagawa, 2001; Maggio & Segal, 2009). In healthy people, stress-induced defects in GABAergic inhibition of the prefrontal cortex include downregulation of GABA following exposure to psychological stressors (Hasler, van der

Veen, Grillon, Drevets, & Shen, 2010). In rats, prefrontal cortical GABA is downregulated following exposure to immobilization stress (Otero Losada, 1988). Moreover, failure of rats to learn to escape in the learned helplessness task (a stressful test of rodents with predictive validity for antidepressant drug action) is associated with downregulation of $GABA_AR$s in the neocortex, hippocampus, and striatum (Drugan et al., 1989; Tan, Zhong, & Yan, 2004). In the hippocampus of rodents, acute stress results in rapid upregulation of neurosteroid-sensitive, tonically active δ subunit-containing $GABA_AR$s (Maguire & Mody, 2007; Serra, Pisu, Mostallino, Sanna, & Biggio, 2008), while chronic stress results in loss or impaired function of PV^+-interneurons (Hu et al., 2010) and downregulation of GABA (Gronli et al., 2007). A wealth of data from analyses of δ subunit KO mice further supports the conclusion that aberrant steroid synthesis and dysregulation of δ-containing $GABA_AR$s contribute to stress-, ovarian cycle-, and pregnancy-related mood disorders (Maguire, Ferando, Simonsen, & Mody, 2009; Maguire & Mody, 2007, 2008; Maguire, Stell, Rafizadeh, & Mody, 2005; reviewed in MacKenzie & Maguire, 2014).

Chronic and early-life stress have detrimental effects on adult hippocampal neurogenesis (Gould & Tanapat, 1999; Karten, Olariu, & Cameron, 2005; Mirescu, Peters, & Gould, 2004). Importantly, neurogenesis is subject to dynamic control by local GABAergic interneurons and $GABA_AR$s that control the mitotic activation of stem cells, the pace of neural maturation, and the synaptic integration of neural progenitor cells (Section 6). Therefore, it is readily conceivable that excessive stress-induced impairments of GABAergic transmission and excess glutamate release contribute to the lasting detrimental effects of stress on hippocampal function. Notably, hippocampal neurogenesis is essential for hippocampus-mediated inhibition of HPA axis function (Schloesser, Manji, & Martinowich, 2009; Snyder, Soumier, Brewer, Pickel, & Cameron, 2011). Moreover, disinhibition of HPA axis function correlates with depressive-like increases in emotional reactivity of mice with impaired hippocampal neurogenesis (Shen et al., 2010; Snyder et al., 2011). Although several studies have reported normal emotional behavior following ablation of hippocampal neurogenesis in mice (reviewed in Petrik, Lagace, & Eisch, 2012), hippocampal neurogenesis is also essential for hippocampus-dependent pattern separation, a cognitive task that is impaired in MDD (Fujii et al., 2014; Leal et al., 2014).

Compared to the hippocampal formation, there is relatively limited information on how chronic stress affects GABA function in the cortex. However, based on data from chronically stressed rats, there is a consensus that such treatment results in reduced expression and function of $GABA_ARs$ (Weizman et al., 1989). Recent analyses of the medial prefrontal cortex of chronically stressed rats further revealed dendritic atrophy of Martinotti cells and reduced expression of GAD67 (Gilabert-Juan, Castillo-Gomez, Guirado, Molto, & Nacher, 2013). These changes are reminiscent of molecular and cellular changes observed in postmortem brain of patients with MDD (Section 2). Importantly, dysfunction of the prefrontal cortex also leads to disinhibition of the HPA axis (Diorio et al., 1993; Sullivan & Gratton, 1999). Thus, chronic stress-induced GABAergic dysfunction in the prefrontal cortex and hippocampus may result in HPA hyperactivity even if the HPA axis *per se* is functioning properly (Shen et al., 2010). The ensuing systemic excessive release of stress hormones is predicted to perpetuate the problem through further impairment of GABAergic mechanisms in hippocampus and cortex. However, this positive feedback loop may be prevented by antidepressants that restore the function of neural circuits in both hippocampus and cortex (Lagace et al., 2010; Snyder et al., 2011; Surget et al., 2011).

6. GABAergic TRANSMISSION IN RELATION TO THE NEUROTROPHIC DEFICIT HYPOTHESIS OF MDD

The neurotrophic deficit hypothesis of MDD is principally based on brain imaging studies that have revealed brain volume reductions in limbic regions implicated in depression, including primary areas of pathology such as the hippocampus and prefrontal cortex. In some studies, volume reductions of the hippocampus were shown to correlate with duration of untreated illness, suggesting that they may represent long-term consequences rather than predict illness onset (Cobb et al., 2013; MacQueen et al., 2003; Sheline, Sanghavi, Mintun, & Gado, 1999). However, a recent meta-analysis of studies investigating first episode depression has found volume losses that are consistent with the neurotrophic deficit hypothesis of MDD (Cole, Costafreda, McGuffin, & Fu, 2011). Similarly in subjects at high familial risk of MDD, volume reductions in areas 24, 25, and 32 of the cortex arise before the onset of illness, suggesting a causal relationship (reviewed in Murray et al., 2011). Moreover, there is evidence that antidepressant drug treatment can protect from brain volume loss (Sheline, Gado, & Kraemer, 2003). Some studies suggest that brain volume reduction in MDD is due to loss of neurons and/or glia (Drevets, Price, & Furey, 2008; Miguel-Hidalgo & Rajkowska, 2002). However, more recent data suggest that brain volume reduction in MDD is predominantly due to reductions in neuropil rather than cell numbers (Cobb et al., 2013).

Low peripheral BDNF levels and downregulation of BDNF and its receptor TrkB in brain are some of the best replicated findings in the field of MDD (Banerjee, Ghosh, Ghosh, Bhattacharyya, & Mondal, 2013; Dwivedi et al., 2003, 2009; Guilloux et al., 2012; Kim et al., 2007; Shimizu et al., 2003). BDNF deficit-induced gene expression changes in mice are highly correlated with gene expression changes of orthologous genes in postmortem brain of depressed patients (Tripp et al., 2012). Moreover, preclinical studies provide a possible causal link between reduced BDNF/TrkB signaling and loss of brain volume in MDD. While stress decreases the expression of BDNF in limbic structures involved in emotion regulation (Gronli et al., 2006; Roth, Lubin, Funk, & Sweatt, 2009), antidepressant drug treatments invariably result in increased expression, and release of hippocampal BDNF (reviewed in Duman & Monteggia, 2006; Russo-Neustadt & Chen, 2005). Knockdown of BDNF in mice selectively in the dentate gyrus and global reduction of neural activity-induced expression of BDNF results in depression-related behavior (Sakata, Jin, & Jha, 2010; Taliaz, Stall, Dar, & Zangen, 2010). Moreover, the behavioral effects of antidepressants are reduced in BDNF- and TrkB-deficient mice (Sairanen et al., 2005).

Despite overwhelming evidence that BDNF plays a key role in regulating depression-related behavior and antidepressant drug mechanisms in animal models, there is no evidence from human GWAS studies for association of BDNF with MDD. However, a role for BDNF in depressive disorders has been proposed based on gene × environment interactions of two allelic versions of the precursor peptide of BDNF, proBDNF. The Met allele of Val66Met proBDNF negatively affects activity-dependent secretion of BDNF (Chen et al., 2004; Egan et al., 2003). Moreover, in a mouse model of this human polymorphism, the Met allele resulted in increased anxiety- and depression-related behavior, increased vulnerability to stress, and reduced responsiveness to the antidepressant fluoxetine (Chen et al., 2006; Yu et al., 2012). Consistent with these preclinical findings, an initial analysis of healthy human volunteers for gene × environment interactions revealed that the proBDNF Met allele conferred greater risk for development of an early-life stress-associated anxiety or depressive syndrome (Gatt et al., 2009). However, more recent studies found that the Met allele is protective rather than a risk factor in that it reduces risk for suicide attempts or depressive symptoms associated with early childhood trauma or stress (Chen, Li, & McGue, 2013; Kim et al., 2010; Perroud et al., 2008) and delays the onset of symptoms and severity of symptoms in patients with schizophrenia (Suchanek et al., 2013 and references therein). Thus, while BDNF may affect emotional reactivity and risk for psychiatric disorders, the directionality of such effects in humans is not predictable, and the mechanism is not specific for MDD. The ambiguity of interpreting the human genetic data of BDNF may have an explanation in the circuit-specific behavioral functions of BDNF. While BDNF mimics behavioral effects of antidepressants when injected into the dentate gyrus of rodents (Shirayama, Chen, Nakagawa, Russell, & Duman, 2002), its effects in the ventral tegmental area (VTA) are the opposite of those in the hippocampus, with BDNF infused into the VTA promoting behavioral despair (Eisch et al., 2003) and knockdown of its gene in the VTA leading to antidepressant-like effects (Berton et al., 2006).

Functional interactions between BDNF signaling, GABAergic transmission, and trophic support of neurons are best illustrated in the context of adult hippocampal neurogenesis, which is increasingly recognized as a cellular substrate for emotion regulation and behavioral antidepressant drug effects. Specifically, while antidepressant drug treatment increases neurogenesis (Malberg, Eisch, Nestler, & Duman, 2000), ablation of neurogenesis by radiation or chemical genetic means abolishes antidepressant-induced behavioral effects (David et al., 2009; Santarelli et al., 2003). Neurogenesis is also essential for antidepressant-mediated protection from detrimental effects of stress (Surget et al., 2011). BDNF deficit-induced defects in the production and maturation of hippocampal granule cells mimic corresponding deficits induced by knockout of $GABA_AR$ subunits (Duveau et al., 2011; Ren et al., 2014; Waterhouse et al., 2012). Moreover, the effects of limiting BDNF on neuronal maturation can be reversed by barbiturates, which is consistent with the earlier notion that BDNF is involved in the maturation of GABAergic input to these neurons (Porcher et al., 2011; Waterhouse et al., 2012; Section 4). Conversely, the fact that hippocampal neurogenesis is critically dependent on normal GABAergic transmission via $GABA_ARs$ suggests that antidepressants (or antidepressant-mediated increases in BDNF) act by modulating GABAergic input to these neurons. Neural activity-dependent release of GABA from PV^+ interneurons delimits the mitotic activation of RGLs in the subgranule cell layer of the dentate gyrus (Song et al., 2012). This mechanism is controlled by $\alpha4\beta\gamma2$ subunit-containing $GABA_ARs$ in RGLs and early progenitors (Duveau et al., 2011; Song et al., 2012). The subsequent maturation of dendrites and dendritic spines of granule cell progenitors is critically dependent on $\alpha2\beta\gamma2$ receptors, but independent of δ-containing receptors, which appear to mediate GABAergic inhibition mainly of mature granule cells (Duveau et al., 2011; Ren et al., 2014).

Interestingly, the neuronal survival promoting function of BDNF is critically dependent on NFATc4, a DNA-binding transcription factor whose nuclear translocation and activity are controlled by BDNF through phospholipase C-mediated activation of calcineurin, which enables nuclear translocation of NFATc4 (Groth & Mermelstein, 2003; Quadrato et al., 2012). Dephosphorylation of NFATc4 by calcineurin requires sustained elevation of intracellular Ca^{2+}. In addition to acting downstream of BDNF, NFATc4 may also promote expression of BDNF. Importantly, NFATc4 promotes hippocampal neurogenesis at least in part by enhancing the transcription of $GABA_AR$ $\alpha2$ and $\alpha4$ subunit genes (Quadrato et al., 2014). This mechanism is consistent with a role of enhanced GABAergic transmission in neurogenesis-dependent antidepressant drug mechanisms.

Proper maturation and synaptic innervation of developing granule cells are also critically dependent on a cell autonomic gradual change of E_{Cl} (Ge, Pradhan, Ming, & Song, 2007). The chloride concentration of immature neurons is elevated initially and renders GABA depolarizing for these cells. GABA-mediated membrane depolarization then enables a signaling cascade that involves $GABA_AR$-mediated depolarization, depolarization-mediated influx of Ca^{2+} (Borodinsky et al., 2003; Fiszman & Schousboe, 2004; Gascon et al., 2006; Maric et al., 2001; Schmidt-Hieber, Jonas, & Bischofberger, 2004; Tashiro, Sandler, Toni, Zhao, & Gage, 2006), and activation of Ca^{2+}-dependent kinases such as Ca^{2+}-calmodulin-dependent kinase II and Protein Kinase C (Shaywitz & Greenberg, 1999). These and other kinases phosphorylate the DNA-binding transcription factor CREB (cAMP-regulated element-binding protein) (Fujioka, Fujioka, & Duman, 2004; Gur et al., 2007; Jagasia et al., 2009; Nakagawa et al., 2002) and enable GABA/$GABA_AR$-mediated activation of CREB target genes, including the genes encoding KCC2 and BDNF (Aguado et al., 2003; Shieh, Hu, Bobb, Timmusk, & Ghosh, 1998; Tao, Finkbeiner, Arnold, Shaywitz, & Greenberg, 1998; reviewed in Fiumelli & Woodin, 2007). Evidence from cultured cortical neurons indicates that independent of these transcriptional responses, $GABA_AR$-depolarization-mediated Ca^{2+}-influx also promotes the cell surface accumulation of $GABA_ARs$ and secretion of BDNF, thereby further facilitating this signaling mechanism (Porcher et al., 2011). Secreted BDNF in turn acts presynaptically to promote GABA (and glutamate) release and to increase the relative excitability of GABAergic versus glutamatergic neurons (Jovanovic, Czernik, Fienberg, Greengard, & Sihra, 2000; Wardle & Poo, 2003). The mechanism of BDNF- and depolarizing GABA-dependent maturation of GABAergic circuits appears to be self-limiting as BDNF-induced expression of KCC2 drives the change from depolarizing GABA in immature neurons to hyperpolarizing and inhibitory actions in mature granule cells (Aguado et al., 2003; Fiumelli & Woodin, 2007), which terminates GABA-depolarization-mediated Ca^{2+} influx.

There is an expansive body of literature demonstrating trophic actions of GABA in the developing brain. It is well established that in embryonic brain and immature neurons, GABAergic excitation provides essential excitatory drive for neurite growth and functional maturation of neurons, preceding glutamatergic synaptic transmission during brain development, and continuing into adulthood in the context of maturation of adult-born neurons (reviewed in Ben-Ari, 2013; Represa & Ben-Ari, 2005). The interplay of neural activity and trophic functions of GABA is pivotal for normal activity-dependent circuit formation but exposes neural circuits to diverse mechanisms of vulnerability by environmental disturbances. Activity-dependent trophic functions of GABA and BDNF also provide a mechanistic rationale for association of MDD with reduced expression of BDNF even in the absence of a discernable genetic basis for MDD.

7. GABAergic TRANSMISSION IN RELATION TO GLUTAMATERGIC ETIOLOGIES OF MDD

The glutamatergic hypothesis locates the primary deficits of MDD in the excitatory component of cognitive–emotional circuits. Brain imaging studies suggest that depression is characterized by an increase in neural activity/excitation in several brain regions believed to be critically involved in the pathology of MDD (Drevets, 2001; Savitz & Drevets, 2009). Elevated glutamate levels measured in blood plasma and postmortem brain tissue suggest altered glutamate transmission in depressed patients (Hashimoto, Sawa, & Iyo, 2007; Lan et al., 2009; Mitani et al., 2006) and postmortem brain tissue analysis revealed alterations in glutamate receptor densities (Deschwanden et al., 2011; Feyissa et al., 2010). Mouse models with genetically modified components of the glutamate system show alterations in depressive-like behavior, while established animal models of depression have altered glutamatergic neurotransmission (Tokita et al., 2012). Stress (Pittenger & Duman, 2008) and current antidepressant treatments (Kucukibrahimoglu et al., 2009; Maes, Verkerk, Vandoolaeghe, Lin, & Scharpe, 1998) are known to impact the glutamate system and, perhaps most informative, the discovery of antidepressant properties of the NMDA receptor antagonist ketamine has stimulated development of what might become a new generation of fast acting antidepressants (Duman & Aghajanian, 2012; Krystal, Sanacora, & Duman, 2013).

Consistent with the aforementioned reductions in brain volume, neural cell densities, and neuropil in postmortem brain of MDD patients (Section 6), excess glutamate release is known to lead to atrophy and cell death. The cytotoxic effects of glutamate are thought to involve increased activation of NMDA receptors (Hardingham & Bading, 2010; Papouin et al., 2012) as a consequence of increased neural activity and neurotransmitter release and corresponding synaptic spillover. Given the intricate relationship between GABAergic and glutamatergic transmitter systems and the pivotal need for neural circuits to balance excitation and inhibition, it is readily conceivable that excitotoxicity following elevated glutamatergic transmission might be a consequence of the aforementioned reductions in markers of GABAergic inhibitory synaptic transmission (Section 2).

Excess activation of NMDARs has also been proposed to involve reduced clearance of glutamate caused by loss of astroglia (Rajkowska & Miguel-Hidalgo, 2007). Such a mechanism is supported by reductions in GFAP immunoreactivity (Si, Miguel-Hidalgo, O'Dwyer, Stockmeier, & Rajkowska, 2004) and glia density in postmortem brain of depressed subjects (Rajkowska & Miguel-Hidalgo, 2007) as well as reports of decreased expression of glutamate transporters and glutamine synthase in several brain areas of depressed patients (Miguel-Hidalgo et al., 2010).

The neurotoxic and proapoptotic effects of glutamate have long been thought to involve selective activation of extrasynaptic NMDARs followed by induction of proapoptotic gene expression programs (reviewed in Hardingham & Bading, 2010). More recently, neurotoxic effects mediated selectively by synaptic NMDARs have also been described (Papouin et al., 2012). As an alternate mechanism, Lee et al. recently showed that elevated glutamate can trigger NMDAR-mediated downregulation of KCC2 in cultured hippocampal neurons, a mechanism that removes the driving force for GABAergic inhibition (Lee et al., 2011). This type of mechanism is predicted to exacerbate and perpetuate the detrimental effects of excess glutamate by removing GABAergic inhibition and thereby further increasing excitability and glutamate release. Moreover, it could be argued that such an ionic mechanism would be faster than transcription-dependent reprogramming of cells and hence contribute upstream of gene expression-dependent proapoptotic processes.

Additional support for a glutamatergic hypothesis of depression comes from the previously mentioned reversal of chronic stress-related features of animal models of depression by treatment with antidepressants (Section 5). Acute and repeated/chronic stressors have different effects that both involve changes in the function of NMDARs. Short-time treatment of rats with corticosterone and acute stress result in a delayed and sustained potentiation of the synaptic response and surface expression of NMDARs, which facilitates working memory function (Yuen et al., 2011). By contrast, repeated stress suppresses glutamate receptor expression and leads to reduced prefrontal cortex NMDA and AMPA receptor expression, and impaired working memory (Yuen et al., 2012). While cortical GABAergic changes under these conditions have not yet been described, the release of stress hormones via HPA axis, which triggers cortical changes in NMDAR function, is under tight control by GABAergic mechanisms, as already detailed in Section 5. Briefly, activity-dependent changes in the function of CRH neurons in the VPN provide striking examples of interaction between GABAergic and glutamatergic synaptic plasticity mechanisms. In adulthood, excessive stress leads to a depolarization shift of E_{Cl} that renders GABAergic inhibition of CRH neurons ineffective, leading to excessive and prolonged HPA axis activation (Hewitt et al., 2009). This shift in E_{CL} is likely due to NMDAR-mediated downregulation of KCC2 (Lee et al., 2011). Thus, the lasting pathology consists of reduced efficacy of GABAergic inputs, while the cause was an NMDAR-dependent change in E_{Cl}. By contrast, in young animals, repeated stress leads to neurosteroid-mediated enhancement of GABAergic inhibition and a compensatory yet stable increase in glutamatergic synapse formation that permanently alters the balance between excitatory and inhibitory inputs to CRH neurons and thereby leads to lifelong hyperexcitability of the HPA axis (Gunn et al., 2013). Thus, the lasting pathology in this case consists of glutamatergic hyperexcitability, and the causal aberration was a developmental and transient change in GABAergic transmission.

A similar scenario can be envisioned for depression-related defects in adult hippocampal neurogenesis. As mentioned in Section 2, the depressive-like phenotype of $GABA_AR$ $\gamma2^{+/-}$ mice includes a marked deficit in dendritic spine maturation of adult-born hippocampal granule cells, indicative of defects in glutamatergic synaptic transmission (Ren et al., 2014) and similar maturational defects were also reported for mice with reduced expression of BDNF (Waterhouse et al., 2012) and mice lacking the $GABA_AR$ $\alpha2$ subunit (Duveau et al., 2011). Importantly, the maturational defects of granule cells in α2 knockout mice could be reversed by systemic treatment with gabapentin, an antagonist of thrombospondin receptors that acts as a neuroleptic by suppressing the formation of glutamatergic synapses (Duveau et al., 2011). This finding suggests that defects in glutamatergic transmission associated with depressive states are a consequence of causal GABAergic deficits and an excessive rather than a reduced drive to form new glutamatergic synapses.

Pharmacological antagonism of the NMDA receptor, notably by a single subanesthetic dose of ketamine, has rapid and lasting antidepressant effects in patients and animal models (Berman et al., 2000; Crane, 1959; Garcia et al., 2009; Trullas & Skolnick, 1990; reviewed in Krystal et al., 2013). More recent studies indicate that GluN2B-specific antagonists (i.e., Ro25-6981, CP101,606) have antidepressant effects similar to ketamine although with shorter duration (Maeng et al., 2008; Preskorn et al., 2008). Compared to GluN2A receptors, GluN2B receptors are enriched in the extrasynaptic membrane and implicated in the cytotoxic effects of glutamate. Although the mechanism of action of ketamine remains poorly understood, its effects in mice depend on GSK3β (Beurel, Song, & Jope, 2011) and they are augmented by inhibitory phosphorylation of GSK3β (Liu et al., 2013). GSK3β is a key kinase in the Wnt signaling pathway that also serves as a pharmacologically relevant target of the mood stabilizer lithium (Voleti & Duman, 2012). Although the relevant targets phosphorylated by GSK3β in this context are largely unknown, lithium-mediated inhibition of GSK3β was recently shown to strengthen GABAergic synaptic transmission (Tyagarajan et al., 2013; Tyagarajan, Ghosh, Yevenes, et al., 2011) as detailed already in Section 4. It stands to reason then that the antidepressant mechanisms of ketamine, similar to monoaminergic antidepressants, might involve potentiation of GABAergic transmission.

In keeping with evidence that MDD involves decreased expression of synapse-related genes and reduced density of glutamatergic synapses (Kang et al., 2012), ketamine-induced antidepressant effects involve an mTOR-dependent mechanism of synaptogenesis, including rapid induction of spine formation and increased function of glutamatergic synapses (reviewed in Duman & Li, 2012; Li et al., 2010, 2011). It follows that in order to maintain the excitation/inhibition balance and to prevent ketamine-induced neurotoxic release of glutamate, ketamine-induced enhancement of glutamatergic transmission must occur in concert with augmentation of GABAergic transmission. By extension, and given the evidence for GABAergic defects in MDD reviewed in Section 2, it should become abundantly clear that depressive brain states involve not only functional deficits in glutamatergic transmission but also reduced GABAergic activity. GABA sets the tempo of brain development, particularly in the cortex and hippocampus (reviewed in Ben-Ari, 2013; Ge et al., 2007). Its function through $GABA_A$Rs is subject to direct modulation by neurosteroids (reviewed in Comenencia-Ortiz, Moss, & Davies, 2014; Gunn et al., 2014; MacKenzie & Maguire, 2014), and the function of $GABA_A$Rs itself is vulnerable to environmental effects on E_{Cl} (Ben-Ari, Khalilov, Kahle, & Cherubini, 2012). The collective evidence suggests that defects in GABergic transmission are causative rather than a consequence of pathological changes in glutamatergic transmission.

8. CONCLUSION

The quest to elucidate the biology underlying MDD is complicated by symptom heterogeneity, the fact that pathological changes are distributed across a multitude of forebrain circuits, and a virtual lack of concrete insights from human GWAS. Nevertheless, there has been remarkable progress in our understanding of the biology of mood disorders over the last two decades. Here, we have summarized increasing evidence for convergence of the GABAergic deficit hypothesis of MDD with other proposed etiologies of this disorder. We attempted to make the case that alterations in markers of GABAergic transmission associated with MDD are not merely epiphenomena but that they are causally involved in the etiology of the disorder. A GABA-centric view of the biology of MDD has provided novel insights into pathological mechanisms of stress, the detrimental effects of stress on cortical and hippocampal deficits associated with MDD as well as functional changes in the HPA axis. Moreover, it is becoming increasingly evident that antidepressant drug effects depend on mechanisms that restore GABAergic inhibition.

A number of aspects of GABAergic transmission in the context of MDD remain unexplained. One major obstacle to the GABAergic hypothesis of MDD is the lack of drug therapies that are based on enhancement of GABA function and show therapeutic efficacy in MDD. In particular, benzodiazepines are largely ineffective as antidepressants, except for the occasional use of alprazolam to augment the efficacy of conventional antidepressant drug therapies (Birkenhager, Moleman, & Nolen, 1995; Petty, Trivedi, Fulton, & Rush, 1995). Possible explanations for this phenomenon include the fact that benzodiazepines induce rapid tolerance and also increase lysosomal degradation of major subtypes of $GABA_ARs$, rendering these drugs unsuitable for long-term therapies (Jacob et al., 2012; Vinkers et al., 2012), although they are highly effective as anxiolytics. However, it is conceivable that novel subtype-specific agonists of $GABA_ARs$ will alleviate this problem (Rudolph & Mohler, 2014).

Reports on the rapid antidepressant action of ketamine and other NMDAR antagonists have sparked considerable excitement regarding their potential application as next-generation antidepressant therapeutics and they point to a key role of altered glutamatergic transmission in the pathology of MDD (Krystal et al., 2013). However, whether altered GABAergic transmission also plays a role in the therapeutic efficacy of NMDAR antagonists is not yet known. Exploring the relationship between GABAergic defects, alterations in glutamatergic transmission, and the antidepressant efficacy of NMDAR antagonists will likely provide valuable new insights into the pathophysiology of MDD.

We have reviewed evidence pointing to a key role of reduced GABAergic inhibition in cortex, hippocampus, and VPN in the cellular, hormonal, and behavioral changes associated with MDD. In addition, GABAergic mechanisms are known to exert powerful control over dopaminergic reward circuits in the VTA (Brown et al., 2012; Tan et al., 2012; Tritsch, Ding, & Sabatini, 2012; van Zessen, Phillips, Budygin, & Stuber, 2012). Moreover, preclinical studies suggest that defects in GABAergic transmission may result in reduced reward seeking behavior thought to represent depressive-like traits (Shen et al., 2010). However, the precise location and nature of corresponding changes in GABAergic transmission remain to be explored.

Lastly, the pathobiology of MDD and other psychiatric conditions is closely linked to inflammatory processes, which might also play a causal role in the etiology of MDD (Dantzer, O'Connor, Freund, Johnson, & Kelley, 2008; Walker, Kavelaars, Heijnen, & Dantzer, 2014). For example, immunotherapies of viral infections with proinflammatory cytokines result in side effects including sickness behavior that resemble depressive-like states (Raison et al., 2009). Moreover, transcriptome analyses of postmortem brains of MDD patients show significant changes in expression of inflammation-related genes (Shelton et al., 2011). However, whether and how inflammation affects GABAergic and glutamatergic transmission or other mechanisms influencing the excitation/inhibition balance is largely unknown. One speculative indication that GABAergic mechanisms might play a role is based on evidence that inflammatory processes can contribute to epilepsy (Zattoni et al., 2011), pointing to disrupted excitation–inhibition balance of neural circuits. Conversely, epilepsy often emerges as a consequence of genetically determined defects in GABAergic inhibition, and the disease is frequently comorbid with depressive disorder. Progress and new insights in all of these areas should allow us to soon draw a more complete picture on the role of altered GABAergic neurotransmission in the etiology of MDD.

CONFLICT OF INTEREST

The authors declare no conflicts of interest.

ACKNOWLEDGMENTS

We thank Matthew Shorey for critical comments on the manuscript. Bernhard Luscher's formative years as a scientist included 9 years as a group leader at the Institute of Pharmacology of the University of Zurich. He is eternally grateful for Hanns Mohler's expert guidance and mentorship during those years. Research in the Luscher Laboratory currently is supported by Grants MH099851, MH097247, and MH083911 from the National Institute of Mental Health (NIMH). The contents of this review are solely the responsibility of the authors and do not necessarily represent the views of the NIMH or the NIH.

REFERENCES

Aguado, F., Carmona, M. A., Pozas, E., Aguilo, A., Martinez-Guijarro, F. J., Alcantara, S., et al. (2003). BDNF regulates spontaneous correlated activity at early developmental stages by increasing synaptogenesis and expression of the K+/Cl− co-transporter KCC2. *Development*, *130*(7), 1267–1280.

Aihara, M., Ida, I., Yuuki, N., Oshima, A., Kumano, H., Takahashi, K., et al. (2007). HPA axis dysfunction in unmedicated major depressive disorder and its normalization by pharmacotherapy correlates with alteration of neural activity in prefrontal cortex and limbic/paralimbic regions. *Psychiatry Research*, *155*(3), 245–256.

Aimone, J. B., Deng, W., & Gage, F. H. (2011). Resolving new memories: A critical look at the dentate gyrus, adult neurogenesis, and pattern separation. *Neuron*, *70*(4), 589–596.

Akana, S. F., Chu, A., Soriano, L., & Dallman, M. F. (2001). Corticosterone exerts site-specific and state-dependent effects in prefrontal cortex and amygdala on regulation of adrenocorticotropic hormone, insulin and fat depots. *Journal of Neuroendocrinology*, *13*(7), 625–637.

American Psychiatric Association. (2000). *Diagnostic and statistical manual of mental disorders* (4th ed.). Washington, DC: American Psychiatric Press, (Text revision).

Angst, J., & Merikangas, K. R. (2001). Multi-dimensional criteria for the diagnosis of depression. *Journal of Affective Disorders*, *62*(1–2), 7–15.

Arnsten, A. F. (2009). Stress signalling pathways that impair prefrontal cortex structure and function. *Nature Reviews. Neuroscience*, *10*(6), 410–422.

Aukes, M. F., Laan, W., Termorshuizen, F., Buizer-Voskamp, J. E., Hennekam, E. A., Smeets, H. M., et al. (2012). Familial clustering of schizophrenia, bipolar disorder, and major depressive disorder. *Genetics in Medicine*, *14*(3), 338–341.

Baldwin, D. S., Evans, D. L., Hirschfeld, R. M., & Kasper, S. (2002). Can we distinguish anxiety from depression? *Psychopharmacology Bulletin, 36*(Suppl. 2), 158–165.

Bale, T. L., Baram, T. Z., Brown, A. S., Goldstein, J. M., Insel, T. R., McCarthy, M. M., et al. (2010). Early life programming and neurodevelopmental disorders. *Biological Psychiatry, 68*(4), 314–319.

Banerjee, R., Ghosh, A. K., Ghosh, B., Bhattacharyya, S., & Mondal, A. C. (2013). Decreased mRNA and protein expression of BDNF, NGF, and their receptors in the hippocampus from suicide: An analysis in human postmortem brain. *Clinical Medicine Insights: Pathology, 6*, 1–11.

Bauer, M., Whybrow, P. C., Angst, J., Versiani, M., Moller, H. J., & World Federation of Societies Biol Psychiat Task Force on Treatment Guidelines for Unipolar Depressive Disorders, (2002). World Federation of Societies of Biological Psychiatry (WFSBP) Guidelines for Biological Treatment of Unipolar Depressive Disorders, Part 1: Acute and continuation treatment of major depressive disorder. *World Journal of Biological Psychiatry, 3*(1), 5–43.

Bell-McGinty, S., Butters, M. A., Meltzer, C. C., Greer, P. J., Reynolds, C. F., 3rd., & Becker, J. T. (2002). Brain morphometric abnormalities in geriatric depression: Long-term neurobiological effects of illness duration. *The American Journal of Psychiatry, 159*(8), 1424–1427.

Ben-Ari, Y. (2013). The developing cortex. In M. L. Olivier Dulac & B. S. Harvey (Eds.), *Handbook of clinical neurology* (pp. 417–426). Elsevier.

Ben-Ari, Y., Khalilov, I., Kahle, K. T., & Cherubini, E. (2012). The GABA excitatory/inhibitory shift in brain maturation and neurological disorders. *The Neuroscientist, 18*(5), 467–486.

Berman, R. M., Cappiello, A., Anand, A., Oren, D. A., Heninger, G. R., Charney, D. S., et al. (2000). Antidepressant effects of ketamine in depressed patients. *Biological Psychiatry, 47*(4), 351–354.

Berton, O., McClung, C. A., Dileone, R. J., Krishnan, V., Renthal, W., Russo, S. J., et al. (2006). Essential role of BDNF in the mesolimbic dopamine pathway in social defeat stress. *Science, 311*(5762), 864–868.

Beurel, E., Song, L., & Jope, R. S. (2011). Inhibition of glycogen synthase kinase-3 is necessary for the rapid antidepressant effect of ketamine in mice. *Molecular Psychiatry, 16*(11), 1068–1070.

Bhagwagar, Z., Wylezinska, M., Taylor, M., Jezzard, P., Matthews, P. M., & Cowen, P. J. (2004). Increased brain GABA concentrations following acute administration of a selective serotonin reuptake inhibitor. *The American Journal of Psychiatry, 161*(2), 368–370.

Bhattacharyya, S., Puri, S., Miledi, R., & Panicker, M. M. (2002). Internalization and recycling of 5-HT2A receptors activated by serotonin and protein kinase C-mediated mechanisms. *Proceedings of the National Academy of Sciences of the United States of America, 99*(22), 14470–14475.

Birkenhager, T. K., Moleman, P., & Nolen, W. A. (1995). Benzodiazepines for depression? A review of the literature. *International Clinical Psychopharmacology, 10*(3), 181–195.

Blackwood, D. H., Pickard, B. J., Thomson, P. A., Evans, K. L., Porteous, D. J., & Muir, W. J. (2007). Are some genetic risk factors common to schizophrenia, bipolar disorder and depression? Evidence from DISC1, GRIK4 and NRG1. *Neurotoxicity Research, 11*(1), 73–83.

Bohacek, J., Farinelli, M., Mirante, O., Steiner, G., Gapp, K., Coiret, G., et al. (2014). Pathological brain plasticity and cognition in the offspring of males subjected to postnatal traumatic stress. *Molecular Psychiatry*. http://dx.doi.org/10.1038/mp.2014.80, Advance online publication.

Borodinsky, L. N., O'Leary, D., Neale, J. H., Vicini, S., Coso, O. A., & Fiszman, M. L. (2003). GABA-induced neurite outgrowth of cerebellar granule cells is mediated by

$GABA_A$ receptor activation, calcium influx and CaMKII and erk1/2 pathways. *Journal of Neurochemistry, 84*(6), 1411–1420.

Bos, R., Sadlaoud, K., Boulenguez, P., Buttigieg, D., Liabeuf, S., Brocard, C., et al. (2013). Activation of 5-HT2A receptors upregulates the function of the neuronal K-Cl cotransporter KCC2. *Proceedings of the National Academy of Sciences of the United States of America, 110*(1), 348–353.

Bremne, J. D., & Vermetten, E. (2001). Stress and development: Behavioral and biological consequences. *Development and Psychopathology, 13*(3), 473–489.

Brickley, S. G., & Mody, I. (2012). Extrasynaptic $GABA_A$ receptors: Their function in the CNS and implications for disease. *Neuron, 73*(1), 23–34.

Brown, M. T., Tan, K. R., O'Connor, E. C., Nikonenko, I., Muller, D., & Luscher, C. (2012). Ventral tegmental area GABA projections pause accumbal cholinergic interneurons to enhance associative learning. *Nature, 492*(7429), 452–456.

Bunney, W. E., Jr., & Davis, J. M. (1965). Norepinephrine in depressive reactions. A review. *Archives of General Psychiatry, 13*(6), 483–494.

Calfa, G., Bussolino, D., & Molina, V. A. (2007). Involvement of the lateral septum and the ventral hippocampus in the emotional sequelae induced by social defeat: Role of glucocorticoid receptors. *Behavioural Brain Research, 181*(1), 23–34.

Carlen, M., Meletis, K., Siegle, J. H., Cardin, J. A., Futai, K., Vierling-Claassen, D., et al. (2012). A critical role for NMDA receptors in parvalbumin interneurons for gamma rhythm induction and behavior. *Molecular Psychiatry, 17*(5), 537–548.

Chen, Z. Y., Jing, D., Bath, K. G., Ieraci, A., Khan, T., Siao, C. J., et al. (2006). Genetic variant BDNF (Val66Met) polymorphism alters anxiety-related behavior. *Science, 314*(5796), 140–143.

Chen, J., Li, X., & McGue, M. (2013). The interacting effect of the BDNF Val66Met polymorphism and stressful life events on adolescent depression is not an artifact of gene–environment correlation: Evidence from a longitudinal twin study. *Journal of Child Psychology and Psychiatry, 54*(10), 1066–1073.

Chen, A. I., Nguyen, C. N., Copenhagen, D. R., Badurek, S., Minichiello, L., Ranscht, B., et al. (2011). TrkB (tropomyosin-related kinase B) controls the assembly and maintenance of GABAergic synapses in the cerebellar cortex. *Journal of Neuroscience, 31*(8), 2769–2780.

Chen, Z. Y., Patel, P. D., Sant, G., Meng, C. X., Teng, K. K., Hempstead, B. L., et al. (2004). Variant brain-derived neurotrophic factor (BDNF) (Met66) alters the intracellular trafficking and activity-dependent secretion of wild-type BDNF in neurosecretory cells and cortical neurons. *Journal of Neuroscience, 24*(18), 4401–4411.

Chew, M. L., Mulsant, B. H., Pollock, B. G., Lehman, M. E., Greenspan, A., Mahmoud, R. A., et al. (2008). Anticholinergic activity of 107 medications commonly used by older adults. *Journal of American Geriatrics Society, 56*(7), 1333–1341.

Choudary, P. V., Molnar, M., Evans, S. J., Tomita, H., Li, J. Z., Vawter, M. P., et al. (2005). Altered cortical glutamatergic and GABAergic signal transmission with glial involvement in depression. *Proceedings of the National Academy of Sciences of the United States of America, 102*(43), 15653–15658.

Clelland, C. D., Choi, M., Romberg, C., Clemenson, G. D., Jr., Fragniere, A., Tyers, P., et al. (2009). A functional role for adult hippocampal neurogenesis in spatial pattern separation. *Science, 325*(5937), 210–213.

Clerc, G. E., Ruimy, P., & Verdeau-Palles, J. (1994). A double-blind comparison of venlafaxine and fluoxetine in patients hospitalized for major depression and melancholia. The Venlafaxine French Inpatient Study Group. *International Clinical Psychopharmacology, 9*(3), 139–143.

Cobb, J. A., Simpson, J., Mahajan, G. J., Overholser, J. C., Jurjus, G. J., Dieter, L., et al. (2013). Hippocampal volume and total cell numbers in major depressive disorder. *Journal of Psychiatric Research, 47*(3), 299–306.

Cole, J., Costafreda, S. G., McGuffin, P., & Fu, C. H. (2011). Hippocampal atrophy in first episode depression: A meta-analysis of magnetic resonance imaging studies. *Journal of Affective Disorders*, *134*(1–3), 483–487.

Comenencia-Ortiz, E., Moss, S. J., & Davies, P. A. (2014). Phosphorylation of $GABA_A$ receptors influences receptor trafficking and neurosteroid actions. *Psychopharmacology*, *231*(17), 3453–3465.

Coppen, A. (1967). The biochemistry of affective disorders. *The British Journal of Psychiatry*, *113*(504), 1237–1264.

Crane, G. E. (1959). Cycloserine as an antidepressant agent. *The American Journal of Psychiatry*, *115*(11), 1025–1026.

Crestani, F., Lorez, M., Baer, K., Essrich, C., Benke, D., Laurent, J. P., et al. (1999). Decreased $GABA_A$-receptor clustering results in enhanced anxiety and a bias for threat cues. *Nature Neuroscience*, *2*(9), 833–839.

Cross-Disorder Group of the Psychiatric Genomics Consortium, Lee, S. H., Ripke, S., Neale, B. M., Faraone, S. V., Purcell, S. M., et al. (2013). Genetic relationship between five psychiatric disorders estimated from genome-wide SNPs. *Nature Genetics*, *45*(9), 984–994.

Cullinan, W. E., Herman, J. P., & Watson, S. J. (1993). Ventral subicular interaction with the hypothalamic paraventricular nucleus: Evidence for a relay in the bed nucleus of the stria terminalis. *Journal of Comparative Neurology*, *332*(1), 1–20.

Curley, A. A., Eggan, S. M., Lazarus, M. S., Huang, Z. J., Volk, D. W., & Lewis, D. A. (2013). Role of glutamic acid decarboxylase 67 in regulating cortical parvalbumin and GABA membrane transporter 1 expression: Implications for schizophrenia. *Neurobiology of Disease*, *50*, 179–186.

D'Addario, C., Dell'Osso, B., Galimberti, D., Palazzo, M. C., Benatti, B., Di Francesco, A., et al. (2013). Epigenetic modulation of BDNF gene in patients with major depressive disorder. *Biological Psychiatry*, *73*(2), e6–7.

Dantzer, R., O'Connor, J. C., Freund, G. G., Johnson, R. W., & Kelley, K. W. (2008). From inflammation to sickness and depression: When the immune system subjugates the brain. *Nature Reviews. Neuroscience*, *9*(1), 46–56.

David, D. J., Samuels, B. A., Rainer, Q., Wang, J. W., Marsteller, D., Mendez, I., et al. (2009). Neurogenesis-dependent and -independent effects of fluoxetine in an animal model of anxiety/depression. *Neuron*, *62*(4), 479–493.

de Boer, T. (1996). The pharmacologic profile of mirtazapine. *Journal of Clinical Psychiatry*, *57*(Suppl. 4), 19–25.

Dejanovic, B., & Schwarz, G. (2014). Neuronal nitric oxide synthase-dependent S-nitrosylation of gephyrin regulates gephyrin clustering at GABAergic synapses. *Journal of Neuroscience*, *34*(23), 7763–7768.

Dejanovic, B., Semtner, M., Ebert, S., Lamkemeyer, T., Neuser, F., Luscher, B., et al. (2014). Palmitoylation of gephyrin controls receptor clustering and plasticity of GABAergic synapses. *PLoS Biology*, *12*(7), e1001908.

Dekeyne, A., Mannoury la Cour, C., Gobert, A., Brocco, M., Lejeune, F., Serres, F., et al. (2008). S32006, a novel 5-HT2C receptor antagonist displaying broad-based antidepressant and anxiolytic properties in rodent models. *Psychopharmacology*, *199*(4), 549–568.

Deschwanden, A., Karolewicz, B., Feyissa, A. M., Treyer, V., Ametamey, S. M., Johayem, A., et al. (2011). Reduced metabotropic glutamate receptor 5 density in major depression determined by [(11)C]ABP688 PET and postmortem study. *The American Journal of Psychiatry*, *168*(7), 727–734.

Dias, B. G., & Ressler, K. J. (2014). Parental olfactory experience influences behavior and neural structure in subsequent generations. *Nature Neuroscience*, *17*(1), 89–96.

Diaz, S. L., Narboux-Neme, N., Trowbridge, S., Scotto-Lomassese, S., Kleine Borgmann, F. B., Jessberger, S., et al. (2013). Paradoxical increase in survival of newborn neurons in the dentate gyrus of mice with constitutive depletion of serotonin. *The European Journal of Neuroscience*, *38*(5), 2650–2658.

Diorio, D., Viau, V., & Meaney, M. J. (1993). The role of the medial prefrontal cortex (cingulate gyrus) in the regulation of hypothalamic-pituitary-adrenal responses to stress. *Journal of Neuroscience*, *13*(9), 3839–3847.

Dong, E., Agis-Balboa, R. C., Simonini, M. V., Grayson, D. R., Costa, E., & Guidotti, A. (2005). Reelin and glutamic acid decarboxylase67 promoter remodeling in an epigenetic methionine-induced mouse model of schizophrenia. *Proceedings of the National Academy of Sciences of the United States of America*, *102*(35), 12578–12583.

Drevets, W. C. (2001). Neuroimaging and neuropathological studies of depression: Implications for the cognitive-emotional features of mood disorders. *Current Opinion in Neurobiology*, *11*(2), 240–249.

Drevets, W. C., Price, J. L., & Furey, M. L. (2008). Brain structural and functional abnormalities in mood disorders: Implications for neurocircuitry models of depression. *Brain Structure and Function*, *213*(1–2), 93–118.

Drugan, R. C., Morrow, A. L., Weizman, R., Weizman, A., Deutsch, S. I., Crawley, J. N., et al. (1989). Stress-induced behavioral depression in the rat is associated with a decrease in GABA receptor-mediated chloride ion flux and brain benzodiazepine receptor occupancy. *Brain Research*, *487*(1), 45–51.

Duman, R. S., & Aghajanian, G. K. (2012). Synaptic dysfunction in depression: Potential therapeutic targets. *Science*, *338*(6103), 68–72.

Duman, R. S., & Li, N. (2012). A neurotrophic hypothesis of depression: Role of synaptogenesis in the actions of NMDA receptor antagonists. *Philosophical Transactions of the Royal Society of London. Series B, Biological Sciences*, *367*(1601), 2475–2484.

Duman, R. S., Malberg, J., & Nakagawa, S. (2001). Regulation of adult neurogenesis by psychotropic drugs and stress. *Journal of Pharmacol and Experimental Therapeutics*, *299*(2), 401–407.

Duman, R. S., & Monteggia, L. M. (2006). A neurotrophic model for stress-related mood disorders. *Biological Psychiatry*, *59*(12), 1116–1127.

Duman, C. H., Schlesinger, L., Kodama, M., Russell, D. S., & Duman, R. S. (2007). A role for MAP kinase signaling in behavioral models of depression and antidepressant treatment. *Biological Psychiatry*, *61*(5), 661–670.

Duric, V., Banasr, M., Licznerski, P., Schmidt, H. D., Stockmeier, C. A., Simen, A. A., et al. (2010). A negative regulator of MAP kinase causes depressive behavior. *Nature Medicine*, *16*(11), 1328–1332.

Duveau, V., Laustela, S., Barth, L., Gianolini, F., Vogt, K. E., Keist, R., et al. (2011). Spatiotemporal specificity of $GABA_A$ receptor-mediated regulation of adult hippocampal neurogenesis. *European Journal of Neuroscience*, *34*(3), 362–373.

Dwivedi, Y., Rizavi, H. S., Conley, R. R., Roberts, R. C., Tamminga, C. A., & Pandey, G. N. (2003). Altered gene expression of brain-derived neurotrophic factor and receptor tyrosine kinase B in postmortem brain of suicide subjects. *Archives of General Psychiatry*, *60*(8), 804–815.

Dwivedi, Y., Rizavi, H. S., Roberts, R. C., Conley, R. C., Tamminga, C. A., & Pandey, G. N. (2001). Reduced activation and expression of ERK1/2 MAP kinase in the post-mortem brain of depressed suicide subjects. *Journal of Neurochemistry*, 77(3), 916–928.

Dwivedi, Y., Rizavi, H. S., Teppen, T., Sasaki, N., Chen, H., Zhang, H., et al. (2007). Aberrant extracellular signal-regulated kinase (ERK) 5 signaling in hippocampus of suicide subjects. *Neuropsychopharmacology*, *32*(11), 2338–2350.

Dwivedi, Y., Rizavi, H. S., Zhang, H., Mondal, A. C., Roberts, R. C., Conley, R. R., et al. (2009). Neurotrophin receptor activation and expression in human postmortem brain: Effect of suicide. *Biological Psychiatry*, *65*(4), 319–328.

Dwivedi, Y., Rizavi, H. S., Zhang, H., Roberts, R. C., Conley, R. R., & Pandey, G. N. (2010). Modulation in activation and expression of phosphatase and tensin homolog on chromosome ten, Akt1, and 3-phosphoinositide-dependent kinase 1: Further evidence demonstrating altered phosphoinositide 3-kinase signaling in postmortem brain of suicide subjects. *Biological Psychiatry*, *67*(11), 1017–1025.

Earnheart, J. C., Schweizer, C., Crestani, F., Iwasato, T., Itohara, S., Mohler, H., et al. (2007). GABAergic control of adult hippocampal neurogenesis in relation to behavior indicative of trait anxiety and depression states. *Journal of Neuroscience*, *27*(14), 3845–3854.

Egan, M. F., Kojima, M., Callicott, J. H., Goldberg, T. E., Kolachana, B. S., Bertolino, A., et al. (2003). The BDNF val66met polymorphism affects activity-dependent secretion of BDNF and human memory and hippocampal function. *Cell*, *112*(2), 257–269.

Eisch, A. J., Bolanos, C. A., de Wit, J., Simonak, R. D., Pudiak, C. M., Barrot, M., et al. (2003). Brain-derived neurotrophic factor in the ventral midbrain-nucleus accumbens pathway: A role in depression. *Biological Psychiatry*, *54*(10), 994–1005.

Essrich, C., Lorez, M., Benson, J., Fritschy, J.-M., & Luscher, B. (1998). Postsynaptic clustering of major $GABA_A$ receptor subtypes requires the gamma2 subunit and gephyrin. *Nature Neuroscience*, *1*(7), 563–571.

Fava, M., Rush, A. J., Alpert, J. E., Balasubramani, G. K., Wisniewski, S. R., Carmin, C. N., et al. (2008). Difference in treatment outcome in outpatients with anxious versus nonanxious depression: A STAR*D report. *The American Journal of Psychiatry*, *165*(3), 342–351.

Feyissa, A. M., Woolverton, W. L., Miguel-Hidalgo, J. J., Wang, Z., Kyle, P. B., Hasler, G., et al. (2010). Elevated level of metabotropic glutamate receptor 2/3 in the prefrontal cortex in major depression. *Progress in Neuropsychopharmacology and Biological Psychiatry*, *34*(2), 279–283.

Fiszman, M. L., & Schousboe, A. (2004). Role of calcium and kinases on the neurotrophic effect induced by gamma-aminobutyric acid. *Journal of Neuroscience Research*, *76*(4), 435–441.

Fiumelli, H., & Woodin, M. A. (2007). Role of activity-dependent regulation of neuronal chloride homeostasis in development. *Current Opinion in Neurobiology*, *17*(1), 81–86.

Flint, J., & Kendler, K. S. (2014). The genetics of major depression. *Neuron*, *81*(3), 484–503.

Franklin, T. B., Russig, H., Weiss, I. C., Graff, J., Linder, N., Michalon, A., et al. (2010). Epigenetic transmission of the impact of early stress across generations. *Biological Psychiatry*, *68*(5), 408–415.

Fritschy, J.-M., & Mohler, H. (1995). $GABA_A$ receptor heterogeneity in the adult rat brain: Differential regional and cellular distribution of seven major subunits. *Journal of Comparative Neurology*, *359*, 154–194.

Fujii, T., Saito, D. N., Yanaka, H. T., Kosaka, H., & Okazawa, H. (2014). Depressive mood modulates the anterior lateral CA1 and DG/CA3 during a pattern separation task in cognitively intact individuals: A functional MRI study. *Hippocampus*, *24*(2), 214–224.

Fujioka, T., Fujioka, A., & Duman, R. S. (2004). Activation of cAMP signaling facilitates the morphological maturation of newborn neurons in adult hippocampus. *Journal of Neuroscience*, *24*(2), 319–328.

Gabbay, V., Mao, X., Klein, R. G., Ely, B. A., Babb, J. S., Panzer, A. M., et al. (2012). Anterior cingulate cortex gamma-aminobutyric acid in depressed adolescents: Relationship to anhedonia. *Archives of General Psychiatry*, *69*(2), 139–149.

Gamez, W., Watson, D., & Doebbeling, B. N. (2007). Abnormal personality and the mood and anxiety disorders: Implications for structural models of anxiety and depression. *Journal of Anxiety Disorders*, *21*(4), 526–539.

Garcia, L. S., Comim, C. M., Valvassori, S. S., Reus, G. Z., Stertz, L., Kapczinski, F., et al. (2009). Ketamine treatment reverses behavioral and physiological alterations induced by chronic mild stress in rats. *Progress in Neuropsychopharmacology and Biological Psychiatry*, *33*(3), 450–455.

Gascon, E., Dayer, A. G., Sauvain, M. O., Potter, G., Jenny, B., De Roo, M., et al. (2006). GABA regulates dendritic growth by stabilizing lamellipodia in newly generated interneurons of the olfactory bulb. *Journal of Neuroscience*, *26*(50), 12956–12966.

Gatt, J. M., Nemeroff, C. B., Dobson-Stone, C., Paul, R. H., Bryant, R. A., Schofield, P. R., et al. (2009). Interactions between BDNF Val66Met polymorphism and early life stress predict brain and arousal pathways to syndromal depression and anxiety. *Molecular Psychiatry*, *14*(7), 681–695.

Ge, S., Pradhan, D. A., Ming, G. L., & Song, H. (2007). GABA sets the tempo for activity-dependent adult neurogenesis. *Trends in Neurosciences*, *30*(1), 1–8.

Gilabert-Juan, J., Castillo-Gomez, E., Guirado, R., Molto, M. D., & Nacher, J. (2013). Chronic stress alters inhibitory networks in the medial prefrontal cortex of adult mice. *Brain Structure and Function*, *218*(6), 1591–1605.

Goren, M. Z., Kucukibrahimoglu, E., Berkman, K., & Terzioglu, B. (2007). Fluoxetine partly exerts its actions through GABA: A neurochemical evidence. *Neurochemical Research*, *32*(9), 1559–1565.

Gorman, J. M. (1996). Comorbid depression and anxiety spectrum disorders. *Depression and Anxiety*, *4*(4), 160–168.

Gould, E., & Tanapat, P. (1999). Stress and hippocampal neurogenesis. *Biological Psychiatry*, *46*(11), 1472–1479.

Gourley, S. L., Wu, F. J., Kiraly, D. D., Ploski, J. E., Kedves, A. T., Duman, R. S., et al. (2008). Regionally specific regulation of ERK MAP kinase in a model of antidepressant-sensitive chronic depression. *Biological Psychiatry*, *63*(4), 353–359.

Gronli, J., Bramham, C., Murison, R., Kanhema, T., Fiske, E., Bjorvatn, B., et al. (2006). Chronic mild stress inhibits BDNF protein expression and CREB activation in the dentate gyrus but not in the hippocampus proper. *Pharmacology, Biochemistry, and Behavior*, *85*(4), 842–849.

Gronli, J., Fiske, E., Murison, R., Bjorvatn, B., Sorensen, E., Ursin, R., et al. (2007). Extracellular levels of serotonin and GABA in the hippocampus after chronic mild stress in rats. A microdialysis study in an animal model of depression. *Behavioural Brain Research*, *181*(1), 42–51.

Gross, C., Zhuang, X., Stark, K., Ramboz, S., Oosting, R., Kirby, L., et al. (2002). Serotonin1A receptor acts during development to establish normal anxiety-like behaviour in the adult. *Nature*, *416*(6879), 396–400.

Groth, R. D., & Mermelstein, P. G. (2003). Brain-derived neurotrophic factor activation of NFAT (nuclear factor of activated T-cells)-dependent transcription: A role for the transcription factor NFATc4 in neurotrophin-mediated gene expression. *Journal of Neuroscience*, *23*(22), 8125–8134.

Guerry, J. D., & Hastings, P. D. (2011). In search of HPA axis dysregulation in child and adolescent depression. *Clinical Child and Family Psychology Review*, *14*(2), 135–160.

Guilloux, J. P., Douillard-Guilloux, G., Kota, R., Wang, X., Gardier, A. M., Martinowich, K., et al. (2012). Molecular evidence for BDNF- and GABA-related dysfunctions in the amygdala of female subjects with major depression. *Molecular Psychiatry*, *17*(11), 1130–1142.

Guillozet-Bongaarts, A. L., Hyde, T. M., Dalley, R. A., Hawrylycz, M. J., Henry, A., Hof, P. R., et al. (2014). Altered gene expression in the dorsolateral prefrontal cortex of individuals with schizophrenia. *Molecular Psychiatry*, *19*(4), 478–485.

Gunn, B. G., Cunningham, L., Cooper, M. A., Corteen, N. L., Seifi, M., Swinny, J. D., et al. (2013). Dysfunctional astrocytic and synaptic regulation of hypothalamic glutamatergic transmission in a mouse model of early-life adversity: Relevance to neurosteroids and programming of the stress response. *Journal of Neuroscience*, *33*(50), 19534–19554.

Gunn, B. G., Cunningham, L., Mitchell, S. G., Swinny, J. D., Lambert, J. J., & Belelli, D. (2014). GABA receptor-acting neurosteroids: A role in the development and regulation of the stress response. *Frontiers in Neuroendocrinology*. http://dx.doi.org/10.1016/j.yfrne.2014.06.001, in press.

Gur, T. L., Conti, A. C., Holden, J., Bechtholt, A. J., Hill, T. E., Lucki, I., et al. (2007). cAMP response element-binding protein deficiency allows for increased neurogenesis and a rapid onset of antidepressant response. *Journal of Neuroscience*, *27*(29), 7860–7868.

Guthrie, C. R., Murray, A. T., Franklin, A. A., & Hamblin, M. W. (2005). Differential agonist-mediated internalization of the human 5-hydroxytryptamine 7 receptor isoforms. *Journal of Pharmacol and Experimental Therapeutics*, *313*(3), 1003–1010.

Hardingham, G. E., & Bading, H. (2010). Synaptic versus extrasynaptic NMDA receptor signalling: Implications for neurodegenerative disorders. *Nature Reviews. Neuroscience*, *11*(10), 682–696.

Hashimoto, K., Sawa, A., & Iyo, M. (2007). Increased levels of glutamate in brains from patients with mood disorders. *Biological Psychiatry*, *62*(11), 1310–1316.

Hashimoto, T., Volk, D. W., Eggan, S. M., Mirnics, K., Pierri, J. N., Sun, Z., et al. (2003). Gene expression deficits in a subclass of GABA neurons in the prefrontal cortex of subjects with schizophrenia. *Journal of Neuroscience*, *23*(15), 6315–6326.

Hasin, D. S., Goodwin, R. D., Stinson, F. S., & Grant, B. F. (2005). Epidemiology of major depressive disorder: Results from the National Epidemiologic Survey on Alcoholism and Related Conditions. *Archives of General Psychiatry*, *62*(10), 1097–1106.

Hasler, G., van der Veen, J. W., Grillon, C., Drevets, W. C., & Shen, J. (2010). Effect of acute psychological stress on prefrontal GABA concentration determined by proton magnetic resonance spectroscopy. *The American Journal of Psychiatry*, *167*(10), 1226–1231.

Hasler, G., van der Veen, J. W., Tumonis, T., Meyers, N., Shen, J., & Drevets, W. C. (2007). Reduced prefrontal glutamate/glutamine and gamma-aminobutyric acid levels in major depression determined using proton magnetic resonance spectroscopy. *Archives of General Psychiatry*, *64*(2), 193–200.

Heisler, L. K., Chu, H. M., Brennan, T. J., Danao, J. A., Bajwa, P., Parsons, L. H., et al. (1998). Elevated anxiety and antidepressant-like responses in serotonin 5-HT1A receptor mutant mice. *Proceedings of the National Academy of Sciences of the United States of America*, *95*(25), 15049–15054.

Hek, K., Demirkan, A., Lahti, J., Terracciano, A., Teumer, A., Cornelis, M. C., et al. (2013). A genome-wide association study of depressive symptoms. *Biological Psychiatry*, *73*(7), 667–678.

Herman, J. P., Cullinan, W. E., Morano, M. I., Akil, H., & Watson, S. J. (1995). Contribution of the ventral subiculum to inhibitory regulation of the hypothalamo-pituitary-adrenocortical axis. *Journal of Neuroendocrinology*, 7(6), 475–482.

Hewitt, S. A., Wamsteeker, J. I., Kurz, E. U., & Bains, J. S. (2009). Altered chloride homeostasis removes synaptic inhibitory constraint of the stress axis. *Nature Neuroscience*, *12*(4), 438–443.

Hickie, I., Naismith, S., Ward, P. B., Turner, K., Scott, E., Mitchell, P., et al. (2005). Reduced hippocampal volumes and memory loss in patients with early- and late-onset depression. *The British Journal of Psychiatry*, *186*, 197–202.

Holm, M. M., Nieto-Gonzalez, J. L., Vardya, I., Henningsen, K., Jayatissa, M. N., Wiborg, O., et al. (2011). Hippocampal GABAergic dysfunction in a rat chronic mild stress model of depression. *Hippocampus*, *21*(4), 422–433.

Holsboer, F. (2001). Stress, hypercortisolism and corticosteroid receptors in depression: Implications for therapy. *Journal of Affective Disorders*, *62*(1–2), 77–91.

Holtzheimer, P. E., & Mayberg, H. S. (2011). Deep brain stimulation for psychiatric disorders. *Annual Review of Neuroscience*, *34*, 289–307.

Hong, E. J., McCord, A. E., & Greenberg, M. E. (2008). A biological function for the neuronal activity-dependent component of Bdnf transcription in the development of cortical inhibition. *Neuron*, *60*(4), 610–624.

Hu, W., Zhang, M., Czeh, B., Flugge, G., & Zhang, W. (2010). Stress impairs GABAergic network function in the hippocampus by activating nongenomic glucocorticoid receptors and affecting the integrity of the parvalbumin-expressing neuronal network. *Neuropsychopharmacology*, *35*(8), 1693–1707.

Huang, Z. J., Kirkwood, A., Pizzorusso, T., Porciatti, V., Morales, B., Bear, M. F., et al. (1999). BDNF regulates the maturation of inhibition and the critical period of plasticity in mouse visual cortex. *Cell*, *98*(6), 739–755.

Inoue, W., Baimoukhametova, D. V., Fuzesi, T., Cusulin, J. I., Koblinger, K., Whelan, P. J., et al. (2013). Noradrenaline is a stress-associated metaplastic signal at GABA synapses. *Nature Neuroscience*, *16*(5), 605–612.

Jackson, K., Soutto, M., Peng, D., Hu, T., Marshal, D., & El-Rifai, W. (2011). Epigenetic silencing of somatostatin in gastric cancer. *Digestive Diseases and Sciences*, *56*(1), 125–130.

Jacob, T. C., Michels, G., Silayeva, L., Haydon, J., Succol, F., & Moss, S. J. (2012). Benzodiazepine treatment induces subtype-specific changes in $GABA_A$ receptor trafficking and decreases synaptic inhibition. *Proceedings of the National Academy of Sciences of the United States of America*, *109*(45), 18595–18600.

Jacob, T. C., Moss, S. J., & Jurd, R. (2008). $GABA_A$ receptor trafficking and its role in the dynamic modulation of neuronal inhibition. *Nature Reviews. Neuroscience*, *9*(5), 331–343.

Jacobson-Pick, S., Audet, M. C., McQuaid, R. J., Kalvapalle, R., & Anisman, H. (2012). Stressor exposure of male and female juvenile mice influences later responses to stressors: Modulation of $GABA_A$ receptor subunit mRNA expression. *Neuroscience*, *215*, 114–126.

Jagasia, R., Steib, K., Englberger, E., Herold, S., Faus-Kessler, T., Saxe, M., et al. (2009). GABA-cAMP response element-binding protein signaling regulates maturation and survival of newly generated neurons in the adult hippocampus. *Journal of Neuroscience*, *29*(25), 7966–7977.

Jovanovic, J. N., Czernik, A. J., Fienberg, A. A., Greengard, P., & Sihra, T. S. (2000). Synapsins as mediators of BDNF-enhanced neurotransmitter release. *Nature Neuroscience*, *3*(4), 323–329.

Jurd, R., Tretter, V., Walker, J., Brandon, N. J., & Moss, S. J. (2010). Fyn kinase contributes to tyrosine phosphorylation of the $GABA_A$ receptor gamma2 subunit. *Molecular and Cellular Neuroscience*, *44*(2), 129–134.

Kang, H. J., Voleti, B., Hajszan, T., Rajkowska, G., Stockmeier, C. A., Licznerski, P., et al. (2012). Decreased expression of synapse-related genes and loss of synapses in major depressive disorder. *Nature Medicine*, *18*(9), 1413–1417.

Karolewicz, B., Maciag, D., O'Dwyer, G., Stockmeier, C. A., Feyissa, A. M., & Rajkowska, G. (2010). Reduced level of glutamic acid decarboxylase-67 kDa in the prefrontal cortex in major depression. *International Journal of Neuropsychopharmacology*, *13*(4), 411–420.

Karpova, N. N. (2014). Role of BDNF epigenetics in activity-dependent neuronal plasticity. *Neuropharmacology*, *76 Pt C*, 709–718.

Karten, Y. J., Olariu, A., & Cameron, H. A. (2005). Stress in early life inhibits neurogenesis in adulthood. *Trends in Neurosciences*, *28*(4), 171–172.

Kathol, R. G., Jaeckle, R. S., Lopez, J. F., & Meller, W. H. (1989). Pathophysiology of HPA axis abnormalities in patients with major depression: An update. *The American Journal of Psychiatry*, *146*(3), 311–317.

Kaufman, R. E., Ostacher, M. J., Marks, E. H., Simon, N. M., Sachs, G. S., Jensen, J. E., et al. (2009). Brain GABA levels in patients with bipolar disorder. *Progress in Neuropsychopharmacology and Biological Psychiatry*, *33*(3), 427–434.

Keller, M. B., McCullough, J. P., Klein, D. N., Arnow, B., Dunner, D. L., Gelenberg, A. J., et al. (2000). A comparison of nefazodone, the cognitive behavioral-analysis system of psychotherapy, and their combination for the treatment of chronic depression. *New England Journal of Medicine*, *342*(20), 1462–1470.

Kendler, K. S., Gatz, M., Gardner, C. O., & Pedersen, N. L. (2006). A Swedish national twin study of lifetime major depression. *The American Journal of Psychiatry*, *163*(1), 109–114.

Kendler, K. S., Karkowski, L. M., & Prescott, C. A. (1999). Causal relationship between stressful life events and the onset of major depression. *The American Journal of Psychiatry*, *156*(6), 837–841.

Kessler, R. C., Berglund, P., Demler, O., Jin, R., Koretz, D., Merikangas, K. R., et al. (2003). The epidemiology of major depressive disorder: Results from the National Comorbidity Survey Replication (NCS-R). *JAMA*, *289*(23), 3095–3105.

Khan, A., Khan, S. R., Walens, G., Kolts, R., & Giller, E. L. (2003). Frequency of positive studies among fixed and flexible dose antidepressant clinical trials: An analysis of the food and drug administration summary basis of approval reports. *Neuropsychopharmacology*, *28*(3), 552–557.

Kheirbek, M. A., Klemenhagen, K. C., Sahay, A., & Hen, R. (2012). Neurogenesis and generalization: A new approach to stratify and treat anxiety disorders. *Nature Neuroscience*, *15*(12), 1613–1620.

Kim, S. J., Cho, S. J., Jang, H. M., Shin, J., Park, P. W., Lee, Y. J., et al. (2010). Interaction between brain-derived neurotrophic factor Val66Met polymorphism and recent negative stressor in harm avoidance. *Neuropsychobiology*, *61*(1), 19–26.

Kim, Y. K., Lee, H. P., Won, S. D., Park, E. Y., Lee, H. Y., Lee, B. H., et al. (2007). Low plasma BDNF is associated with suicidal behavior in major depression. *Progress in Neuropsychopharmacology and Biological Psychiatry*, *31*(1), 78–85.

Kittler, J. T., Chen, G., Kukhtina, V., Vahedi-Faridi, A., Gu, Z., Tretter, V., et al. (2008). Regulation of synaptic inhibition by phospho-dependent binding of the AP2 complex to a YECL motif in the $GABA_A$ receptor gamma2 subunit. *Proceedings of the National Academy of Sciences of the United States of America*, *105*(9), 3616–3621.

Klempan, T. A., Sequeira, A., Canetti, L., Lalovic, A., Ernst, C., ffrench-Mullen, J., et al. (2009). Altered expression of genes involved in ATP biosynthesis and GABAergic neurotransmission in the ventral prefrontal cortex of suicides with and without major depression. *Molecular Psychiatry*, *14*(2), 175–189.

Klempin, F., Babu, H., De Pietri Tonelli, D., Alarcon, E., Fabel, K., & Kempermann, G. (2010). Oppositional effects of serotonin receptors 5-HT1a, 2, and 2c in the regulation of adult hippocampal neurogenesis. *Frontiers in Molecular Neuroscience*, *3*(14).

Klempin, F., Beis, D., Mosienko, V., Kempermann, G., Bader, M., & Alenina, N. (2013). Serotonin is required for exercise-induced adult hippocampal neurogenesis. *Journal of Neuroscience*, *33*(19), 8270–8275.

Klumpers, U. M., Veltman, D. J., Drent, M. L., Boellaard, R., Comans, E. F., Meynen, G., et al. (2010). Reduced parahippocampal and lateral temporal $GABA_A$-[^{11}C]flumazenil binding in major depression: Preliminary results. *European Journal of Nuclear Medicine and Molecular Imaging*, *37*(3), 565–574.

Kneussel, M., Brandstätter, J. H., Laube, B., Stahl, S., Müller, U., & Betz, H. (1999). Loss of postsynaptic $GABA_A$ receptor clustering in gephyrin-deficient mice. *Journal of Neuroscience*, *19*(21), 9289–9297.

Kohara, K., Kitamura, A., Adachi, N., Nishida, M., Itami, C., Nakamura, S., et al. (2003). Inhibitory but not excitatory cortical neurons require presynaptic brain-derived neurotrophic factor for dendritic development, as revealed by chimera cell culture. *Journal of Neuroscience*, *23*(14), 6123–6131.

Koolschijn, P. C., van Haren, N. E., Lensvelt-Mulders, G. J., Hulshoff Pol, H. E., & Kahn, R. S. (2009). Brain volume abnormalities in major depressive disorder: A meta-analysis of magnetic resonance imaging studies. *Human Brain Mapping*, *30*(11), 3719–3735.

Krishnan, V., & Nestler, E. J. (2008). The molecular neurobiology of depression. *Nature*, *455*(7215), 894–902.
Krystal, J. H., Sanacora, G., & Duman, R. S. (2013). Rapid-acting glutamatergic antidepressants: The path to ketamine and beyond. *Biological Psychiatry*, *73*(12), 1133–1141.
Kucukibrahimoglu, E., Saygin, M. Z., Caliskan, M., Kaplan, O. K., Unsal, C., & Goren, M. Z. (2009). The change in plasma GABA, glutamine and glutamate levels in fluoxetine- or S-citalopram-treated female patients with major depression. *European Journal of Clinical Pharmacology*, *65*(6), 571–577.
Lagace, D. C., Donovan, M. H., DeCarolis, N. A., Farnbauch, L. A., Malhotra, S., Berton, O., et al. (2010). Adult hippocampal neurogenesis is functionally important for stress-induced social avoidance. *Proceedings of the National Academy of Sciences of the United States of America*, *107*(9), 4436–4441.
Lan, M. J., McLoughlin, G. A., Griffin, J. L., Tsang, T. M., Huang, J. T., Yuan, P., et al. (2009). Metabonomic analysis identifies molecular changes associated with the pathophysiology and drug treatment of bipolar disorder. *Molecular Psychiatry*, *14*(3), 269–279.
Leal, S. L., Tighe, S. K., Jones, C. K., & Yassa, M. A. (2014). Pattern separation of emotional information in hippocampal dentate and CA3. *Hippocampus*, *24*(9), 1146–1155.
Lee, H. H., Deeb, T. Z., Walker, J. A., Davies, P. A., & Moss, S. J. (2011). NMDA receptor activity downregulates KCC2 resulting in depolarizing $GABA_A$ receptor-mediated currents. *Nature Neuroscience*, *14*(6), 736–743.
Lee, V., Sarkar, J., & Maguire, J. (2014). Loss of Gabrd in CRH neurons blunts the corticosterone response to stress and diminishes stress-related behaviors. *Psychoneuroendocrinology*, *41*, 75–88.
Levinson, A. J., Fitzgerald, P. B., Favalli, G., Blumberger, D. M., Daigle, M., & Daskalakis, Z. J. (2010). Evidence of cortical inhibitory deficits in major depressive disorder. *Biological Psychiatry*, *67*(5), 458–464.
Li, N., Lee, B., Liu, R. J., Banasr, M., Dwyer, J. M., Iwata, M., et al. (2010). mTOR-dependent synapse formation underlies the rapid antidepressant effects of NMDA antagonists. *Science*, *329*(5994), 959–964.
Li, N., Liu, R. J., Dwyer, J. M., Banasr, M., Lee, B., Son, H., et al. (2011). Glutamate N-methyl-D-aspartate receptor antagonists rapidly reverse behavioral and synaptic deficits caused by chronic stress exposure. *Biological Psychiatry*, *69*(8), 754–761.
Li, H. Y., & Sawchenko, P. E. (1998). Hypothalamic effector neurons and extended circuitries activated in "neurogenic" stress: A comparison of footshock effects exerted acutely, chronically, and in animals with controlled glucocorticoid levels. *Journal of Comparative Neurology*, *393*(2), 244–266.
Lichtenstein, P., Yip, B. H., Bjork, C., Pawitan, Y., Cannon, T. D., Sullivan, P. F., et al. (2009). Common genetic determinants of schizophrenia and bipolar disorder in Swedish families: A population-based study. *Lancet*, *373*(9659), 234–239.
Liu, R. J., Fuchikami, M., Dwyer, J. M., Lepack, A. E., Duman, R. S., & Aghajanian, G. K. (2013). GSK-3 inhibition potentiates the synaptogenic and antidepressant-like effects of subthreshold doses of ketamine. *Neuropsychopharmacology*, *38*(11), 2268–2277.
Low, K., Crestani, F., Keist, R., Benke, D., Brunig, I., Benson, J. A., et al. (2000). Molecular and neuronal substrate for the selective attenuation of anxiety. *Science*, *290*(5489), 131–134.
Luscher, B., & Fuchs, T. (2013). Adult hippocampal neurogenesis in the absence of serotonin (Commentary on Diaz et al.). *European Journal of Neuroscience*, *38*(5), 2649.
Luscher, B., Fuchs, T., & Kilpatrick, C. L. (2011). $GABA_A$ receptor trafficking-mediated plasticity of inhibitory synapses. *Neuron*, *70*(3), 385–409.
Luscher, B., Shen, Q., & Sahir, N. (2011). The GABAergic deficit hypothesis of major depressive disorder. *Molecular Psychiatry*, *16*(4), 383–406.
Maciag, D., Hughes, J., O'Dwyer, G., Pride, Y., Stockmeier, C. A., Sanacora, G., et al. (2010). Reduced density of calbindin immunoreactive GABAergic neurons in the

occipital cortex in major depression: Relevance to neuroimaging studies. *Biological Psychiatry, 67*(5), 465–470.

MacKenzie, G., & Maguire, J. (2014). The role of ovarian hormone-derived neurosteroids on the regulation of $GABA_A$ receptors in affective disorders. *Psychopharmacology, 231*(17), 3333–3342.

MacQueen, G. M., Campbell, S., McEwen, B. S., Macdonald, K., Amano, S., Joffe, R. T., et al. (2003). Course of illness, hippocampal function, and hippocampal volume in major depression. *Proceedings of the National Academy of Sciences of the United States of America, 100*(3), 1387–1392.

Maeng, S., Zarate, C. A., Jr., Du, J., Schloesser, R. J., McCammon, J., Chen, G., et al. (2008). Cellular mechanisms underlying the antidepressant effects of ketamine: Role of alpha-amino-3-hydroxy-5-methylisoxazole-4-propionic acid receptors. *Biological Psychiatry, 63*(4), 349–352.

Maes, M., Verkerk, R., Vandoolaeghe, E., Lin, A., & Scharpe, S. (1998). Serum levels of excitatory amino acids, serine, glycine, histidine, threonine, taurine, alanine and arginine in treatment-resistant depression: Modulation by treatment with antidepressants and prediction of clinical responsivity. *Acta Psychiatrica Scandinavica, 97*(4), 302–308.

Maggio, N., & Segal, M. (2009). Differential corticosteroid modulation of inhibitory synaptic currents in the dorsal and ventral hippocampus. *Journal of Neuroscience, 29*(9), 2857–2866.

Maguire, J., Ferando, I., Simonsen, C., & Mody, I. (2009). Excitability changes related to $GABA_A$ receptor plasticity during pregnancy. *Journal of Neuroscience, 29*(30), 9592–9601.

Maguire, J., & Mody, I. (2007). Neurosteroid synthesis-mediated regulation of $GABA_A$ receptors: Relevance to the ovarian cycle and stress. *Journal of Neuroscience, 27*(9), 2155–2162.

Maguire, J., & Mody, I. (2008). $GABA_AR$ plasticity during pregnancy: Relevance to postpartum depression. *Neuron, 59*(2), 207–213.

Maguire, J. L., Stell, B. M., Rafizadeh, M., & Mody, I. (2005). Ovarian cycle-linked changes in $GABA_A$ receptors mediating tonic inhibition alter seizure susceptibility and anxiety. *Nature Neuroscience, 8*(6), 797–804.

Major Depressive Disorder Working Group of the Psychiatric GWAS Consortium, Ripke, S., Wray, N. R., Lewis, C. M., Hamilton, S. P., & Weissman, M. M. (2013). A mega-analysis of genome-wide association studies for major depressive disorder. *Molecular Psychiatry, 18*(4), 497–511.

Malberg, J. E., Eisch, A. J., Nestler, E. J., & Duman, R. S. (2000). Chronic antidepressant treatment increases neurogenesis in adult rat hippocampus. *Journal of Neuroscience, 20*(24), 9104–9110.

Maric, D., Liu, Q. Y., Maric, I., Chaudry, S., Chang, Y. H., Smith, S. V., et al. (2001). GABA expression dominates neuronal lineage progression in the embryonic rat neocortex and facilitates neurite outgrowth via $GABA_A$ autoreceptor/Cl− channels. *Journal of Neuroscience, 21*(7), 2343–2360.

Martisova, E., Solas, M., Horrillo, I., Ortega, J. E., Meana, J. J., Tordera, R. M., et al. (2012). Long lasting effects of early-life stress on glutamatergic/GABAergic circuitry in the rat hippocampus. *Neuropharmacology, 62*(5–6), 1944–1953.

Matussek, N. (1969). Die catecholamin- und serotoninhypothese der depression. In H. Hippius & H. Seebach (Eds.), *Das depressive syndrome*. München, Berlin, Wien: Urban & Schwarzenberg.

Merali, Z., Du, L., Hrdina, P., Palkovits, M., Faludi, G., Poulter, M. O., et al. (2004). Dysregulation in the suicide brain: mRNA expression of corticotropin-releasing hormone receptors and $GABA_A$ receptor subunits in frontal cortical brain region. *Journal of Neuroscience, 24*(6), 1478–1485.

Miguel-Hidalgo, J. J., & Rajkowska, G. (2002). Morphological brain changes in depression: Can antidepressants reverse them? *CNS Drugs, 16*(6), 361–372.

Miguel-Hidalgo, J. J., Waltzer, R., Whittom, A. A., Austin, M. C., Rajkowska, G., & Stockmeier, C. A. (2010). Glial and glutamatergic markers in depression, alcoholism, and their comorbidity. *Journal of Affective Disorders*, *127*(1–3), 230–240.

Mihalek, R. M., Banerjee, P. K., Korpi, E. R., Quinlan, J. J., Firestone, L. L., Mi, Z. P., et al. (1999). Attenuated sensitivity to neuroactive steroids in gamma-aminobutyrate type A receptor delta subunit knockout mice. *Proceedings of the National Academy of Sciences of the United States of America*, *96*(22), 12905–12910.

Mineur, Y. S., Obayemi, A., Wigestrand, M. B., Fote, G. M., Calarco, C. A., Li, A. M., et al. (2013). Cholinergic signaling in the hippocampus regulates social stress resilience and anxiety- and depression-like behavior. *Proceedings of the National Academy of Sciences of the United States of America*, *110*(9), 3573–3578.

Mirescu, C., Peters, J. D., & Gould, E. (2004). Early life experience alters response of adult neurogenesis to stress. *Nature Neuroscience*, 7(8), 841–846.

Mitani, H., Shirayama, Y., Yamada, T., Maeda, K., Ashby, C. R., Jr., & Kawahara, R. (2006). Correlation between plasma levels of glutamate, alanine and serine with severity of depression. *Progress in Neuropsychopharmacology and Biological Psychiatry*, *30*(6), 1155–1158.

Mizoguchi, Y., Kanematsu, T., Hirata, M., & Nabekura, J. (2003). A rapid increase in the total number of cell surface functional $GABA_A$ receptors induced by brain-derived neurotrophic factor in rat visual cortex. *Journal of Biological Chemistry*, *278*(45), 44097–44102.

Mohler, H. (2012). The GABA system in anxiety and depression and its therapeutic potential. *Neuropharmacology*, *62*(1), 42–53.

Morris, H. M., Hashimoto, T., & Lewis, D. A. (2008). Alterations in somatostatin mRNA expression in the dorsolateral prefrontal cortex of subjects with schizophrenia or schizoaffective disorder. *Cerebral Cortex*, *18*(7), 1575–1587.

Murray, C. J., & Lopez, A. D. (1996). Evidence-based health policy—Lessons from the Global Burden of Disease Study. *Science*, *274*(5288), 740–743.

Murray, E. A., Wise, S. P., & Drevets, W. C. (2011). Localization of dysfunction in major depressive disorder: Prefrontal cortex and amygdala. *Biological Psychiatry*, *69*(12), e43–54.

Nakagawa, S., Kim, J. E., Lee, R., Malberg, J. E., Chen, J., Steffen, C., et al. (2002). Regulation of neurogenesis in adult mouse hippocampus by cAMP and the cAMP response element-binding protein. *Journal of Neuroscience*, *22*(9), 3673–3682.

Nibuya, M., Morinobu, S., & Duman, R. S. (1995). Regulation of BDNF and trkB mRNA in rat brain by chronic electroconvulsive seizure and antidepressant drug treatments. *Journal of Neuroscience*, *15*(11), 7539–7547.

Okamoto, H., Voleti, B., Banasr, M., Sarhan, M., Duric, V., Girgenti, M. J., et al. (2010). Wnt2 expression and signaling is increased by different classes of antidepressant treatments. *Biological Psychiatry*, *68*(6), 521–527.

Ongur, D., Prescot, A. P., McCarthy, J., Cohen, B. M., & Renshaw, P. F. (2010). Elevated gamma-aminobutyric acid levels in chronic schizophrenia. *Biological Psychiatry*, *68*(7), 667–670.

Otero Losada, M. E. (1988). Changes in central GABAergic function following acute and repeated stress. *British Journal of Pharmacology*, *93*(3), 483–490.

Papouin, T., Ladepeche, L., Ruel, J., Sacchi, S., Labasque, M., Hanini, M., et al. (2012). Synaptic and extrasynaptic NMDA receptors are gated by different endogenous coagonists. *Cell*, *150*(3), 633–646.

Parks, C. L., Robinson, P. S., Sibille, E., Shenk, T., & Toth, M. (1998). Increased anxiety of mice lacking the serotonin1A receptor. *Proceedings of the National Academy of Sciences of the United States of America*, *95*(18), 10734–10739.

Paul, I. A., & Skolnick, P. (2003). Glutamate and depression: Clinical and preclinical studies. *Annals of the New York Academy of Sciences*, *1003*, 250–272.

Perroud, N., Courtet, P., Vincze, I., Jaussent, I., Jollant, F., Bellivier, F., et al. (2008). Interaction between BDNF Val66Met and childhood trauma on adult's violent suicide attempt. *Genes, Brain, and Behavior*, *7*(3), 314–322.

Perry, P. J. (1996). Pharmacotherapy for major depression with melancholic features: Relative efficacy of tricyclic versus selective serotonin reuptake inhibitor antidepressants. *Journal of Affective Disorders*, *39*(1), 1–6.

Petrik, D., Lagace, D. C., & Eisch, A. J. (2012). The neurogenesis hypothesis of affective and anxiety disorders: Are we mistaking the scaffolding for the building? *Neuropharmacology*, *62*(1), 21–34.

Petty, F., Trivedi, M. H., Fulton, M., & Rush, A. J. (1995). Benzodiazepines as antidepressants: Does GABA play a role in depression? *Biological Psychiatry*, *38*(9), 578–591.

Pinna, G., Costa, E., & Guidotti, A. (2006). Fluoxetine and norfluoxetine stereospecifically and selectively increase brain neurosteroid content at doses that are inactive on 5-HT reuptake. *Psychopharmacology*, *186*(3), 362–372.

Pinna, G., Costa, E., & Guidotti, A. (2009). SSRIs act as selective brain steroidogenic stimulants (SBSSs) at low doses that are inactive on 5-HT reuptake. *Current Opinion in Pharmacology*, *9*(1), 24–30.

Pittenger, C., & Duman, R. S. (2008). Stress, depression, and neuroplasticity: A convergence of mechanisms. *Neuropsychopharmacology*, *33*(1), 88–109.

Porcher, C., Hatchett, C., Longbottom, R. E., McAinch, K., Sihra, T. S., Moss, S. J., et al. (2011). Positive feedback regulation between gamma-aminobutyric acid type A ($GABA_A$) receptor signaling and brain-derived neurotrophic factor (BDNF) release in developing neurons. *Journal of Biological Chemistry*, *286*(24), 21667–21677.

Poulter, M. O., Du, L., Weaver, I. C., Palkovits, M., Faludi, G., Merali, Z., et al. (2008). $GABA_A$ receptor promoter hypermethylation in suicide brain: Implications for the involvement of epigenetic processes. *Biological Psychiatry*, *64*(8), 645–652.

Preskorn, S. H., Baker, B., Kolluri, S., Menniti, F. S., Krams, M., & Landen, J. W. (2008). An innovative design to establish proof of concept of the antidepressant effects of the NR2B subunit selective N-methyl-D-aspartate antagonist, CP-101,606, in patients with treatment-refractory major depressive disorder. *Journal of Clinical Psychopharmacology*, *28*(6), 631–637.

Quadrato, G., Benevento, M., Alber, S., Jacob, C., Floriddia, E. M., Nguyen, T., et al. (2012). Nuclear factor of activated T cells (NFATc4) is required for BDNF-dependent survival of adult-born neurons and spatial memory formation in the hippocampus. *Proceedings of the National Academy of Sciences of the United States of America*, *109*(23), E1499–1508.

Quadrato, G., Elnaggar, M. Y., Duman, C., Sabino, A., Forsberg, K., & Di Giovanni, S. (2014). Modulation of $GABA_A$ receptor signaling increases neurogenesis and suppresses anxiety through NFATc4. *Journal of Neuroscience*, *34*(25), 8630–8645.

Radley, J. J., Gosselink, K. L., & Sawchenko, P. E. (2009). A discrete GABAergic relay mediates medial prefrontal cortical inhibition of the neuroendocrine stress response. *Journal of Neuroscience*, *29*(22), 7330–7340.

Raison, C. L., Borisov, A. S., Majer, M., Drake, D. F., Pagnoni, G., Woolwine, B. J., et al. (2009). Activation of central nervous system inflammatory pathways by interferon-alpha: Relationship to monoamines and depression. *Biological Psychiatry*, *65*(4), 296–303.

Rajkowska, G., & Miguel-Hidalgo, J. J. (2007). Gliogenesis and glial pathology in depression. *CNS & Neurological Disorders Drug Targets*, *6*(3), 219–233.

Rajkowska, G., O'Dwyer, G., Teleki, Z., Stockmeier, C. A., & Miguel-Hidalgo, J. J. (2007). GABAergic neurons immunoreactive for calcium binding proteins are reduced in the prefrontal cortex in major depression. *Neuropsychopharmacology*, *32*(2), 471–482.

Ren, Z., Sahir, N., Murakami, S., Luellen, B. A., Earnheart, J. C., Lal, R., et al. (2014). Defects in dendrite and spine maturation and synaptogenesis associated with an anxious-depressive-like phenotype of $GABA_A$ receptor-deficient mice. *Neuropharmacology*, http://dx.doi.org/10.1016/j.neuropharm.2014.07.019, (Epub ahead of print).

Represa, A., & Ben-Ari, Y. (2005). Trophic actions of GABA on neuronal development. *Trends in Neurosciences*, *28*(6), 278–283.

Riad, M., Zimmer, L., Rbah, L., Watkins, K. C., Hamon, M., & Descarries, L. (2004). Acute treatment with the antidepressant fluoxetine internalizes 5-HT1A autoreceptors and reduces the in vivo binding of the PET radioligand [^{18}F]MPPF in the nucleus raphe dorsalis of rat. *Journal of Neuroscience*, *24*(23), 5420–5426.

Rico, B., Xu, B., & Reichardt, L. F. (2002). TrkB receptor signaling is required for establishment of GABAergic synapses in the cerebellum. *Nature Neuroscience*, *5*(3), 225–233.

Roose, S. P., Glassman, A. H., Attia, E., & Woodring, S. (1994). Comparative efficacy of selective serotonin reuptake inhibitors and tricyclics in the treatment of melancholia. *The American Journal of Psychiatry*, *151*(12), 1735–1739.

Rosenzweig-Lipson, S., Sabb, A., Stack, G., Mitchell, P., Lucki, I., Malberg, J. E., et al. (2007). Antidepressant-like effects of the novel, selective, 5-HT2C receptor agonist WAY-163909 in rodents. *Psychopharmacology*, *192*(2), 159–170.

Rossignol, E. (2011). Genetics and function of neocortical GABAergic interneurons in neurodevelopmental disorders. *Neural Plasticity*, *2011*, 649325.

Roth, T. L., Lubin, F. D., Funk, A. J., & Sweatt, J. D. (2009). Lasting epigenetic influence of early-life adversity on the BDNF gene. *Biological Psychiatry*, *65*(9), 760–769.

Rowland, L. M., Kontson, K., West, J., Edden, R. A., Zhu, H., Wijtenburg, S. A., et al. (2013). In vivo measurements of glutamate, GABA, and NAAG in schizophrenia. *Schizophrenia Bulletin*, *39*(5), 1096–1104.

Rubio, A., Sanchez-Mut, J. V., Garcia, E., Velasquez, Z. D., Oliver, J., Esteller, M., et al. (2012). Epigenetic control of somatostatin and cortistatin expression by beta amyloid peptide. *Journal of Neuroscience Research*, *90*(1), 13–20.

Rudolph, U., & Mohler, H. (2014). $GABA_A$ receptor subtypes: Therapeutic potential in down syndrome, affective disorders, schizophrenia, and autism. *Annual Review of Pharmacology and Toxicology*, *54*, 483–507.

Ruhe, H. G., Mason, N. S., & Schene, A. H. (2007). Mood is indirectly related to serotonin, norepinephrine and dopamine levels in humans: A meta-analysis of monoamine depletion studies. *Molecular Psychiatry*, *12*(4), 331–359.

Russo, S. J., & Nestler, E. J. (2013). The brain reward circuitry in mood disorders. *Nature Reviews. Neuroscience*, *14*(9), 609–625.

Russo-Neustadt, A. A., & Chen, M. J. (2005). Brain-derived neurotrophic factor and antidepressant activity. *Current Pharmaceutical Design*, *11*(12), 1495–1510.

Saarelainen, T., Hendolin, P., Lucas, G., Koponen, E., Sairanen, M., MacDonald, E., et al. (2003). Activation of the TrkB neurotrophin receptor is induced by antidepressant drugs and is required for antidepressant-induced behavioral effects. *Journal of Neuroscience*, *23*(1), 349–357.

Sacher, J., Neumann, J., Funfstuck, T., Soliman, A., Villringer, A., & Schroeter, M. L. (2012). Mapping the depressed brain: A meta-analysis of structural and functional alterations in major depressive disorder. *Journal of Affective Disorders*, *140*(2), 142–148.

Sahay, A., Scobie, K. N., Hill, A. S., O'Carroll, C. M., Kheirbek, M. A., Burghardt, N. S., et al. (2011). Increasing adult hippocampal neurogenesis is sufficient to improve pattern separation. *Nature*, *472*(7344), 466–470.

Sairanen, M., Lucas, G., Ernfors, P., Castren, M., & Castren, E. (2005). Brain-derived neurotrophic factor and antidepressant drugs have different but coordinated effects on neuronal turnover, proliferation, and survival in the adult dentate gyrus. *Journal of Neuroscience*, *25*(5), 1089–1094.

Sakata, K., Jin, L., & Jha, S. (2010). Lack of promoter IV-driven BDNF transcription results in depression-like behavior. *Genes, Brain, and Behavior*, *9*(7), 712–721.

Sakata, K., Woo, N. H., Martinowich, K., Greene, J. S., Schloesser, R. J., Shen, L., et al. (2009). Critical role of promoter IV-driven BDNF transcription in GABAergic transmission and synaptic plasticity in the prefrontal cortex. *Proceedings of the National Academy of Sciences of the United States of America, 106*(14), 5942–5947.

Samuels, B. A., & Hen, R. (2011). Neurogenesis and affective disorders. *European Journal of Neuroscience, 33*(6), 1152–1159.

Samuels, B. A., Leonardo, E. D., Gadient, R., Williams, A., Zhou, J., David, D. J., et al. (2011). Modeling treatment-resistant depression. *Neuropharmacology, 61*(3), 408–413.

Sanacora, G., Gueorguieva, R., Epperson, C. N., Wu, Y. T., Appel, M., Rothman, D. L., et al. (2004). Subtype-specific alterations of gamma-aminobutyric acid and glutamate in patients with major depression. *Archives of General Psychiatry, 61*(7), 705–713.

Sanacora, G., Mason, G. F., Rothman, D. L., Behar, K. L., Hyder, F., Petroff, O. A., et al. (1999). Reduced cortical gamma-aminobutyric acid levels in depressed patients determined by proton magnetic resonance spectroscopy. *Archives of General Psychiatry, 56*(11), 1043–1047.

Sanacora, G., Mason, G. F., Rothman, D. L., Hyder, F., Ciarcia, J. J., Ostroff, R. B., et al. (2003). Increased cortical GABA concentrations in depressed patients receiving ECT. *The American Journal of Psychiatry, 160*(3), 577–579.

Sanacora, G., Mason, G. F., Rothman, D. L., & Krystal, J. H. (2002). Increased occipital cortex GABA concentrations in depressed patients after therapy with selective serotonin reuptake inhibitors. *The American Journal of Psychiatry, 159*(4), 663–665.

Santarelli, L., Saxe, M., Gross, C., Surget, A., Battaglia, F., Dulawa, S., et al. (2003). Requirement of hippocampal neurogenesis for the behavioral effects of antidepressants. *Science, 301*(5634), 805–809.

Sapolsky, R. M., Krey, L. C., & McEwen, B. S. (1984). Glucocorticoid-sensitive hippocampal neurons are involved in terminating the adrenocortical stress response. *Proceedings of the National Academy of Sciences of the United States of America, 81*(19), 6174–6177.

Sarkar, J., Wakefield, S., MacKenzie, G., Moss, S. J., & Maguire, J. (2011). Neurosteroidogenesis is required for the physiological response to stress: Role of neurosteroid-sensitive $GABA_A$ receptors. *Journal of Neuroscience, 31*(50), 18198–18210.

Savitz, J. B., & Drevets, W. C. (2009). Imaging phenotypes of major depressive disorder: Genetic correlates. *Neuroscience, 164*(1), 300–330.

Schildkraut, J. (1965). The catecholamine hypothesis of affective disorders: A review of supporting evidence. *The American Journal of Psychiatry, 122*, 509–522.

Schildkraut, J. J. (1995). The catecholamine hypothesis of affective disorders: A review of supporting evidence. 1965. *Journal of Neuropsychiatry and Clinical Neurosciences*, 7(4), 524–533 (Discussion 523–524).

Schlag, B. D., Lou, Z., Fennell, M., & Dunlop, J. (2004). Ligand dependency of 5-hydroxytryptamine 2C receptor internalization. *Journal of Pharmacol and Experimental Therapeutics, 310*(3), 865–870.

Schloesser, R. J., Manji, H. K., & Martinowich, K. (2009). Suppression of adult neurogenesis leads to an increased hypothalamo–pituitary–adrenal axis response. *Neuroreport, 20*(6), 553–557.

Schmidt, H. D., & Duman, R. S. (2007). The role of neurotrophic factors in adult hippocampal neurogenesis, antidepressant treatments and animal models of depressive-like behavior. *Behavioral Pharmacology, 18*(5–6), 391–418.

Schmidt-Hieber, C., Jonas, P., & Bischofberger, J. (2004). Enhanced synaptic plasticity in newly generated granule cells of the adult hippocampus. *Nature, 429*(6988), 184–187.

Schreiber, G., & Avissar, S. (2007). Regulators of G-protein-coupled receptor-G-protein coupling: Antidepressants mechanism of action. *Expert Review of Neurotherapeutics*, 7(1), 75–84.

Sequeira, A., Klempan, T., Canetti, L., ffrench-Mullen, J., Benkelfat, C., Rouleau, G. A., et al. (2007). Patterns of gene expression in the limbic system of suicides with and without major depression. *Molecular Psychiatry*, *12*(7), 640–655.

Sequeira, A., Mamdani, F., Ernst, C., Vawter, M. P., Bunney, W. E., Lebel, V., et al. (2009). Global brain gene expression analysis links glutamatergic and GABAergic alterations to suicide and major depression. *PLoS One*, *4*(8), e6585.

Serra, M., Pisu, M. G., Mostallino, M. C., Sanna, E., & Biggio, G. (2008). Changes in neuroactive steroid content during social isolation stress modulate $GABA_A$ receptor plasticity and function. *Brain Research Reviews*, *57*(2), 520–530.

Shaywitz, A. J., & Greenberg, M. E. (1999). CREB: A stimulus-induced transcription factor activated by a diverse array of extracellular signals. *Annual Review of Biochemistry*, *68*, 821–861.

Sheline, Y. I., Gado, M. H., & Kraemer, H. C. (2003). Untreated depression and hippocampal volume loss. *The American Journal of Psychiatry*, *160*(8), 1516–1518.

Sheline, Y. I., Sanghavi, M., Mintun, M. A., & Gado, M. H. (1999). Depression duration but not age predicts hippocampal volume loss in medically healthy women with recurrent major depression. *Journal of Neuroscience*, *19*(12), 5034–5043.

Shelton, R. C., Claiborne, J., Sidoryk-Wegrzynowicz, M., Reddy, R., Aschner, M., Lewis, D. A., et al. (2011). Altered expression of genes involved in inflammation and apoptosis in frontal cortex in major depression. *Molecular Psychiatry*, *16*, 751–762.

Shen, Q., Lal, R., Luellen, B. A., Earnheart, J. C., Andrews, A. M., & Luscher, B. (2010). gamma-Aminobutyric acid-type A receptor deficits cause hypothalamic-pituitary-adrenal axis hyperactivity and antidepressant drug sensitivity reminiscent of melancholic forms of depression. *Biological Psychiatry*, *68*(6), 512–520.

Shieh, P. B., Hu, S. C., Bobb, K., Timmusk, T., & Ghosh, A. (1998). Identification of a signaling pathway involved in calcium regulation of BDNF expression. *Neuron*, *20*(4), 727–740.

Shimizu, E., Hashimoto, K., Okamura, N., Koike, K., Komatsu, N., Kumakiri, C., et al. (2003). Alterations of serum levels of brain-derived neurotrophic factor (BDNF) in depressed patients with or without antidepressants. *Biological Psychiatry*, *54*(1), 70–75.

Shirayama, Y., Chen, A. C., Nakagawa, S., Russell, D. S., & Duman, R. S. (2002). Brain-derived neurotrophic factor produces antidepressant effects in behavioral models of depression. *Journal of Neuroscience*, *22*(8), 3251–3261.

Si, X., Miguel-Hidalgo, J. J., O'Dwyer, G., Stockmeier, C. A., & Rajkowska, G. (2004). Age-dependent reductions in the level of glial fibrillary acidic protein in the prefrontal cortex in major depression. *Neuropsychopharmacology*, *29*(11), 2088–2096.

Sibille, E., Morris, H. M., Kota, R. S., & Lewis, D. A. (2011). GABA-related transcripts in the dorsolateral prefrontal cortex in mood disorders. *International Journal of Neuropsychopharmacology*, *14*(6), 721–734.

Smith, K. S., & Rudolph, U. (2012). Anxiety and depression: Mouse genetics and pharmacological approaches to the role of $GABA_A$ receptor subtypes. *Neuropharmacology*, *62*(1), 54–62.

Snyder, J. S., Soumier, A., Brewer, M., Pickel, J., & Cameron, H. A. (2011). Adult hippocampal neurogenesis buffers stress responses and depressive behaviour. *Nature*, *476*(7361), 458–461.

Song, J., Zhong, C., Bonaguidi, M. A., Sun, G. J., Hsu, D., Gu, Y., et al. (2012). Neuronal circuitry mechanism regulating adult quiescent neural stem-cell fate decision. *Nature*, *489*(7414), 150–154.

Stander, V. A., Thomsen, C. J., & Highfill-McRoy, R. M. (2014). Etiology of depression comorbidity in combat-related PTSD: A review of the literature. *Clinical Psychology Review*, *34*(2), 87–98.

Steffens, D. C., Byrum, C. E., McQuoid, D. R., Greenberg, D. L., Payne, M. E., Blitchington, T. F., et al. (2000). Hippocampal volume in geriatric depression. *Biological Psychiatry*, *48*(4), 301–309.

Stell, B. M., Brickley, S. G., Tang, C. Y., Farrant, M., & Mody, I. (2003). Neuroactive steroids reduce neuronal excitability by selectively enhancing tonic inhibition mediated by delta subunit-containing $GABA_A$ receptors. *Proceedings of the National Academy of Sciences of the United States of America*, *100*(24), 14439–14444.

Suchanek, R., Owczarek, A., Paul-Samojedny, M., Kowalczyk, M., Kucia, K., & Kowalski, J. (2013). BDNF val66met polymorphism is associated with age at onset and intensity of symptoms of paranoid schizophrenia in a Polish population. *Journal of Neuropsychiatry and Clinical Neurosciences*, *25*(1), 88–94.

Sullivan, R. M., & Gratton, A. (1999). Lateralized effects of medial prefrontal cortex lesions on neuroendocrine and autonomic stress responses in rats. *Journal of Neuroscience*, *19*(7), 2834–2840.

Sullivan, P. F., Kendler, K. S., & Neale, M. C. (2003). Schizophrenia as a complex trait: Evidence from a meta-analysis of twin studies. *Archives of General Psychiatry*, *60*(12), 1187–1192.

Surget, A., Tanti, A., Leonardo, E. D., Laugeray, A., Rainer, Q., Touma, C., et al. (2011). Antidepressants recruit new neurons to improve stress response regulation. *Molecular Psychiatry*, *16*(12), 1177–1188.

Swartz, C. M., & Guadagno, G. (1998). Melancholia with onset during treatment with SSRIs. *Annals of Clinical Psychiatry*, *10*(4), 177–179.

Taliaz, D., Stall, N., Dar, D. E., & Zangen, A. (2010). Knockdown of brain-derived neurotrophic factor in specific brain sites precipitates behaviors associated with depression and reduces neurogenesis. *Molecular Psychiatry*, *15*(1), 80–92.

Tan, K. R., Yvon, C., Turiault, M., Mirzabekov, J. J., Doehner, J., Labouebe, G., et al. (2012). GABA neurons of the VTA drive conditioned place aversion. *Neuron*, *73*(6), 1173–1183.

Tan, H., Zhong, P., & Yan, Z. (2004). Corticotropin-releasing factor and acute stress prolongs serotonergic regulation of GABA transmission in prefrontal cortical pyramidal neurons. *Journal of Neuroscience*, *24*(21), 5000–5008.

Tao, X., Finkbeiner, S., Arnold, D. B., Shaywitz, A. J., & Greenberg, M. E. (1998). Ca^{2+} influx regulates BDNF transcription by a CREB family transcription factor-dependent mechanism. *Neuron*, *20*(4), 709–726.

Tashiro, A., Sandler, V. M., Toni, N., Zhao, C., & Gage, F. H. (2006). NMDA-receptor-mediated, cell-specific integration of new neurons in adult dentate gyrus. *Nature*, *442*(7105), 929–933.

Tayoshi, S., Nakataki, M., Sumitani, S., Taniguchi, K., Shibuya-Tayoshi, S., Numata, S., et al. (2010). GABA concentration in schizophrenia patients and the effects of antipsychotic medication: A proton magnetic resonance spectroscopy study. *Schizophrenia Research*, *117*(1), 83–91.

Tokita, K., Yamaji, T., & Hashimoto, K. (2012). Roles of glutamate signaling in preclinical and/or mechanistic models of depression. *Pharmacology, Biochemistry, and Behavior*, *100*(4), 688–704.

Toth, M. (2003). 5-HT1A receptor knockout mouse as a genetic model of anxiety. *European Journal of Pharmacology*, *463*(1–3), 177–184.

Tretter, V., Revilla-Sanchez, R., Houston, C., Terunuma, M., Havekes, R., Florian, C., et al. (2009). Deficits in spatial memory correlate with modified {gamma}-aminobutyric acid type A receptor tyrosine phosphorylation in the hippocampus. *Proceedings of the National Academy of Sciences of the United States of America*, *106*(47), 20039–20044.

Tripp, A., Kota, R. S., Lewis, D. A., & Sibille, E. (2011). Reduced somatostatin in subgenual anterior cingulate cortex in major depression. *Neurobiology of Disease*, *42*(1), 116–124.

Tripp, A., Oh, H., Guilloux, J. P., Martinowich, K., Lewis, D. A., & Sibille, E. (2012). Brain-derived neurotrophic factor signaling and subgenual anterior cingulate cortex dysfunction in major depressive disorder. *The American Journal of Psychiatry*, *169*(11), 1194–1202.

Tritsch, N. X., Ding, J. B., & Sabatini, B. L. (2012). Dopaminergic neurons inhibit striatal output through non-canonical release of GABA. *Nature*, *490*(7419), 262–266.

Trullas, R., & Skolnick, P. (1990). Functional antagonists at the NMDA receptor complex exhibit antidepressant actions. *European Journal of Pharmacology*, *185*(1), 1–10.

Tyagarajan, S. K., & Fritschy, J. M. (2014). Gephyrin: A master regulator of neuronal function? *Nature Reviews. Neuroscience*, *15*(3), 141–156.

Tyagarajan, S. K., Ghosh, H., Harvey, K., & Fritschy, J. M. (2011). Collybistin splice variants differentially interact with gephyrin and Cdc42 to regulate gephyrin clustering at GABAergic synapses. *Journal of Cell Science*, *124*(Pt 16), 2786–2796.

Tyagarajan, S. K., Ghosh, H., Yevenes, G. E., Imanishi, S. Y., Zeilhofer, H. U., Gerrits, B., et al. (2013). Extracellular signal-regulated kinase and glycogen synthase kinase 3beta regulate gephyrin postsynaptic aggregation and GABAergic synaptic function in a calpain-dependent mechanism. *Journal of Biological Chemistry*, *288*(14), 9634–9647.

Tyagarajan, S. K., Ghosh, H., Yevenes, G. E., Nikonenko, I., Ebeling, C., Schwerdel, C., et al. (2011). Regulation of GABAergic synapse formation and plasticity by GSK3 beta-dependent phosphorylation of gephyrin. *Proceedings of the National Academy of Sciences of the United States of America*, *108*(1), 379–384.

Uzunov, D. P., Cooper, T. B., Costa, E., & Guidotti, A. (1996). Fluoxetine-elicited changes in brain neurosteroid content measured by negative ion mass fragmentography. *Proceedings of the National Academy of Sciences of the United States of America*, *93*(22), 12599–12604.

Uzunova, V., Sheline, Y., Davis, J. M., Rasmusson, A., Uzunov, D. P., Costa, E., et al. (1998). Increase in the cerebrospinal fluid content of neurosteroids in patients with unipolar major depression who are receiving fluoxetine or fluvoxamine. *Proceedings of the National Academy of Sciences of the United States of America*, *95*(6), 3239–3244.

van Zessen, R., Phillips, J. L., Budygin, E. A., & Stuber, G. D. (2012). Activation of VTA GABA neurons disrupts reward consumption. *Neuron*, *73*(6), 1184–1194.

Verkuyl, J. M., Hemby, S. E., & Joels, M. (2004). Chronic stress attenuates GABAergic inhibition and alters gene expression of parvocellular neurons in rat hypothalamus. *The European Journal of Neuroscience*, *20*(6), 1665–1673.

Vicini, S., Losi, G., & Homanics, G. E. (2002). $GABA_A$ receptor delta subunit deletion prevents neurosteroid modulation of inhibitory synaptic currents in cerebellar neurons. *Neuropharmacology*, *43*(4), 646–650.

Vinkers, C. H., van Oorschot, R., Nielsen, E. O., Cook, J. M., Hansen, H. H., Groenink, L., et al. (2012). $GABA_A$ receptor alpha subunits differentially contribute to diazepam tolerance after chronic treatment. *PLoS One*, 7(8), e43054.

Vithlani, M., Hines, R. M., Zhong, P., Terunuma, M., Hines, D. J., Revilla-Sanchez, R., et al. (2013). The ability of BDNF to modify neurogenesis and depressive-like behaviors is dependent upon phosphorylation of tyrosine residues 365/367 in the $GABA_A$-receptor gamma2 subunit. *Journal of Neuroscience*, *33*(39), 15567–15577.

Voleti, B., & Duman, R. S. (2012). The roles of neurotrophic factor and Wnt signaling in depression. *Clinical Pharmacology and Therapeutics*, *91*(2), 333–338.

Vollenweider, I., Smith, K. S., Keist, R., & Rudolph, U. (2011). Antidepressant-like properties of alpha2-containing $GABA_A$ receptors. *Behavioural Brain Research*, *217*(1), 77–80.

Walker, A. K., Kavelaars, A., Heijnen, C. J., & Dantzer, R. (2014). Neuroinflammation and comorbidity of pain and depression. *Pharmacological Reviews*, *66*(1), 80–101.

Wardle, R. A., & Poo, M. M. (2003). Brain-derived neurotrophic factor modulation of GABAergic synapses by postsynaptic regulation of chloride transport. *Journal of Neuroscience*, *23*(25), 8722–8732.

Waterhouse, E. G., An, J. J., Orefice, L. L., Baydyuk, M., Liao, G. Y., Zheng, K., et al. (2012). BDNF promotes differentiation and maturation of adult-born neurons through GABAergic transmission. *Journal of Neuroscience*, *32*(41), 14318–14330.

Weizman, R., Weizman, A., Kook, K. A., Vocci, F., Deutsch, S. I., & Paul, S. M. (1989). Repeated swim stress alters brain benzodiazepine receptors measured in vivo. *Journal of Pharmacology and Experimental Therapeutics*, *249*(3), 701–707.

World Health Organization. (2008). *The global burden of disease: 2004 update.* Available at, http://www.who.int/healthinfo/global_burden_disease/2004_report_update/en/.2014.

Young, E. A., Altemus, M., Lopez, J. F., Kocsis, J. H., Schatzberg, A. F., DeBattista, C., et al. (2004). HPA axis activation in major depression and response to fluoxetine: A pilot study. *Psychoneuroendocrinology*, *29*(9), 1198–1204.

Yu, H., Wang, D. D., Wang, Y., Liu, T., Lee, F. S., & Chen, Z. Y. (2012). Variant brain-derived neurotrophic factor Val66Met polymorphism alters vulnerability to stress and response to antidepressants. *Journal of Neuroscience*, *32*(12), 4092–4101.

Yuen, E. Y., Liu, W., Karatsoreos, I. N., Ren, Y., Feng, J., McEwen, B. S., et al. (2011). Mechanisms for acute stress-induced enhancement of glutamatergic transmission and working memory. *Molecular Psychiatry*, *16*(2), 156–170.

Yuen, E. Y., Wei, J., Liu, W., Zhong, P., Li, X., & Yan, Z. (2012). Repeated stress causes cognitive impairment by suppressing glutamate receptor expression and function in prefrontal cortex. *Neuron*, *73*(5), 962–977.

Zattoni, M., Mura, M. L., Deprez, F., Schwendener, R. A., Engelhardt, B., Frei, K., et al. (2011). Brain infiltration of leukocytes contributes to the pathophysiology of temporal lobe epilepsy. *Journal of Neuroscience*, *31*(11), 4037–4050.

Zhang, T. Y., Hellstrom, I. C., Bagot, R. C., Wen, X., Diorio, J., & Meaney, M. J. (2010). Maternal care and DNA methylation of a glutamic acid decarboxylase 1 promoter in rat hippocampus. *Journal of Neuroscience*, *30*(39), 13130–13137.

Zhao, C., Wang, F., Pun, F. W., Mei, L., Ren, L., Yu, Z., et al. (2012). Epigenetic regulation on GABRB2 isoforms expression: Developmental variations and disruptions in psychotic disorders. *Schizophrenia Research*, *134*(2–3), 260–266.

CHAPTER SIX

Mechanisms of Fast Desensitization of $GABA_B$ Receptor-Gated Currents

Adi Raveh*, Rostislav Turecek*,†, Bernhard Bettler*,[1]
*Department of Biomedicine, Institute of Physiology, Pharmazentrum, University of Basel, Basel, Switzerland
†Department of Auditory Neuroscience, Institute of Experimental Medicine, ASCR, Prague, Czech Republic
[1]Corresponding author: e-mail address: bernhard.bettler@unibas.ch

Contents

Abstract

$GABA_B$ receptors ($GABA_BRs$) regulate the excitability of most neurons in the central nervous system by modulating the activity of enzymes and ion channels. In the sustained presence of the neurotransmitter γ-aminobutyric acid, $GABA_BRs$ exhibit a time-dependent decrease in the receptor response—a phenomenon referred to as homologous desensitization. Desensitization prevents excessive receptor influences on neuronal activity. Much work focused on the mechanisms of $GABA_BR$ desensitization that operate at the receptor and control receptor expression at the plasma membrane. Over the past few years, it became apparent that $GABA_BRs$ additionally evolved mechanisms for faster desensitization. These mechanisms operate at the G protein rather than at the receptor and inhibit G protein signaling within seconds of agonist exposure. The mechanisms for fast desensitization are ideally suited to regulate receptor-activated ion channel responses, which influence neuronal activity on a faster timescale than effector enzymes. Here, we provide an update on the mechanisms for fast desensitization of $GABA_BR$ responses and discuss physiological and pathophysiological implications.

Advances in Pharmacology, Volume 73
ISSN 1054-3589
http://dx.doi.org/10.1016/bs.apha.2014.11.004

ABBREVIATIONS

GABA γ-aminobutyric acid
$GABA_BRs$ $GABA_B$ receptors
GAP GTPase-activating protein
GIRK/Kir3 channel G protein-activated inwardly rectifying K^+ channel
GPCR G protein-coupled receptor
KCTD K^+ channel tetramerization domain
PKA cAMP-dependent protein kinase
RGS regulator of G protein signaling

1. INTRODUCTION

γ-Aminobutyric acid B receptors ($GABA_BRs$) are the G protein-coupled receptors (GPCRs) for GABA (γ-aminobutyric acid), the main inhibitory neurotransmitter in the central nervous system (Benke, Zemoura, & Maier, 2012; Chalifoux & Carter, 2011a; Gassmann & Bettler, 2012). $GABA_BRs$ are obligate heteromers composed of principal $GABA_{B1}$ and $GABA_{B2}$ subunits (Marshall, Jones, Kaupmann, & Bettler, 1999; Mohler & Fritschy, 1999). $GABA_BRs$ additionally associate with auxiliary KCTD subunits (named after their K^+ channel tetramerization domain) (Bartoi et al., 2010; Schwenk et al., 2010). While the KCTDs exert little allosteric influence on agonist affinity, they substantially influence biophysical properties of the receptor response (Rajalu et al., 2014; Schwenk et al., 2010; Seddik et al., 2012). $GABA_BRs$ activate heterotrimeric $G\alpha_i$- and $G\alpha_o$-type G proteins by catalyzing the exchange of GDP for GTP at Gα, leading to a partial or complete dissociation of Gα·GTP from Gβγ (Lambert, 2008). Dissociated Gα and Gβγ subunits independently signal to effector enzymes and ion channels. It is well established that $G\alpha_{o/i}$ inhibits adenylyl cyclase while Gβγ gates G protein-activated inwardly rectifying K^+ (GIRK/Kir3) channels (Betke, Wells, & Hamm, 2012; Luscher & Slesinger, 2010) and voltage-activated Ca^{2+} channels (Betke et al., 2012; Chalifoux & Carter, 2011b; Perez-Garci, Larkum, & Nevian, 2013; Tedford & Zamponi, 2006). $GABA_BRs$ are expressed by the vast majority of neurons in the central nervous system (Mohler, Benke, & Fritschy, 2001). The two prevalent subunit isoforms $GABA_{B1a}$ and $GABA_{B1b}$ target $GABA_BRs$ to pre- and postsynaptic sites, respectively (Biermann et al., 2010; Perez-Garci, Gassmann, Bettler, & Larkum, 2006; Shaban et al., 2006; Vigot et al., 2006). Presynaptic $GABA_BRs$ are expressed at inhibitory and excitatory terminals where they inhibit neurotransmitter release by downregulating the activity of voltage-activated Ca^{2+} channels and inhibiting adenylyl cyclase (Betke et al., 2012; Thompson & Gahwiler, 1992). Postsynaptic $GABA_B$Rs are well known to activate Kir3 channels, which inhibit neuronal activity by shunting excitatory transmission and by generating slow postsynaptic inhibitory potentials that hyperpolarize the membrane (Luscher & Slesinger, 2010). Postsynaptic $GABA_B$Rs also inhibit voltage-gated Ca^{2+} channels, which may impact synaptic transmission and plasticity (Chalifoux & Carter, 2011b; Perez-Garci et al., 2006, 2013).

The activity of $GABA_B$Rs at pre- and postsynaptic sites influences numerous neuronal processes and behavioral states. For example, $GABA_B$Rs regulate the expression of long-term potentiation (Davies, Starkey, Pozza, & Collingridge, 1991; Shaban et al., 2006; Vigot et al., 2006), a synaptic plasticity phenomenon implicated in learning and memory processes. An emerging physiological function of $GABA_B$Rs is the control of spontaneous network activity, which is fundamental for cognitive processes (Beenhakker & Huguenard, 2010; Craig, Mayne, Bettler, Paulsen, & McBain, 2013; Isaacson & Scanziani, 2011; Kohl & Paulsen, 2010; Mann, Kohl, & Paulsen, 2009; Scanziani, 2000). The importance of $GABA_B$Rs for the proper functioning of the nervous system is highlighted in studies with genetically modified mice. Knockout mice with a complete or partial lack of $GABA_B$Rs show a wide variety of neurological and behavioral phenotypes, including epileptiform activity, impaired memory, hyperalgesia, hypothermia, contextual hyperactivity, altered stress responses, and increased anxiety (Gassmann et al., 2004; Jacobson, Kelly, Bettler, Kaupmann, & Cryan, 2006; Mombereau et al., 2005; O'Leary et al., 2014; Prosser et al., 2001; Quéva et al., 2003; Schuler et al., 2001; Shaban et al., 2006; Vienne, Bettler, Franken, & Tafti, 2010; Vigot et al., 2006). It is therefore not surprising that deficits in $GABA_B$R-mediated signaling have long been associated with neurological and psychiatric conditions, most notably with epilepsy, anxiety, depression, schizophrenia, obsessive compulsive disorder, bipolar disorder, addiction, and pain (Bowery et al., 2002; Enna & Bowery, 2004; Gassmann & Bettler, 2012; Richter et al., 2012; Vacher & Bettler, 2003).

Because $GABA_B$Rs are important for the regulation of a wide variety of physiological processes in the nervous system, the activity of $GABA_B$Rs is under stringent temporal control. To avoid excessive effects on neuronal activity in the prolonged presence of agonist, $GABA_B$Rs usually exhibit a time-dependent decrease in receptor-activated $G\alpha$ and $G\beta\gamma$ signaling, a process referred to as homologous desensitization (Couve et al., 2002; Cruz et al., 2004; Gjoni & Urwyler, 2009; Gonzalez-Maeso, Wise, Green, & Koenig, 2003; Perroy, Adam, Qanbar, Chenier, & Bouvier, 2003; Sickmann & Alzheimer, 2003; Sodickson & Bean, 1996; Tosetti et al., 2004; Wetherington & Lambert, 2002). Of note, $GABA_BRs$ can also be cross-desensitized by the activity of other receptors (Benke et al., 2012; Gassmann & Bettler, 2012; Guetg et al., 2010; Terunuma, Pangalos, & Moss, 2010), which is referred to as heterologous desensitization (Chuang, Iacovelli, Sallese, & DeBlasi, 1996).

Many GPCRs exhibit a slow form of homologous desensitization that operates at the agonist-occupied receptor and involves phosphorylation at serine and/or threonine residues by G protein-coupled receptor kinases (GRKs), recruitment of arrestins, uncoupling of G proteins from the receptor, receptor internalization via clathrin-dependent endocytosis, followed by either recycling or degradation of the receptors (Evron, Daigle, & Caron, 2012; Kelly, Bailey, & Henderson, 2008; Tsao, Cao, & von Zastrow, 2001). It is meanwhile well established that $GABA_BRs$ are not phosphorylated by GRKs (Fairfax et al., 2004; Perroy et al., 2003) and do not undergo this canonical form of desensitization (Benke et al., 2012; Sudo et al., 2012; Terunuma et al., 2010). Here, we review alternative mechanisms involved in homologous desensitization of $GABA_BR$ responses, with emphasis on mechanisms that enable a fast desensitization (also referred to as "acute" or "short-term" desensitization) of receptor-gated current responses within seconds of agonist exposure.

2. HOMOLOGOUS DESENSITIZATION OPERATING AT THE RECEPTOR

While $GABA_BRs$ are not internalized through typical agonist-induced receptor phosphorylation by the GRKs, the surface availability of $GABA_BRs$ is nevertheless regulated by receptor activity and phosphorylation, albeit in a rather unusual manner. Couve and colleagues demonstrated that cAMP-dependent protein kinase (PKA) phosphorylates serine-892 in the C-terminal domain of the $GABA_{B2}$ subunit, which stabilizes receptors at the cell surface by slowing endocytosis-independent degradation (Couve et al., 2002; Fairfax et al., 2004). Since activation of $GABA_BRs$ inhibits adenylyl cyclase and thus cAMP-dependent phosphorylation by PKA, receptor activity is expected to reduce serine-892 phosphorylation and to facilitate desensitization of the receptor response.

In keeping with such a mechanism, cultured hippocampal neurons of knock-in mice with a nonphosphorylatable mutation of serine-892 to alanine exhibit an accelerated desensitization of $GABA_BR$-activated K^+ currents (Adelfinger et al., 2014). Serine-892 was found to be heavily phosphorylated in the brain (Adelfinger et al., 2014; Couve et al., 2002), which supports that tonic PKA activity normally limits desensitization of $GABA_BR$ responses. Kinases that are not classical effectors of $GABA_BRs$, such as Ca^{2+}/calmodulin-dependent protein kinase, protein kinase C, and 5′AMP-dependent protein kinase, also phosphorylate $GABA_BRs$ and regulate desensitization of the receptor response in a heterologous manner (Dutar & Nicoll, 1988; Guetg et al., 2010; Kuramoto et al., 2007; Pontier et al., 2006; Thompson & Gahwiler, 1992). Phosphorylation-dependent mechanisms of $GABA_BR$ desensitization have been recently reviewed (Benke et al., 2012; Gassmann & Bettler, 2012; Terunuma et al., 2010) and therefore will not be discussed in more detail here. Of note, phosphorylation-dependent mechanisms of desensitization usually reduce surface receptor numbers through second- messenger-mediated multistep processes. Phosphorylation-dependent mechanisms of desensitization are therefore relatively slow and typically operate on timescales of several minutes to hours.

3. HOMOLOGOUS DESENSITIZATION OPERATING AT THE α AND $\beta\gamma$ SUBUNITS OF THE G PROTEIN

It is emerging that $GABA_BRs$ have evolved mechanisms for desensitization that operate on a faster timescale than phosphorylation-dependent mechanisms. Fast mechanisms of desensitization likely evolved to control receptor signaling to effector channels, which regulate neuronal activity on a faster timescale than effector enzymes. Desensitization within seconds of agonist exposure has been reported with $GABA_BR$-gated K^+ or Ca^{2+} currents (Ehrengruber et al., 1997; Isaacson, Solis, & Nicoll, 1993; Sickmann & Alzheimer, 2002, 2003; Sodickson & Bean, 1996). In contrast to the slower mechanisms of desensitization that operate at the receptor, the fast mechanisms operate at the G protein. Fast desensitization is to some extent determined by the kinetic properties of the G protein activation/deactivation cycle, i.e., the rates of GDP–GTP exchange and GTP hydrolysis at Gα (Chuang, Yu, Jan, & Jan, 1998; Leaney et al., 2004; Sickmann & Alzheimer, 2003), which limits the amount of heterotrimeric

G protein that can be activated during prolonged agonist exposure. The G protein cycle therefore contributes to a basal component of fast desensitization that occurs with GPCR-gated currents in general (Chuang et al., 1998). In addition, several regulatory proteins have been identified that directly influence G protein signaling to enhance fast desensitization. At this point, these proteins include the regulator of G protein signaling (RGS) proteins, GRKs, KCTD12, and KCTD12b. Below, we summarize current knowledge on how these regulatory proteins desensitize $GABA_BR$ responses.

3.1. RGS-induced fast desensitization

RGS proteins are a family of cytosolic proteins that contain a conserved RGS domain of approximately 120 amino acids (De Vries, Zheng, Fischer, Elenko, & Farquhar, 2000; Hollinger & Hepler, 2002; Ross & Wilkie, 2000; Willars, 2006). The RGS domain directly binds to activated $G\alpha$ and modulates the timecourse of G protein signaling by acting as a GTPase-activating protein (GAP) (Berman, Wilkie, & Gilman, 1996; Hunt, Fields, Casey, & Peralta, 1996; Watson, Linder, Druey, Kehrl, & Blumer, 1996). GAP activity accelerates the rate of GTP hydrolysis and thus speeds up deactivation of the G protein. The influence of RGS proteins on receptor-gated currents has been analyzed using Kir3 currents as a readout. It is now well established that coexpression of GPCRs with various RGS proteins accelerates activation and deactivation of G protein-gated K^+ currents (Doupnik, Davidson, Lester, & Kofuji, 1997; Doupnik, Jaen, & Zhang, 2004; Lambert et al., 2010; Maurice et al., 2010; Saitoh, Kubo, Miyatani, Asano, & Nakata, 1997). These kinetic effects mostly result from the GAP activity of RGS proteins (Lambert et al., 2010), yet alternative explanations have also been advanced (Garzon, Rodriguez-Munoz, de la Torre-Madrid, & Sanchez-Blazquez, 2005; Hollinger & Hepler, 2002; Zerangue & Jan, 1998).

Fast activation and deactivation kinetics of $GABA_BR$-activated K^+ currents in rat cerebellar granule cells suggested early on that native $GABA_BRs$ are subject to modulation by RGS proteins (Doupnik et al., 2004). Likewise, experiments with the nonhydrolysable GTP analogue GTPγS or mutant $G\alpha$ subunits that either sequester endogenous RGS proteins or render $G\alpha$ insensitive to GAP activity suggested that endogenous RGS proteins regulate $GABA_BR$ responses in human embryonic kidney 293T cells (Mutneja,

Berton, Suen, Luscher, & Slesinger, 2005). Additional experiments showed that RGS4 was responsible for enhancing the fast component of $GABA_B$R-activated K^+ current desensitization in 293T cells. In keeping with these findings, RGS4 was later shown to associate with $GABA_B$Rs (Kim et al., 2014), presumably as part of a complex that also includes the Kir3 channel and the G protein (Ciruela et al., 2010; David et al., 2006; Fowler, Aryal, Suen, & Slesinger, 2007) (Fig. 1A).

The exact mechanism by which RGS proteins induce fast desensitization of K^+ current responses is still unclear. Several processes may contribute to RGS-induced desensitization. First, the GAP activity of RGS proteins is expected to accelerate the G protein cycle, which speeds up use-dependent processes (for example, GTP depletion; Zholos & Bolton, 1996) and thus contributes to a basal desensitization of effector K^+ currents. Consistent with this hypothesis, GDPβS, which binds to but does not activate G proteins, reduces fast K^+ current desensitization (Leaney et al., 2004; Sickmann & Alzheimer, 2003). Second, RGS-induced desensitization may also arise from a shift in G protein activity, which is more synchronized at the peak of the response than at steady state (Chuang et al., 1998). Third, RGS proteins may associate with G proteins in an activity-dependent manner and sequester Gα from receptors or effectors, thus limiting the number of G proteins that can be activated with time (Tesmer, Berman, Gilman, & Sprang, 1997).

Several members of the RGS family, including RGS2, RGS6, RGS7, and RGS8, were reported to influence $GABA_B$Rs-induced K^+ current activation and/or deactivation, however without clearly affecting agonist-induced desensitization (Benians, Nobles, Hosny, & Tinker, 2005; Labouebe et al., 2007; Maity et al., 2012). In many of these studies, the effects of RGS proteins on the fast component of current desensitization may have been missed because of the use of slow drug delivery systems. Specifically, RGS2 was reported to reduce the $GABA_B$R/Kir3 coupling efficiency in dopaminergic neurons of the ventral tegmental area (Labouebe et al., 2007). RGS6 was shown to form a complex with Gβ5 in cerebellar granule cells and to accelerate $GABA_B$R-activated K^+ current deactivation (Maity et al., 2012). Gβ5 also associates with RGS7, which accelerates both activation and deactivation of $GABA_B$R-activated K^+ currents in hippocampal neurons (Fajardo-Serrano et al., 2013; Ostrovskaya et al., 2014; Xie et al., 2010). Likewise, RGS8-enhanced activation and deactivation of receptor-activated Kir3 current (Benians et al., 2005).

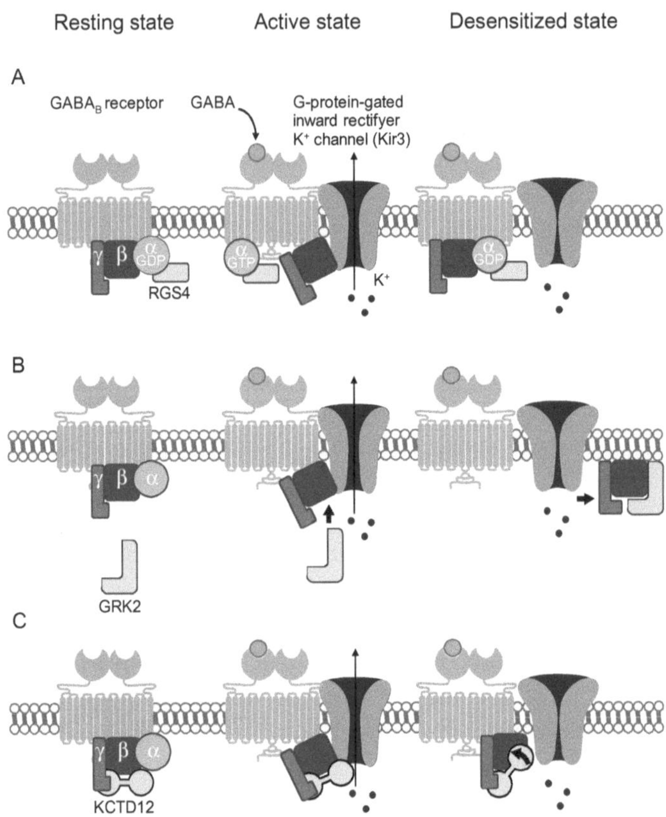

Figure 1 Scheme illustrating mechanisms for fast desensitization of GABA$_B$R-activated Kir3-type K$^+$ currents. These mechanisms operate at α and βγ subunits of the G protein. Resting (left), active (middle), and desensitized (right) states of the current response are shown. (A) RGS4 constitutively assembles with GABA$_B$Rs, the G protein, and Kir3 channels into a signaling complex. The GAP activity of RGS4 synchronizes G proteins in the inactive (α · GDP) heterotrimeric state prior to receptor activation (resting state). Phasic activation of highly synchronized G proteins generates a maximal amount of activated G protein (α · GTP) and thus a maximal Gβγ-induced current response (active state). During prolonged receptor activity, the G proteins cycle less synchronously and a higher proportion of G protein is inactive (α · GDP, desensitized state). RGS4 proteins may also sequester G protein α subunits from the receptor and accelerate local depletion of GTP (not shown). (B) In the absence of receptor activity, GRK2 is localized in the cytoplasm (inactive state). Upon receptor activation, GRK2 is recruited to the plasma membrane (active state). Binding of GRK2 to Gβγ titrates Gβγ away from effector Kir3 channels and induces current desensitization (desensitized state). (C) KCTD12 constitutively assembles with GABA$_B$Rs and the G protein into a signaling complex (inactive state). Constitutive binding of KCTD12 to activated Gβγ allows for a transient activation of Kir3 channels (active state). An activity-dependent rearrangement of KCTD12 at Gβγ leads to a shielding of the Kir3-binding site on Gβγ by KCTD12 and induces current desensitization (desensitized state).

3.2. GRK-induced fast desensitization

GRKs are protein kinases that normally induce GPCR desensitization by phosphorylating agonist-activated receptors, thus promoting uncoupling of the G protein and clathrin-mediated receptor endocytosis (Pitcher, Freedman, & Lefkowitz, 1998). While GRKs do not phosphorylate $GABA_BRs$ (Fairfax et al., 2004; Perroy et al., 2003), it was surprisingly shown that GRK4 and GRK5 still induce homologous desensitization of the $GABA_BR$ response (Kanaide et al., 2007; Perroy et al., 2003). One study reported that a kinase-deficient GRK4 mutant desensitizes $GABA_BR$-stimulated [^{35}S]GTPγS binding in cerebellar granule cells and in heterologous cells (Perroy et al., 2003). While the precise mechanism of GRK4-induced desensitization was not revealed, GRK4 was shown to directly associate with $GABA_BRs$. Since GRK4 expression in the brain is restricted to cerebellar granule cells (Sallese et al., 2000; Virlon et al., 1998), this form of $GABA_BR$ desensitization appears to be highly cell specific. A subsequent study showed that GRK4 and GRK5, but not GRK2, GRK3, or GRK6, desensitize $GABA_BR$-activated Kir3 currents in heterologous cells (Kanaide et al., 2007). The mechanism for desensitization again involves assembly of GRK4 and GRK5 with $GABA_BRs$, more specifically with the $GABA_{B2}$ subunit (Kanaide et al., 2007). Because GRK4 and GRK5 compete with the G protein for binding at $GABA_{B2}$ (Havlickova et al., 2002; Robbins et al., 2001), GRKs may induce desensitization by uncoupling the G protein from the receptor (Benke et al., 2012). A follow-up study showed that GRK4- and GRK5-induced desensitization in heterologous cells is partially suppressed in the presence of ketamine, an NMDA receptor antagonist (Ando et al., 2011). Ketamine was reported to inhibit the association of the GRKs with $GABA_BRs$, but the underlying mechanism was not revealed.

It was recently reported that noncanonical GRK2-induced desensitization of GPCR responses can arise from the sequestration of free Gβγ (Raveh, Cooper, Guy-David, & Reuveny, 2010) (Fig. 1B). Following activation of the receptor, cytosolic GRK2 is recruited to the receptor at the plasma membrane, where the C-terminal pleckstrin homology domain of GRK2 binds to Gβγ (Tesmer, Kawano, Shankaranarayanan, Kozasa, & Tesmer, 2005). Binding of Gβγ by GRK2 desensitizes K^+ current responses by competitively titrating Gβγ away from effector Kir3 channels (Raveh et al., 2010; Shui, Khan, Tsuga, Haga, & Boyett, 1998). This mechanism was shown to desensitize GPCR-activated K^+ currents in heterologous cells and cultured hippocampal neurons within seconds of agonist exposure (Raveh et al., 2010; Turecek et al., 2014). Desensitization is also induced by a kinase-deficient GRK2 mutant, clearly showing that receptor phosphorylation is not required for desensitization (Raveh et al., 2010). Since this noncanonical form of GRK2-induced desensitization operates downstream of the receptor, no receptor specificity is to be expected. Accordingly, in transfected cells, GRK2 desensitizes Kir3 current responses of a variety of GPCRs, including adenosine A1 receptors (Raveh et al., 2010), μ-opioid receptors (Raveh et al., 2010), muscarinic acetylcholine M2 receptors (Shui et al., 1998), and $GABA_BRs$ (Turecek et al., 2014). However, GRK2 failed to desensitize the responses of metabotropic glutamate receptors (mGlu2) and muscarinic acetylcholine receptors (M4), suggesting that GRK2 nonetheless exhibits preference for certain GPCRs (Raveh et al., 2010). Since phosphorylation-independent GRK2 desensitization is induced at the G protein and exhibits limited receptor specificity, this mechanism of desensitization may also desensitize native $GABA_BR$ responses in a heterologous manner when $GABA_BRs$ share G proteins with other receptors.

Noncanonical GRK2-induced (Raveh et al., 2010) and GRK4-induced desensitization (Kanaide et al., 2007; Perroy et al., 2003) depends on distinct GRK domains. GRK4 completely lacks the pleckstrin homology domain that is required for scavenging Gβγ by GRK2. However, the RGS domain of GRKs was shown to bind to Gα and to efficiently block the G protein from interacting with downstream effectors (Lodowski, Pitcher, Capel, Lefkowitz, & Tesmer, 2003). It is therefore conceivable that the RGS domain of GRK4 binds to Gα and keeps the G protein at bay, until receptor signaling is terminated (Lodowski et al., 2003). GRK2 and GRK4 may therefore regulate desensitization through different domains but related mechanisms.

3.3. KCTD12-induced fast desensitization

KCTD proteins associate as tetrameric complexes with $GABA_BRs$ and are examples of auxiliary GPCR subunits (Ivankova et al., 2013; Schwenk et al., 2010; Turecek et al., 2014). During biogenesis, the cytoplasmic KCTD proteins assemble with the receptor at the outer membrane of the endoplasmic reticulum and thereafter remain associated with the receptor throughout its lifecycle (Ivankova et al., 2013). Tyrosine-902 in the intracellular C-terminal domain of the $GABA_{B2}$ subunit is mandatory for binding of the KCTDs to the receptor (Correale et al., 2013; Ivankova et al., 2013; Schwenk et al., 2010). A recent study showed that the KCTDs, like the RGS and GRK proteins, bind to G proteins (Turecek et al., 2014). The KCTDs therefore stabilize a constitutive multiprotein complex comprising $GABA_BRs$, KCTD tetramers, and G protein subunits (Turecek et al., 2014). Two of the KCTD proteins, KCTD12 and 12b, induce a fast and pronounced desensitization of the gating of Ca^{2+} and K^+ currents (Adelfinger et al., 2014; Rajalu et al., 2015; Schwenk et al., 2010; Seddik et al., 2012; Turecek et al., 2014). The KCTDs represent previously unknown regulators of G protein signaling. A recent study has addressed how KCTD12 desensitizes $GABA_BR$-activated K^+ currents (Turecek et al., 2014). Biochemical and electrophysiological experiments revealed that KCTD12 binds to Gβγ and shields its channel binding site, thereby interfering with efficient gating of effector channels (Fig. 1C). This mechanism of desensitization is reminiscent of the nonenzymatic GRK2-induced desensitization described above (Raveh et al., 2010). However, while both KCTD12 and GRK2 induce fast desensitization by uncoupling the Gβγ from the channel, the mechanisms of desensitization differ in several aspects. Scavenging of Gβγ by GRK2 is a poorly reversible process that does not enable recovery of desensitized $GABA_BR$-activated K^+ currents within a 10-min period (Turecek et al., 2014). In contrast, KCTD12-induced K^+ current desensitization is fully reversible within minutes. GRK2 binds at the interface of Gβγ with Gα (Lodowski et al., 2003; Tesmer et al., 2005), which effectively sequesters the subunits of the heterotrimeric G protein (Tesmer et al., 2005) and scavenges free Gβγ away from channels (Raveh et al., 2010). In contrast, KCTD12 remains constitutively associated with the receptor and blocks the channel binding site of Gβγ, thus allowing for fast recovery of the receptor response (Turecek et al., 2014). GRK2, which extends the refractory period of the G protein, therefore appears to be better suited to induce fast and long-lasting desensitization while KCTD12 may be more important for fast and short-lasting desensitization, for example, during repetitive synaptic activity.

Interestingly, KCTD12 not only controls desensitization but also onset, rise time, and amplitude of receptor-activated K^+ currents (Ivankova et al., 2013; Turecek et al., 2014). Pulling together receptor and G protein may overcome the slow diffusion-limited association between receptor and G protein, and thus accelerate G protein activation (Fowler et al., 2007; Ross, 2008). Faster kinetics of $GABA_BR$-activated K^+ currents in the presence of KCTD12 may speed up use-dependent processes and increase basal current desensitization (Chuang et al., 1998), which may contribute to the overall increase in desensitization. In the adult brain, KCTD12 is almost exclusively associated with $GABA_BRs$ (Turecek et al., 2014). This renders KCTD12-induced desensitization highly $GABA_BR$-specific. However, when expressed in excess of $GABA_BRs$, KCTD12 also desensitizes K^+ channel responses of other GPCRs, both *in vitro* and *in vivo* (Schwenk et al., 2010; Turecek et al., 2014). KCTD12 was also shown to disinhibit the Gβγ-mediated inhibition of Ca^{2+} currents (Schwenk et al., 2010). It is likely that KCTD12 exerts this disinhibition by interfering with Gβγ binding to the Ca^{2+} channel, analogous to its mode of action at K^+ channels.

Interestingly, the *KCTD12* gene has been linked to various neuropsychiatric disorders. A genome-wide association study of bipolar I disorder patients identified polymorphisms in *KCTD12* and the Ca^{2+} channel subunit β2 gene *CACNB2* associated with disorder prevalence (Lee et al., 2011). Since $GABA_BRs$, KCTD12, and voltage-gated Ca^{2+} channels are mechanistically linked and constituents of the same signaling complex (Muller et al., 2010; Schwenk et al., 2010), the association of KCTD12 and voltage-gated Ca^{2+} channels with bipolar disorder 1 is intriguing. Moreover, microarray studies using peripheral blood mononuclear cells of schizophrenia patients revealed significantly reduced levels of KCTD12 transcripts (Glatt et al., 2005). In contrast *post mortem* KCTD12 transcripts were reported to be increased in schizophrenia patients (Benes, 2010). A microarray study in major depressive disorder patients identified *KCTD12* as one of several genes exhibiting increased expression in the amygdala (Sibille et al., 2009). A dysregulation of KCTD12 protein is expected to alter current kinetics in the brain and therefore may constitute a risk factor for neuropsychiatric disorders. Of note, a downregulation of KCTD12 protein in disease is expected to specifically affect $GABA_BR$ signaling, since virtually all KCTD12 protein in the adult brain is associated with $GABA_BRs$ (Turecek et al., 2014). In contrast, an upregulation of KCTD12 protein in excess of $GABA_BRs$ may lead to a promiscuous association of KCTD12 with Gβγ and affect the signaling of many GPCRs (Turecek et al., 2014).

4. SLOW AND FAST MECHANISMS OF DESENSITIZATION INFLUENCE EACH OTHER

The close proximity of the serine-892 phosphorylation site and the tyrosine-902 KCTD12 binding site in $GABA_{B2}$ sparked experiments addressing whether the slow and fast mechanisms of desensitization can influence each other (Adelfinger et al., 2014). Bioresonance energy transfer and electrophysiological experiments indicate that serine-892 phosphorylation rearranges KCTD12 at the receptor and slows KCTD12-induced desensitization, both in heterologous cells and in hippocampal neurons. Experiments with KCTD12 knockout mice and knock-in mice carrying a serine-892-to-alanine mutation further demonstrate that tonic serine-892 phosphorylation normally limits KCTD12-induced desensitization in hippocampal neurons. Conversely, assembly of receptors with KCTD12 promotes serine-892 phosphorylation and attenuates slow desensitization. This cross-regulation of serine-892 phosphorylation and KCTD12 activity sharpens the K^+ current response during repeated receptor activation.

5. CONCLUSION

It is becoming apparent that the agonist-induced desensitization of $GABA_BR$ responses in neurons integrates distinct mechanistic underpinnings. Slow homologous desensitization involves reduced PKA-mediated receptor phosphorylation followed by receptor degradation. This mechanism of desensitization may be particularly effective in response to ambient levels of GABA. In addition, faster regulatory mechanisms have evolved to desensitize receptor-gated K^+ and Ca^{2+} current responses. At this point, the fast mechanisms of desensitization all operate at the receptor-activated G protein, which directly gates effector channels and therefore represents an ideal target for the inhibition of channel gating.

The mechanisms for fast desensitization of $GABA_BR$ responses have mostly been studied in heterologous cells. While the mechanisms regulating fast desensitization of $GABA_BR$ responses in identified neurons are still largely unexplored, some progress has recently been made in addressing this issue. For example, KCTD12 was shown to be highly expressed in pyramidal neurons of the hippocampus (Metz, Gassmann, Fakler, Schaeren-Wiemers, & Bettler, 2011), which has provided the opportunity to address the role of KCTD12 in K^+ current desensitization *in vivo*. Electrophysiological analysis of hippocampal neurons derived from KCTD12 knockout mice revealed a significantly reduced desensitization of $GABA_BR$-activated K^+ currents, thus confirming that native KCTD12 contributes to K^+ current desensitization in these neurons (Turecek et al., 2014). Similarly, downregulation of GRK2 in hippocampal neurons reduced fast desensitization of adenosine-activated K^+ currents (Raveh et al., 2010). Dopaminergic neurons in the ventral tegmental area undergo a pronounced desensitization of $GABA_BR$-activated K^+ currents (Cruz et al., 2004; Labouebe et al., 2007). Interestingly, the desensitization in dopaminergic neurons is altered neither by serine-892 phosphorylation nor by deletion of KCTD12 in knockout mice (E. Perez-Garci, unpublished). This reinforces that the mechanisms for desensitization of $GABA_BR$ responses differ between neuronal populations and suggests that additional mechanisms of desensitization remain to be discovered. For example, the subunit composition of Kir3 channels differs between neuronal populations (Cruz et al., 2004; Labouebe et al., 2007) and may influence current desensitization (Bender et al., 2001). Other mechanisms that could be involved in fast desensitization of receptor-gated current responses in the brain are, for example, a receptor-mediated reverse action of G protein that releases GTP from Gα (Hommers, Klenk, Dees, & Bunemann, 2010) and a reduction in electrochemical driving force for K^+ during intensive Kir3 activity (Bender et al., 2004). Moreover, proteins that purify with native $GABA_BR$ complexes from the brain (Schwenk et al., 2010) may induce desensitization of the receptor responses via mechanisms for which there is no precedent.

The relevance of fast desensitization for $GABA_BR$ signaling in the brain remains an interesting topic for future exploration. It is, for example, unclear whether the mechanisms for fast desensitization can alter slow IPSCs in the brain, which typically only last 100–500 ms. Possibly, fast desensitization is physiologically most relevant when $GABA_BR$s are repeatedly activated during synchronous activity of interneurons. $GABA_BR$s have an important role in the control of cortical network activity and in the transmission of information across brain areas (Isaacson & Scanziani, 2011; Scanziani, 2000). It will be interesting to see whether KCTD12 knockout mice or mice overexpressing KCTD12 exhibit perturbed network activity, which may provide a mechanistic explanation for the association of KCTD12 with psychiatric disorders.

CONFLICT OF INTEREST

The authors have no conflict of interest to declare.

ACKNOWLEDGMENTS

We thank Martin Gassmann for help with the manuscript. Our work is supported by grants of the National Center for Competences in Research (NCCR) "Synapsy, Synaptic Bases of Mental Health Disease" and the Swiss National Science Foundation (31003A_152970) to B. B.

REFERENCES

Adelfinger, L., Turecek, R., Ivankova, K., Jensen, A. A., Moss, S. J., Gassmann, M., et al. (2014). $GABA_B$ receptor phosphorylation regulates KCTD12-induced K^+ current desensitization. *Biochemical Pharmacology*, *91*(3), 369–379.

Ando, Y., Hojo, M., Kanaide, M., Takada, M., Sudo, Y., Shiraishi, S., et al. (2011). S(+)-ketamine suppresses desensitization of γ-aminobutyric acid type B receptor-mediated signaling by inhibition of the interaction of γ-aminobutyric acid type B receptors with G protein-coupled receptor kinase 4 or 5. *Anesthesiology*, *114*(2), 401–411.

Bartoi, T., Rigbolt, K. T., Du, D., Kohr, G., Blagoev, B., & Kornau, H. C. (2010). $GABA_B$ receptor constituents revealed by tandem affinity purification from transgenic mice. *The Journal of Biological Chemistry*, *285*(27), 20625–20633.

Beenhakker, M. P., & Huguenard, J. R. (2010). Astrocytes as gatekeepers of $GABA_B$ receptor function. *The Journal of Neuroscience*, *30*(45), 15262–15276.

Bender, K., Wellner-Kienitz, M. C., Bosche, L. I., Rinne, A., Beckmann, C., & Pott, L. (2004). Acute desensitization of GIRK current in rat atrial myocytes is related to K^+ current flow. *The Journal of Physiology*, *561*(2), 471–483.

Bender, K., Wellner-Kienitz, M. C., Inanobe, A., Meyer, T., Kurachi, Y., & Pott, L. (2001). Overexpression of monomeric and multimeric GIRK4 subunits in rat atrial myocytes removes fast desensitization and reduces inward rectification of muscarinic K^+ current ($I_{K(ACh)}$). Evidence for functional homomeric GIRK4 channels. *The Journal of Biological Chemistry*, *276*(31), 28873–28880.

Benes, F. M. (2010). Amygdalocortical circuitry in schizophrenia: From circuits to molecules. *Neuropsychopharmacology*, *35*(1), 239–257.

Benians, A., Nobles, M., Hosny, S., & Tinker, A. (2005). Regulators of G-protein signaling form a quaternary complex with the agonist, receptor, and G-protein. A novel explanation for the acceleration of signaling activation kinetics. *The Journal of Biological Chemistry*, *280*(14), 13383–13394.

Benke, D., Zemoura, K., & Maier, P. J. (2012). Modulation of cell surface $GABA_B$ receptors by desensitization, trafficking and regulated degradation. *World Journal of Biological Chemistry*, *3*(4), 61–72.

Berman, D. M., Wilkie, T. M., & Gilman, A. G. (1996). GAIP and RGS4 are GTPase-activating proteins for the G_i subfamily of G protein α subunits. *Cell*, *86*(3), 445–452.

Betke, K. M., Wells, C. A., & Hamm, H. E. (2012). GPCR mediated regulation of synaptic transmission. *Progress in Neurobiology*, *96*(3), 304–321.

Biermann, B., Ivankova-Susankova, K., Bradaia, A., Abdel Aziz, S., Besseyrias, V., Kapfhammer, J. P., et al. (2010). The Sushi domains of $GABA_B$ receptors function as axonal targeting signals. *The Journal of Neuroscience*, *30*(4), 1385–1394.

Bowery, N. G., Bettler, B., Froestl, W., Gallagher, J. P., Marshall, F., Raiteri, M., et al. (2002). International Union of Pharmacology. XXXIII. Mammalian γ-aminobutyric acidB receptors: Structure and function. *Pharmacological Reviews*, *54*(2), 247–264.

Chalifoux, J. R., & Carter, A. G. (2011a). $GABA_B$ receptor modulation of synaptic function. *Current Opinion in Neurobiology*, *21*(2), 339–344.

Chalifoux, J. R., & Carter, A. G. (2011b). $GABA_B$ receptor modulation of voltage-sensitive calcium channels in spines and dendrites. *The Journal of Neuroscience*, *31*(11), 4221–4232.

Chuang, T. T., Iacovelli, L., Sallese, M., & DeBlasi, A. (1996). G protein-coupled receptors: Heterologous regulation of homologous desensitization and its implications. *Trends in Pharmacological Sciences*, *17*(11), 416–421.

Chuang, H. H., Yu, M., Jan, Y. N., & Jan, L. Y. (1998). Evidence that the nucleotide exchange and hydrolysis cycle of G proteins causes acute desensitization of G-protein gated inward rectifier K^+ channels. *Proceedings of the National Academy of Sciences of the United States of America*, *95*(20), 11727–11732.

Ciruela, F., Fernandez-Duenas, V., Sahlholm, K., Fernandez-Alacid, L., Nicolau, J. C., Watanabe, M., et al. (2010). Evidence for oligomerization between $GABA_B$ receptors and GIRK channels containing the GIRK1 and GIRK3 subunits. *The European Journal of Neuroscience*, *32*(8), 1265–1277.

Correale, S., Esposito, C., Pirone, L., Vitagliano, L., Gaetano, S. D., & Pedone, E. (2013). A biophysical characterization of the folded domains of KCTD12: Insights into interaction with the $GABA_{B2}$ receptor. *Journal of Molecular Recognition*, *26*(10), 488–495.

Couve, A., Thomas, P., Calver, A. R., Hirst, W. D., Pangalos, M. N., Walsh, F. S., et al. (2002). Cyclic AMP-dependent protein kinase phosphorylation facilitates $GABA_B$ receptor-effector coupling. *Nature Neuroscience*, *5*(5), 415–424.

Craig, M. T., Mayne, E. W., Bettler, B., Paulsen, O., & McBain, C. J. (2013). Distinct roles of $GABA_{B1a}$- and $GABA_{B1b}$-containing $GABA_B$ receptors in spontaneous and evoked termination of persistent cortical activity. *The Journal of Physiology*, *591*(4), 835–843.

Cruz, H. G., Ivanova, T., Lunn, M. L., Stoffel, M., Slesinger, P. A., & Luscher, C. (2004). Bi-directional effects of $GABA_B$ receptor agonists on the mesolimbic dopamine system. *Nature Neuroscience*, 7(2), 153–159.

David, M., Richer, M., Mamarbachi, A. M., Villeneuve, L. R., Dupre, D. J., & Hebert, T. E. (2006). Interactions between $GABA_{B1}$ receptors and Kir3 inwardly rectifying potassium channels. *Cellular Signalling*, *18*(12), 2172–2181.

Davies, C. H., Starkey, S. J., Pozza, M. F., & Collingridge, G. L. (1991). GABA autoreceptors regulate the induction of LTP. *Nature*, *349*(6310), 609–611.

De Vries, L., Zheng, B., Fischer, T., Elenko, E., & Farquhar, M. G. (2000). The regulator of G protein signaling family. *Annual Review of Pharmacology and Toxicology*, *40*, 235–271.

Doupnik, C. A., Davidson, N., Lester, H. A., & Kofuji, P. (1997). RGS proteins reconstitute the rapid gating kinetics of $G\beta\gamma$-activated inwardly rectifying K^+ channels. *Proceedings of the National Academy of Sciences of the United States of America*, *94*(19), 10461–10466.

Doupnik, C. A., Jaen, C., & Zhang, Q. (2004). Measuring the modulatory effects of RGS proteins on GIRK channels. *Methods in Enzymology*, *389*, 131–154.

Dutar, P., & Nicoll, R. A. (1988). Pre- and postsynaptic $GABA_B$ receptors in the hippocampus have different pharmacological properties. *Neuron*, *1*(7), 585–591.

Ehrengruber, M. U., Doupnik, C. A., Xu, Y., Garvey, J., Jasek, M. C., Lester, H. A., et al. (1997). Activation of heteromeric G protein-gated inward rectifier K^+ channels overexpressed by adenovirus gene transfer inhibits the excitability of hippocampal neurons. *Proceedings of the National Academy of Sciences of the United States of America*, *94*(13), 7070–7075.

Enna, S. J., & Bowery, N. G. (2004). $GABA_B$ receptor alterations as indicators of physiological and pharmacological function. *Biochemical Pharmacology*, *68*(8), 1541–1548.

Evron, T., Daigle, T. L., & Caron, M. G. (2012). GRK2: Multiple roles beyond G protein-coupled receptor desensitization. *Trends in Pharmacological Sciences*, *33*(3), 154–164.

Fairfax, B. P., Pitcher, J. A., Scott, M. G., Calver, A. R., Pangalos, M. N., Moss, S. J., et al. (2004). Phosphorylation and chronic agonist treatment atypically modulate $GABA_B$ receptor cell surface stability. *The Journal of Biological Chemistry*, *279*(13), 12565–12573.

Fajardo-Serrano, A., Wydeven, N., Young, D., Watanabe, M., Shigemoto, R., Martemyanov, K. A., et al. (2013). Association of RGS7/$G\beta5$ complexes with GIRK channels and $GABA_B$ receptors in hippocampal CA1 pyramidal neurons. *Hippocampus*, *23*(12), 1231–1245.

Fowler, C. E., Aryal, P., Suen, K. F., & Slesinger, P. A. (2007). Evidence for association of $GABA_B$ receptors with Kir3 channels and regulators of G protein signalling (RGS4) proteins. *The Journal of Physiology*, *580*(1), 51–65.

Garzon, J., Rodriguez-Munoz, M., de la Torre-Madrid, E., & Sanchez-Blazquez, P. (2005). Effector antagonism by the regulators of G protein signalling (RGS) proteins causes desensitization of μ-opioid receptors in the CNS. *Psychopharmacology*, *180*(1), 1–11.

Gassmann, M., & Bettler, B. (2012). Regulation of neuronal $GABA_B$ receptor functions by subunit composition. *Nature Reviews. Neuroscience*, *13*(6), 380–394.
Gassmann, M., Shaban, H., Vigot, R., Sansig, G., Haller, C., Barbieri, S., et al. (2004). Redistribution of $GABA_{B1}$ protein and atypical $GABA_B$ responses in $GABA_{B2}$-deficient mice. *The Journal of Neuroscience*, *24*(27), 6086–6097.
Gjoni, T., & Urwyler, S. (2009). Changes in the properties of allosteric and orthosteric $GABA_B$ receptor ligands after a continuous, desensitizing agonist pretreatment. *The European Journal of Pharmacology*, *603*(1–3), 37–41.
Glatt, S. J., Everall, I. P., Kremen, W. S., Corbeil, J., Sasik, R., Khanlou, N., et al. (2005). Comparative gene expression analysis of blood and brain provides concurrent validation of SELENBP1 up-regulation in schizophrenia. *Proceedings of the National Academy of Sciences of the United States of America*, *102*(43), 15533–15538.
Gonzalez-Maeso, J., Wise, A., Green, A., & Koenig, J. A. (2003). Agonist-induced desensitization and endocytosis of heterodimeric $GABA_B$ receptors in CHO-K1 cells. *The European Journal of Pharmacology*, *481*(1), 15–23.
Guetg, N., Abdel Aziz, S., Holbro, N., Turecek, R., Rose, T., Seddik, R., et al. (2010). NMDA receptor-dependent $GABA_B$ receptor internalization via CaMKII phosphorylation of serine 867 in $GABA_{B1}$. *Proceedings of the National Academy of Sciences of the United States of America*, *107*(31), 13924–13929.
Havlickova, M., Prezeau, L., Duthey, B., Bettler, B., Pin, J. P., & Blahos, J. (2002). The intracellular loops of the GB2 subunit are crucial for G-protein coupling of the heteromeric γ-aminobutyrate B receptor. *Molecular Pharmacology*, *62*(2), 343–350.
Hollinger, S., & Hepler, J. R. (2002). Cellular regulation of RGS proteins: Modulators and integrators of G protein signaling. *Pharmacological Reviews*, *54*(3), 527–559.
Hommers, L. G., Klenk, C., Dees, C., & Bunemann, M. (2010). G proteins in reverse mode: Receptor-mediated GTP release inhibits G protein and effector function. *The Journal of Biological Chemistry*, *285*(11), 8227–8233.
Hunt, T. W., Fields, T. A., Casey, P. J., & Peralta, E. G. (1996). RGS10 is a selective activator of $G\alpha_i$ GTPase activity. *Nature*, *383*(6596), 175 177.
Isaacson, J. S., & Scanziani, M. (2011). How inhibition shapes cortical activity. *Neuron*, *72*(2), 231–243.
Isaacson, J. S., Solis, J. M., & Nicoll, R. A. (1993). Local and diffuse synaptic actions of GABA in the hippocampus. *Neuron*, *10*(2), 165–175.
Ivankova, K., Turecek, R., Fritzius, T., Seddik, R., Prezeau, L., Comps-Agrar, L., et al. (2013). Up-regulation of $GABA_B$ receptor signaling by constitutive assembly with the K^+ channel tetramerization domain-containing protein 12 (KCTD12). *The Journal of Biological Chemistry*, *288*(34), 24848–24856.
Jacobson, L. H., Kelly, P. H., Bettler, B., Kaupmann, K., & Cryan, J. F. (2006). $GABA_{B1}$ receptor isoforms differentially mediate the acquisition and extinction of aversive taste memories. *The Journal of Neuroscience*, *26*(34), 8800–8803.
Kanaide, M., Uezono, Y., Matsumoto, M., Hojo, M., Ando, Y., Sudo, Y., et al. (2007). Desensitization of $GABA_B$ receptor signaling by formation of protein complexes of $GABA_{B2}$ subunit with GRK4 or GRK5. *Journal of Cellular Physiology*, *210*(1), 237–245.
Kelly, E., Bailey, C. P., & Henderson, G. (2008). Agonist-selective mechanisms of GPCR desensitization. *British Journal of Pharmacology*, *153*(Suppl. 1), S379–S388.
Kim, G., Jung, S., Son, H., Kim, S., Choi, J., Lee, D. H., et al. (2014). The $GABA_B$ receptor associates with regulators of G-protein signaling 4 protein in the mouse prefrontal cortex and hypothalamus. *BMB Reports*, *47*(6), 324–329.
Kohl, M. M., & Paulsen, O. (2010). The roles of $GABA_B$ receptors in cortical network activity. *Advances in Pharmacology*, *58*, 205–229.
Kuramoto, N., Wilkins, M. E., Fairfax, B. P., Revilla-Sanchez, R., Terunuma, M., Tamaki, K., et al. (2007). Phospho-dependent functional modulation of $GABA_B$ receptors by the metabolic sensor AMP-dependent protein kinase. *Neuron*, *53*(2), 233–247.

Labouebe, G., Lomazzi, M., Cruz, H. G., Creton, C., Lujan, R., Li, M., et al. (2007). RGS2 modulates coupling between $GABA_B$ receptors and GIRK channels in dopamine neurons of the ventral tegmental area. *Nature Neuroscience*, *10*(12), 1559–1568.

Lambert, N. A. (2008). Dissociation of heterotrimeric G proteins in cells. *Science Signaling*, *1*(25), re5.

Lambert, N. A., Johnston, C. A., Cappell, S. D., Kuravi, S., Kimple, A. J., Willard, F. S., et al. (2010). Regulators of G-protein signaling accelerate GPCR signaling kinetics and govern sensitivity solely by accelerating GTPase activity. *Proceedings of the National Academy of Sciences of the United States of America*, *107*(15), 7066–7071.

Leaney, J. L., Benians, A., Brown, S., Nobles, M., Kelly, D., & Tinker, A. (2004). Rapid desensitization of G protein-gated inwardly rectifying K^+ currents is determined by G protein cycle. *American Journal of Physiology. Cell Physiology*, *287*(1), C182–C191.

Lee, M. T., Chen, C. H., Lee, C. S., Chen, C. C., Chong, M. Y., Ouyang, W. C., et al. (2011). Genome-wide association study of bipolar I disorder in the Han Chinese population. *Molecular Psychiatry*, *16*(5), 548–556.

Lodowski, D. T., Pitcher, J. A., Capel, W. D., Lefkowitz, R. J., & Tesmer, J. J. (2003). Keeping G proteins at bay: A complex between G protein-coupled receptor kinase 2 and $G\beta\gamma$. *Science*, *300*(5623), 1256–1262.

Luscher, C., & Slesinger, P. A. (2010). Emerging roles for G protein-gated inwardly rectifying potassium (GIRK) channels in health and disease. *Nature Reviews. Neuroscience*, *11*(5), 301–315.

Maity, B., Stewart, A., Yang, J., Loo, L., Sheff, D., Shepherd, A. J., et al. (2012). Regulator of G protein signaling 6 (RGS6) protein ensures coordination of motor movement by modulating $GABA_B$ receptor signaling. *The Journal of Biological Chemistry*, *287*(7), 4972–4981.

Mann, E. O., Kohl, M. M., & Paulsen, O. (2009). Distinct roles of $GABA_A$ and $GABA_B$ receptors in balancing and terminating persistent cortical activity. *The Journal of Neuroscience*, *29*(23), 7513–7518.

Marshall, F. H., Jones, K. A., Kaupmann, K., & Bettler, B. (1999). $GABA_B$ receptors—The first 7TM heterodimers. *Trends in Pharmacological Sciences*, *20*(10), 396–399.

Maurice, P., Daulat, A. M., Turecek, R., Ivankova-Susankova, K., Zamponi, F., Kamal, M., et al. (2010). Molecular organization and dynamics of the melatonin MT_1 receptor/RGS20/G_i protein complex reveal asymmetry of receptor dimers for RGS and G_i coupling. *The EMBO Journal*, *29*(21), 3646–3659.

Metz, M., Gassmann, M., Fakler, B., Schaeren-Wiemers, N., & Bettler, B. (2011). Distribution of the auxiliary $GABA_B$ receptor subunits KCTD8, 12, 12b, and 16 in the mouse brain. *The Journal of Comparative Neurology*, *519*(8), 1435–1454.

Mohler, H., Benke, D., & Fritschy, J. M. (2001). $GABA_B$-receptor isoforms molecular architecture and distribution. *Life Sciences*, *68*(19–20), 2297–2300.

Mohler, H., & Fritschy, J. M. (1999). $GABA_B$ receptors make it to the top—As dimers. *Trends in Pharmacological Sciences*, *20*(3), 87–89.

Mombereau, C., Kaupmann, K., Gassmann, M., Bettler, B., van der Putten, H., & Cryan, J. F. (2005). Altered anxiety and depression-related behaviour in mice lacking $GABA_{B2}$ receptor subunits. *Neuroreport*, *16*(3), 307–310.

Muller, C. S., Haupt, A., Bildl, W., Schindler, J., Knaus, H. G., Meissner, M., et al. (2010). Quantitative proteomics of the Cav2 channel nano-environments in the mammalian brain. *Proceedings of the National Academy of Sciences of the United States of America*, *107*(34), 14950–14957.

Mutneja, M., Berton, F., Suen, K. F., Luscher, C., & Slesinger, P. A. (2005). Endogenous RGS proteins enhance acute desensitization of $GABA_B$ receptor-activated GIRK currents in HEK-293T cells. *Pflügers Archiv: European Journal of Physiology.*, *450*(1), 61–73.

O'Leary, O. F., Felice, D., Galimberti, S., Savignac, H. M., Bravo, J. A., Crowley, T., et al. (2014). $GABA_{B1}$ receptor subunit isoforms differentially regulate stress resilience. *Proceedings of the National Academy of Sciences of the United States of America, 111*(42), 15232–15237.

Ostrovskaya, O., Xie, K., Masuho, I., Fajardo-Serrano, A., Lujan, R., Wickman, K., et al. (2014). RGS7/Gβ5/R7BP complex regulates synaptic plasticity and memory by modulating hippocampal $GABA_BR$-GIRK signaling. *eLife, 3*, e02053.

Perez-Garci, E., Gassmann, M., Bettler, B., & Larkum, M. E. (2006). The $GABA_{B1b}$ isoform mediates long-lasting inhibition of dendritic Ca^{2+} spikes in layer 5 somatosensory pyramidal neurons. *Neuron, 50*(4), 603–616.

Perez-Garci, E., Larkum, M. E., & Nevian, T. (2013). Inhibition of dendritic Ca^{2+} spikes by $GABA_B$ receptors in cortical pyramidal neurons is mediated by a direct $G_{i/o}$-β-subunit interaction with Cav1 channels. *The Journal of Physiology, 591*(7), 1599–1612.

Perroy, J., Adam, L., Qanbar, R., Chenier, S., & Bouvier, M. (2003). Phosphorylation-independent desensitization of $GABA_B$ receptor by GRK4. *The EMBO Journal, 22*(15), 3816–3824.

Pitcher, J. A., Freedman, N. J., & Lefkowitz, R. J. (1998). G protein-coupled receptor kinases. *Annual Review of Biochemistry, 67*, 653–692.

Pontier, S. M., Lahaie, N., Ginham, R., St-Gelais, F., Bonin, H., Bell, D. J., et al. (2006). Coordinated action of NSF and PKC regulates $GABA_B$ receptor signaling efficacy. *The EMBO Journal, 25*(12), 2698–2709.

Prosser, H. M., Gill, C. H., Hirst, W. D., Grau, E., Robbins, M., Calver, A., et al. (2001). Epileptogenesis and enhanced prepulse inhibition in $GABA_{B1}$-deficient mice. *Molecular and Cellular Neurosciences, 17*(6), 1059–1070.

Quéva, C., Bremner-Danielsen, M., Edlund, A., Ekstrand, A. J., Elg, S., Erickson, S., et al. (2003). Effects of GABA agonists on body temperature regulation in $GABA_{B1}{}^{-/-}$ mice. *British Journal of Pharmacology, 140*(2), 315–322.

Rajalu, M., Fritzius, T., Adelfinger, L., Jacquier, V., Besseyrias, V., Gassmann, M., et al. (2015). Pharmacological characterization of $GABA_B$ receptor subtypes assembled with auxiliary KCTD subunits. *Neuropharmacology, 88*, 145–154.

Raveh, A., Cooper, A., Guy-David, L., & Reuveny, E. (2010). Nonenzymatic rapid control of GIRK channel function by a G protein-coupled receptor kinase. *Cell, 143*(5), 750–760.

Richter, M. A., de Jesus, D. R., Hoppenbrouwers, S., Daigle, M., Deluce, J., Ravindran, L. N., et al. (2012). Evidence for cortical inhibitory and excitatory dysfunction in obsessive compulsive disorder. *Neuropsychopharmacology, 37*(5), 1144–1151.

Robbins, M. J., Calver, A. R., Filippov, A. K., Hirst, W. D., Russell, R. B., Wood, M. D., et al. (2001). $GABA_{B2}$ is essential for G-protein coupling of the $GABA_B$ receptor heterodimer. *The Journal of Neuroscience, 21*(20), 8043–8052.

Ross, E. M. (2008). Coordinating speed and amplitude in G-protein signaling. *Current Biology, 18*(17), 777–783.

Ross, E. M., & Wilkie, T. M. (2000). GTPase-activating proteins for heterotrimeric G proteins: Regulators of G protein signaling (RGS) and RGS-like proteins. *Annual Review of Biochemistry, 69*, 795–827.

Saitoh, O., Kubo, Y., Miyatani, Y., Asano, T., & Nakata, H. (1997). RGS8 accelerates G-protein-mediated modulation of K^+ currents. *Nature, 390*(6659), 525–529.

Sallese, M., Salvatore, L., D'Urbano, E., Sala, G., Storto, M., Launey, T., et al. (2000). The G-protein-coupled receptor kinase GRK4 mediates homologous desensitization of metabotropic glutamate receptor 1. *FASEB Journal, 14*(15), 2569–2580.

Scanziani, M. (2000). GABA spillover activates postsynaptic $GABA_B$ receptors to control rhythmic hippocampal activity. *Neuron, 25*(3), 673–681.

Schuler, V., Luscher, C., Blanchet, C., Klix, N., Sansig, G., Klebs, K., et al. (2001). Epilepsy, hyperalgesia, impaired memory, and loss of pre- and postsynaptic $GABA_B$ responses in mice lacking $GABA_{B1}$. *Neuron, 31*(1), 47–58.

Schwenk, J., Metz, M., Zolles, G., Turecek, R., Fritzius, T., Bildl, W., et al. (2010). Native $GABA_B$ receptors are heteromultimers with a family of auxiliary subunits. *Nature*, *465*(7295), 231–235.

Seddik, R., Jungblut, S. P., Silander, O. K., Rajalu, M., Fritzius, T., Besseyrias, V., et al. (2012). Opposite effects of KCTD subunit domains on $GABA_B$ receptor-mediated desensitization. *The Journal of Biological Chemistry*, *287*(47), 39869–39877.

Shaban, H., Humeau, Y., Herry, C., Cassasus, G., Shigemoto, R., Ciocchi, S., et al. (2006). Generalization of amygdala LTP and conditioned fear in the absence of presynaptic inhibition. *Nature Neuroscience*, *9*(8), 1028–1035.

Shui, Z., Khan, I. A., Tsuga, H., Haga, T., & Boyett, M. R. (1998). Role of receptor kinase in short-term desensitization of cardiac muscarinic K^+ channels expressed in Chinese hamster ovary cells. *The Journal of Physiology*, *507*(2), 325–334.

Sibille, E., Wang, Y., Joeyen-Waldorf, J., Gaiteri, C., Surget, A., Oh, S., et al. (2009). A molecular signature of depression in the amygdala. *The American Journal of Psychiatry*, *166*(9), 1011–1024.

Sickmann, T., & Alzheimer, C. (2002). Agonist-specific maturation of GIRK current responses in acutely isolated pyramidal neurons of rat neocortex. *Brain Research*, *929*(2), 166–174.

Sickmann, T., & Alzheimer, C. (2003). Short-term desensitization of G-protein-activated, inwardly rectifying K^+ (GIRK) currents in pyramidal neurons of rat neocortex. *Journal of Neurophysiology*, *90*(4), 2494–2503.

Sodickson, D. L., & Bean, B. P. (1996). $GABA_B$ receptor-activated inwardly rectifying potassium current in dissociated hippocampal CA3 neurons. *The Journal of Neuroscience*, *16*(20), 6374–6385.

Sudo, Y., Hojo, M., Ando, Y., Takada, M., Murata, H., Kurata, S., et al. (2012). $GABA_B$ receptors do not internalize after baclofen treatment, possibly due to a lack of β-arrestin association: Study with a real-time visualizing assay. *Synapse*, *66*(9), 759–769.

Tedford, H. W., & Zamponi, G. W. (2006). Direct G protein modulation of Cav2 calcium channels. *Pharmacological Reviews*, *58*(4), 837–862.

Terunuma, M., Pangalos, M. N., & Moss, S. J. (2010). Functional modulation of $GABA_B$ receptors by protein kinases and receptor trafficking. *Advances in Pharmacology*, *58*, 113–122.

Tesmer, J. J., Berman, D. M., Gilman, A. G., & Sprang, S. R. (1997). Structure of RGS4 bound to AlF_4^--activated $G_{i\alpha1}$: Stabilization of the transition state for GTP hydrolysis. *Cell*, *89*(2), 251–261.

Tesmer, V. M., Kawano, T., Shankaranarayanan, A., Kozasa, T., & Tesmer, J. J. (2005). Snapshot of activated G proteins at the membrane: The $G\alpha_q$-GRK2-$G\beta\gamma$ complex. *Science*, *310*(5754), 1686–1690.

Thompson, S. M., & Gahwiler, B. H. (1992). Comparison of the actions of baclofen at presynaptic and postsynaptic receptors in the rat hippocampus in vitro. *The Journal of Physiology*, *451*(1), 329–345.

Tosetti, P., Bakels, R., Colin-Le Brun, I., Ferrand, N., Gaiarsa, J. L., & Caillard, O. (2004). Acute desensitization of presynaptic $GABA_B$-mediated inhibition and induction of epileptiform discharges in the neonatal rat hippocampus. *The European Journal of Neuroscience*, *19*(12), 3227–3234.

Tsao, P., Cao, T., & von Zastrow, M. (2001). Role of endocytosis in mediating downregulation of G-protein-coupled receptors. *Trends in Pharmacological Sciences*, *22*(2), 91–96.

Turecek, R., Schwenk, J., Fritzius, T., Ivankova, K., Zolles, G., Adelfinger, L., et al. (2014). Auxiliary $GABA_B$ receptor subunits uncouple G protein βγ subunits from effector channels to induce desensitization. *Neuron*, *82*(5), 1032–1044.

Vacher, C. M., & Bettler, B. (2003). $GABA_B$ receptors as potential therapeutic targets. *Current Drug Targets. CNS and Neurological Disorders*, *2*(4), 248–259.
Vienne, J., Bettler, B., Franken, P., & Tafti, M. (2010). Differential effects of $GABA_B$ receptor subtypes, γ-hydroxybutyric acid, and baclofen on EEG activity and sleep regulation. *The Journal of Neuroscience*, *30*(42), 14194–14204.
Vigot, R., Barbieri, S., Brauner-Osborne, H., Turecek, R., Shigemoto, R., Zhang, Y. P., et al. (2006). Differential compartmentalization and distinct functions of $GABA_B$ receptor variants. *Neuron*, *50*(4), 589–601.
Virlon, B., Firsov, D., Cheval, L., Reiter, E., Troispoux, C., Guillou, F., et al. (1998). Rat G protein-coupled receptor kinase GRK4: Identification, functional expression, and differential tissue distribution of two splice variants. *Endocrinology*, *139*(6), 2784–2795.
Watson, N., Linder, M. E., Druey, K. M., Kehrl, J. H., & Blumer, K. J. (1996). RGS family members: GTPase-activating proteins for heterotrimeric G-protein α-subunits. *Nature*, *383*(6596), 172–175.
Wetherington, J. P., & Lambert, N. A. (2002). $GABA_B$ receptor activation desensitizes postsynaptic $GABA_B$ and A_1 adenosine responses in rat hippocampal neurones. *The Journal of Physiology*, *544*(2), 459–467.
Willars, G. B. (2006). Mammalian RGS proteins: Multifunctional regulators of cellular signalling. *Seminars in Cell & Developmental Biology*, *17*(3), 363–376.
Xie, K., Allen, K. L., Kourrich, S., Colon-Saez, J., Thomas, M. J., Wickman, K., et al. (2010). Gβ5 recruits R7 RGS proteins to GIRK channels to regulate the timing of neuronal inhibitory signaling. *Nature Neuroscience*, *13*(6), 661–663.
Zerangue, N., & Jan, L. Y. (1998). G-protein signaling: Fine-tuning signaling kinetics. *Current Biology*, *8*(9), 313–316.
Zholos, A. V., & Bolton, T. B. (1996). A novel GTP-dependent mechanism of ileal muscarinic metabotropic channel desensitization. *British Journal of Pharmacology*, *119*(5), 997–1005.

CHAPTER SEVEN

Allosteric Ligands and Their Binding Sites Define γ-Aminobutyric Acid (GABA) Type A Receptor Subtypes

Richard W. Olsen[1]
Department of Molecular & Medical Pharmacology, David Geffen School of Medicine at UCLA, Los Angeles, California, USA
[1]Corresponding author: e-mail address: rolsen@mednet.ucla.edu

Contents

Abstract

$GABA_A$ receptors ($GABA_ARs$) mediate rapid inhibitory transmission in the brain. $GABA_ARs$ are ligand-gated chloride ion channel proteins and exist in about a dozen or more heteropentameric subtypes exhibiting variable age and brain regional localization and thus participation in differing brain functions and diseases. $GABA_ARs$ are also subject to modulation by several chemotypes of allosteric ligands that help define structure and function, including subtype definition. The channel blocker picrotoxin identified a noncompetitive channel blocker site in $GABA_ARs$. This ligand site is located in the transmembrane channel pore, whereas the GABA agonist site is in the extracellular domain at subunit interfaces, a site useful for low energy coupled conformational changes of the functional channel domain. Two classes of pharmacologically important allosteric modulatory ligand binding sites reside in the extracellular domain at modified agonist sites at other subunit interfaces: the benzodiazepine site and the high-affinity, relevant to intoxication, ethanol site. The benzodiazepine site is specific for certain $GABA_AR$ subtypes, mainly synaptic, while the ethanol site is found at a modified benzodiazepine site

Advances in Pharmacology, Volume 73
ISSN 1054-3589
http://dx.doi.org/10.1016/bs.apha.2014.11.005

on different, extrasynaptic, subtypes. In the transmembrane domain are allosteric modulatory ligand sites for diverse chemotypes of general anesthetics: the volatile and intravenous agents, barbiturates, etomidate, propofol, long-chain alcohols, and neurosteroids. The last are endogenous positive allosteric modulators. X-ray crystal structures of prokaryotic and invertebrate pentameric ligand-gated ion channels, and the mammalian $GABA_AR$ protein, allow homology modeling of $GABA_AR$ subtypes with the various ligand sites located to suggest the structure and function of these proteins and their pharmacological modulation.

ABBREVIATIONS

CNS central nervous system
ECD extracellular domain
GABA γ-aminobutyric acid
$GABA_AR$ $GABA_A$ receptor(s)
LGIC ligand-gated ion channel(s)
M1 transmembrane helix #1
M2 transmembrane helix #2
M3 transmembrane helix #3
M4 transmembrane helix #4
PAM positive allosteric modulator(s)
SDS–PAGE sodium dodecyl sulfate–polyacrylamide gel electrophoresis
TMD transmembrane domain

1. INTRODUCTION

1.1. γ-Aminobutyric acid

γ-Aminobutyric acid (GABA) is the major inhibitory neurotransmitter in the nervous system (Curtis & Johnston, 1974; Roberts, Chase, & Tower, 1975). GABA mediates most fast inhibitory neurotransmission, both synaptic and extrasynaptic, via the $GABA_A$ receptors ($GABA_AR$s) (Barnard et al., 1998; Enna, 1983; Olsen & Sieghart, 2008). $GABA_AR$s are ligand-gated chloride channels, members of the ligand-gated ion channel (LGIC) superfamily that includes the pentameric cys-loop receptors such as nicotinic acetylcholine receptors, inhibitory glycine receptors, and $5HT_3$ receptors (Dutertre, Becker, & Betz, 2012; Galzi & Changeux, 1994; Sigel & Steinmann, 2012; Smart & Paoletti, 2012). GABA also mediates slower, modulatory neurotransmission involving $GABA_B$ receptors, G protein-coupled receptors (GPCRs) that regulate other receptors and voltage-gated ion channels involved in neurotransmission, some involving protein phosphorylation (Bowery et al., 2002). The two types of GABA receptor are not related to each other in structure or function (except both are mostly, but not entirely, inhibitory) and differ in their brain regional as well as subcellular localization.

Finally, there were "non-A, non-B" responses to GABA, insensitive to baclofen and bicuculline, sometimes called $GABA_C$ receptors, although they were likely a mixture of several discrete receptors (Johnston, 1997). Although insensitive to baclofen and not GPCRs, these responses were similar to those of $GABA_A$Rs in being chloride ion dependent, and they have significantly different pharmacology including a less extended, more bent, conformation of GABA as the receptor-specific conformation, thus showing selectivity for a series of GABA analogues that are not potent on classical GABAR. Further, the $GABA_C$R have little or no sensitivity to a number of $GABA_A$R PAMs (benzodiazepines (BZs), barbiturates, and neurosteroids) (Ng et al., 2011). The modern definition of receptor families is based on gene and thus protein sequence: several subunits (named ρ) that produce homopentameric chloride channels with function similar to $GABA_A$Rs but $GABA_C$ pharmacology had homology with, and were thus classified as a subset of, $GABA_A$Rs by the receptor nomenclature committee of the International Union of Pharmacology (Barnard et al., 1998; Olsen & Sieghart, 2008).

1.2. Brief history of function/pharmacophysiology and binding of $GABA_A$ receptors

Sites for GABA, BZ, and picrotoxin allow identification and characterization of $GABA_A$ receptors ($GABA_A$Rs). Historical pharmacology/ physiology/ neurochemistry on different organs/central nervous system (CNS) regions and circuits demonstrate subtypes. Selective agonists or antagonists, including those for allosteric ligand sites (medicinal chemistry) were developed for tools: radioligand binding, affinity labeling, and affinity columns.

$GABA_A$Rs, the subject of this chapter, are abundant in the mammalian CNS where they participate in all circuits and thus functions and behaviors, as well as most neuropsychiatric disorders (Rudolph & Möhler, 2014). $GABA_A$Rs are most known for a role in, at least, epilepsy, anxiety and sleep disorders, substance abuse, and problems in learning and memory and sensorimotor processing (Olsen, 2014; Olsen et al., 1992; Rudolph & Möhler, 2004). Fortunately, there are about 2 dozen subtypes of $GABA_A$Rs (Olsen & Sieghart, 2008) made from the heteromeric assembly of 19 highly homologous but distinct subunit genes (Lüddens, Korpi, & Seeburg, 1995; Rudolph & Möhler, 2004; Schofield et al., 1987; Whiting,

McKernan, & Wafford, 1995). Coupled with the fact that $GABA_A$Rs are arguably the major molecular target for drugs acting on the brain, mostly positive allosteric modulators, PAM (Barnard et al., 1998; Christopoulos et al., 2014; Lüddens et al., 1995; Olsen, 1982, 2014; Olsen & Sieghart, 2008; Rudolph & Möhler, 2006; Wick et al., 1998), the multiple subtypes afford the likelihood that these subtypes might be specifically modulated by drugs and thus function- and disease-specific pharmacological agents could be developed, if not already produced (Möhler, 2012; Whiting, 2003).

$GABA_A$Rs were identified with radioligand binding assays for the three major drug sites on the protein: the GABA sites (agonist/antagonist), the picrotoxin sites (channel blocker), and the BZ sites (PAM). These three ligand sites will be covered as briefly as possible so I can concentrate on two other classes of site that I have worked on extensively: the anesthetic sites discovered biochemically by our development of the radioactive picrotoxin binding assay, and the high-affinity EtOH sites discovered as novel BZ binding sites on the δ subunit-containing $GABA_A$R. $GABA_A$R studies required an assay defining receptor-specific binding using a radioligand that is specific for the receptor being studied, or a nonradioactive (cold) ligand that is specific for the receptor so that whatever binding of a nonspecific ligand that is displaced by the specific ligand can be used as the "receptor-specific" binding. For example, acetylcholine, or GABA, the neurotransmitters themselves, will bind to many (>7) proteins in brain/muscle, but one can count only the portion of the binding of acetylcholine, or a nonmetabolizable analogue (either one a nonspecific ligand), that is inhibited by the receptor-specific blocker α-bungarotoxin (Galzi & Changeux, 1994), or the fraction of GABA binding that is measured in sodium-free buffer (to eliminate transporter activity) and inhibited by the receptor-specific agonist muscimol or the receptor-specific antagonist bicuculline (Enna & Snyder, 1975; Greenlee, Van Ness, & Olsen, 1978; Krogsgaard-Larsen & Johnston, 1978; Zukin, Young, & Snyder, 1974). Of course, one can alternatively use radiolabeled α-bungarotoxin or a closely related snake neurotoxin (Meunier et al., 1971) for the nicotinic acetylcholine receptor, or radiolabeled muscimol (specific agonist) (Beaumont, Chilton, Yamamura, & Enna, 1978) or bicuculline (specific antagonist) (Möhler & Okada, 1977b; Olsen, Snowhill, & Wamsley, 1984) for the $GABA_A$R, if one can obtain such a ligand. The appropriate radioligand was often the tool that allowed receptor studies to proceed, e.g., isosteric agonists/antagonists for neurotransmitter receptors (Pert & Snyder, 1973; Yamamura, Enna, & Kuhar, 1978), allosteric PAMs like BZ

(Braestrup & Squires, 1977; Möhler & Okada, 1977a), and noncompetitive antagonists like picrotoxin (Möhler & Okada, 1977a; Ticku, Ban, & Olsen, 1978).

1.2.1 The GABA sites

In addition, the determination of receptor-specific binding required the comparison of potencies for a series of chemically related compounds (analogues) with restricted conformational flexibility/increased rigidity, both naturally occurring and specifically synthesized as tools, especially the isoxazole series of Krogsgaard-Larsen, including muscimol and THIP (gaboxadol) (Johnston, Allan, Kennedy, & Twitchin, 1979; Krogsgaard-Larsen, Frolund, Liljefors, & Ebert, 2004; Krogsgaard-Larsen, Honore, & Thyssen, 1979; Meera, Wallner, & Otis, 2011; Olsen et al., 1981); on neuronal inhibitory transmission measured by electrophysiology, with their potency on the receptor binding assay (Egebjerg, Schousboe, & Krogsgaard-Larsen, 2002; Johnston, 2000); likewise keeping in mind the species and tissue specificity associated with the neurotransmitter actions. The *in vivo* pharmacology of GABA and analogues is not very important for a variety of reasons including rapid metabolism and poor blood–brain barrier penetration, but what is known about analogue potencies must be consistent between *in vivo* and *in vitro* activities.

1.2.2 The BZ sites

BZ drugs, historically some of the most prescribed drugs ever, were shown to be CNS depressants and eventually to act as PAM on $GABA_AR$ (Costa & Guidotti, 1979; Haefely, 1982).

BZ receptors were identified by radioligand binding and shown to exhibit *receptor-specific* pharmacology (Braestrup & Squires, 1977; Möhler & Okada, 1977a). BZ binding was found to be enhanced in the test tube by GABA (Karobath, Placheta, Lippitsch, & Krogsgaard-Larsen, 1979; Tallman, Paul, Skolnick, & Gallager, 1980), indicating that the BZ sites are present on the GABAR proteins. This was verified by solubilizing and purifying one protein with GABA, BZ, and modulatory sites (Gavish & Snyder, 1980; King, Nielsen, Stauber, & Olsen, 1987; Sigel & Barnard, 1984; Supavilai & Karobath, 1984), as well as cloning of the BZ site-containing GABAR (Schofield et al., 1987).

A spectrum of partial agonists was found for the BZ sites (Braestrup, Nielsen, Krogsgaard-Larsen, & Falch, 1979). Also, pharmacological antagonists were discovered (Ro15-1788, AKA flumazenil) (Hunkeler et al.,

1981) and even ligands with the opposite efficacy as classical BZs (CNS depressants). These excitatory agents (anxiogenic, proconvulsant) were named *inverse agonists*; they included BZ analogues, plus variants like imidazobenzodiazepines (Ro15-4513) and β-carbolines, such as β-CCE and DMCM (Braestrup, Schmiechen, Neef, Nielsen, & Petersen, 1982). Braestrup, Nielsen, and Olsen (1980) discovered a putative endogenous ligand β-CCE in human urine and brain for the brain BZ binding sites, which exhibited nanomolar affinity and inverse agonist efficacy. GABA inhibited BZ inverse agonist ligand binding (Braestrup & Nielsen, 1981) including β-CCE (remember it enhanced traditional BZ agonist binding). However, the molecule β-CCE is a possible metabolite/breakdown product of tryptophan or serotonin, and this ligand is not actually in the brain but produced by chemical modification of brain ingredients during extraction and characterization. Other putative endogenous ligands for the brain BZ sites, both small molecules, termed "endozepines," and peptides, notably "diazepam binding inhibitor" DBI, have been described (Papadopoulos, Berkovich, Krueger, Costa, & Guidotti, 1991). Some of them clearly arise from the environment. Whether any of them is biologically active remains controversial (Möhler, 2014). One class of natural products with potent efficacy on BZ receptors are the *flavonoids*, herbal medicine ingredients from kudzu tree, e.g., daidzein, genistein (Lukas et al., 2013), and *Hovenia*, e.g., dihydromyricetin (Shen et al., 2012). These molecules have low micromolar affinity for BZ sites and variable efficacy and $GABA_AR$ subtype selectivity, the most common being a BZ agonist-like activity (Kahnberg et al., 2002). Several flavonoids have been described as neuroprotectants, antioxidants, and promoters of alcohol elimination (Arolfo et al., 2009); it is not certain if these activities, while interesting, involve the BZ sites on $GABA_ARs$ or have physiological relevance.

Heterogeneity of binding suggesting pharmacological subtypes is confirmed by cloning: The BZ receptor sites were identified immediately as showing heterogeneity in several characteristics including pharmacology but also brain regional localization. Affinities of many ligands, e.g., zolpidem (Lloyd & Zivkovic, 1988) and triazolopyridazines (Squires et al., 1979), varied with brain region; second, the binding densities of BZ, GABA, and picrotoxin site ligands did not agree across brain regions (Lo, Niehoff, Kuhar, & Snyder, 1983; Olsen, McCabe, & Wamsley, 1990). Ligands for BZ and picrotoxin sites also allosterically modulate each other (Olsen, 1982; Supavilai, Mannonen, & Karobath, 1982), consistent with multiple allosterically coupled sites on a single protein complex, the $GABA_AR$-chloride channel.

This helped establish that these drugs acted on the brain via PAM action on $GABA_A$Rs. This was of major interest because it suggested presumably circuitry and behavioral distinctions consistent with behavioral pharmacological subtypes, and the potential development of new drugs with subtype selectivity and hopefully improved clinical profiles. Molecular cloning of a family of 19 homologous subunit genes yields several dozens of likely heteropentameric subtypes in nature, of which maybe one dozen is sufficiently abundant to contribute significantly to function and pharmacology (Barnard et al., 1998; Hevers & Lüddens, 1998; Olsen & Sieghart, 2008; Rudolph & Möhler, 2004; Schofield et al., 1987; Whiting et al., 1995). Years of work and clever techniques such as genetic engineering were required to sort out the functional and pharmacological subtypes of $GABA_A$R, but the work has been well worth it.

Numerous *global knockout mice* were engineered for several of the $GABA_A$R subunit genes, e.g., γ2 (Günther et al., 1995), β3 (Homanics et al., 1997), δ (Mihalek et al., 1999), α4 (Chandra et al., 2006), and α5 (Collinson et al., 2002). Many such knockout studies generated informative phenotypes with valuable insights into the function of the gene under study, but a disappointing number gave either lethal or uninformative phenotypes due mostly to multiple compensatory changes in expression of other gene products (Boehm et al., 2004; Olsen & Homanics, 2000; Rudolph & Möhler, 2004). This prompted new technology development, such as the regional- and/or age-specific (*conditional*) *knockouts* (see, e.g., Engin et al., 2014; McHugh et al., 2007) or cell type-specific *gene expression knock-down* using virus-transfected siRNA expressed under control of cell-specific promoters (e.g., Dixon, Walker, King, & Stephens, 2012; Liu et al., 2011; Nie, Rewal, Gill, Ron, & Janak, 2011; Rewal et al., 2009).

A brilliant refinement of this preliminary understanding of α subunits and pharmacological subtypes was provided by the mouse α subunit knockins, first by Möhler and colleagues (Rudolph et al., 1999) and then by the Merck team of Whiting (McKernan et al., 2000). This approach used the α H101 residue (in α1, rat) that was R in the non-BZ binding α4 and α6 (Hevers & Lüddens, 1998). The H was switched to R by genetic engineering in mouse to give a point-mutated α subunit which did not show any change in expression and localization, $GABA_A$R channel function, or pharmacology, except for the BZ sensitivity, which was eliminated. Then the animals could be tested for sensitivity to BZ in any behavioral assay of interest. The point-mutated α1 subunit resulted in a large reduction in the sedative, but not anxiolytic actions of BZ in the mouse knockin (McKernan et al., 2000;

Rudolph et al., 1999). However, the α2 H101R knockin lost the BZ anxiolytic effect but not the sedative action (Low et al., 2000). This strategy was able to elegantly dissect the molecular targets of BZ action on various behaviors, but even more importantly, it gave strong clues to the function of the GABAR subtypes in behaviors and neuropsychiatric disorders (Rudolph & Möhler, 2004) and could be correlated to the brain regional expression and importance of the receptor subtypes (Barnard et al., 1998; Benke et al., 2004; Whiting et al., 1995). The approach was used to study other subunits of $GABA_AR$ like β, indicating molecular roles in other phenomena such as general anesthesia (Rudolph & Antkowiak, 2004).

Finally, now that the tremendous importance of $GABA_ARs$ in virtually all brain functions is established, and the ability to sort out which subtypes mediate which circuit-dependent behaviors, we gain greater understanding of normal brain function and the corresponding neuropsychiatric diseases (Rudolph & Möhler, 2006). In addition to baseline functions, it is also evident that many disorders result from environmentally induced plastic changes in GABAR subunit composition in critical regions (e.g., Kumar et al., 2009; Liang et al., 2006; for alcohol-treated animals) and insights into these mechanisms may be necessary for understanding brain function. The BZ sites are discussed in detail in the chapter "From $GABA_A$ receptors to CNS drugs: Past, present, and future" by Möhler.

1.2.3 The picrotoxinin sites

Picrotoxinin is a non-nitrogenous natural product from plants that possesses potent and specific ability to block $GABA_AR$ function (Curtis & Johnston, 1974; Olsen, 1982); it does so noncompetitively, by binding at a site within the transmembrane domain (TMD) channel (Chen, Durkin, & Casida, 2006; Hibbs & Gouaux, 2011; Olsen, 2014) and nowhere near the GABA binding site, which has been identified on the extracellular domain (ECD) of the protein (Fig. 1) (Miller & Aricescu, 2014; Olsen et al., 2014). Picrotoxin, as isolated from plants, is a molecular pair of two isomers: the much more active picrotoxinin and the less active picrotin. The $GABA_AR$ channel blocking activity makes picrotoxin a dangerous convulsant. Its pharmacology is shared by some close structural analogues found in related plants (Jarboe, Poerter, & Buckler, 1968; Olsen, 2014).

The Olsen lab synthesized a radiolabeled analogue $[^3H]$dihydropicrotoxinin and demonstrated specific receptor binding sites in membrane homogenates of crayfish muscle and mammalian brain (Olsen, 1982; Ticku et al., 1978). The binding was inhibited by appropriate concentrations

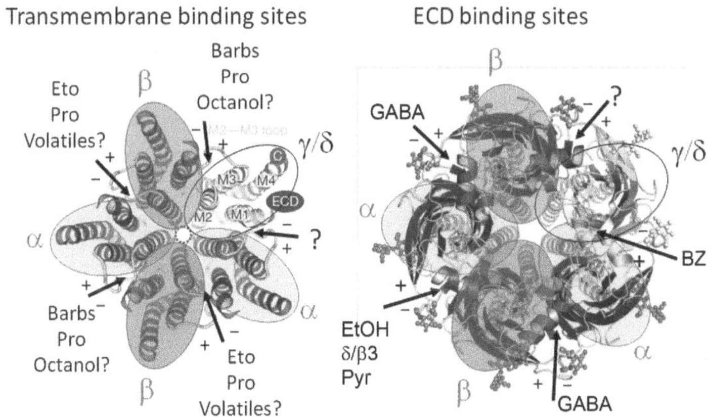

Figure 1 GABA$_A$R ligand sites at subunit interfaces identified by mutagenesis and/or affinity labeling. The left panel shows the transmembrane binding sites, while the right panel shows the ECD binding sites. The protein structures are taken from the X-ray crystallography-derived structure of the recombinant mammalian homomeric β3 GABA$_A$R (Miller & Aricescu, 2014), on which is displayed a homologous native GABA$_A$R comprised of an α–β–α–β–γ/δ heteropentamer (actual subunits arbitrary, no specific sequence implied although they are all homologous to β3). The protein is viewed looking from the extracellular face, perpendicular to the cell membrane/synapse. Thus, the ECD pentamer on right would actually be positioned directly on top of the TMD pentamer at left. Both portions indicate locations of ligand binding sites found in the β3 homomer structure. The two α subunits are indicated by the green (or dark gray in the print version)-shaded oval, the two β subunits by the pink (or light gray in the print version)-shaded ovals, and the one γ/δ subunit indicated by the clear (white in the print version) oval. An example C-terminus is indicated by a small red (dark gray in the print version) circled "C" at the bottom of the TMD of the γ/δ subunit; the M1,2,3,4 domains are also labeled in this example subunit, and the N-terminus of the TMD of each subunit would attach to its ECD at the position indicated by the small blue (or black in the print version) oval "ECD." Ligand binding sites for the compounds listed (in shorthand) are indicated by arrows (cf. Table 1). Note that the heteropentameric proteins show several different but homologous subunit interfaces (Table 1), so the pharmacological specificity varies with GABA$_A$R subtype. The ligands named are BZ (benzodiazepines) and GABA, EtOH, and Pyr (pyrazoloquinolines, see Table 1) in the ECD. In the TMD, Eto (etomidate), Pro (propofol), octanol, volatiles, and barbs (barbiturates), binding sites are located.

of appropriate picrotoxin analogues, and also by several "cage convulsants" (Ticku & Olsen, 1979), previously demonstrated to be GABA$_A$R noncompetitive inhibitors (Bowery, Collins, & Hill, 1976). These cage convulsants (trioxabicyclo-octanes) were synthesized as a systematic attempt to make invertebrate-selective pesticides acting on the nervous system (Casida, 1993; Chen et al., 2006), including the highly toxic *t*-butyl bicyclophosphorothionate (TBPS). This ligand was radiolabeled and made

Table 1 Summary of $GABA_AR$ ligand sites at subunit interfaces identified by mutagenesis and/or affinity labeling (cf. Fig. 1)

Interface	Extracellular domain	Transmembrane domain
β+/α- (two copies)	GABA (1,2)	Etomidate (3–6)
		Propofol (3,6–10)[a]
		Volatiles? (4,10)
α+/β-	EtOH (on δ) (11)	Barbiturates (12)
	Imidazo-BZ (on δ) (11)	Propofol (9)
	Pyrazoloquinolines (13)	Octanol? (9)
γ+/β-	?	Barbiturates (12)
		Propofol (9)
		Octanol? (9)
α+/γ-	BZ (14–16)	?
α+/δ-	?	?
4δ+/β-	?	Barbiturates? (8)
		Propofol? (8)
		Octanol? (7)
?		Neurosteroids (17)[b,c]

[a]A residue in the TMD (M3) but not at subunit interfaces was affinity labeled by a propofol analogue (Yip et al., 2013).
[b]Four residues in TMD (TM 1,2,3,4) were identified as critical for neurosteroid action (Hosie, Wilkins, da Silva, & Smart, 2006).
[c]A different residue in the TMD (M3) but not at subunit interfaces was affinity labeled by a neurosteroid (Chen et al., 2012).
(1) Amin and Weiss (1993); (2) Smith and Olsen (1994); (3) Li et al. (2006); (4) Mihic et al. (1997); (5) Belelli, Lambert, Peters, Wafford, and Whiting (1997); (6) Jurd et al. (2003); (7) Bali and Akabas (2004); (8) Krasowski, Nishikawa, Nikolaeva, Lin, and Harrison (2001); (9) Jayakar et al. (2014); (10) Li, Chiara, Cohen, and Olsen (2010); (11) Wallner, Hanchar, and Olsen (2014); (12) Chiara et al. (2013); (13) Ramerstorfer et al. (2011); (14) Duncalfe, Carpenter, Smillie, Martin, and Dunn (1996); (15) Smith and Olsen (1995); (16) Sawyer, Chiara, Olsen, and Cohen (2002); (17) Li, Chiara, Cohen, and Olsen (2009).

a superior affinity ligand tool for the picrotoxin site (Squires, Casida, Richardson, & Saederup, 1983); the binding activity was found to be allosterically inhibited by GABA agonists and shown to be located on the $GABA_AR$ protein (King et al., 1987; Sigel & Barnard, 1984; Supavilai & Karobath, 1984). There is no conclusive evidence for any significant difference in sensitivity to picrotoxin site ligands for subtypes of $GABA_AR$, except for perhaps the insecticides.

Many insecticides, e.g., lindane and dieldrin, were found to inhibit the picrotoxin/TBPS site in $GABA_AR$, including Casida's synthetic cage convulsants (Casida, 1993), as well as some new-generation potent compounds such as fipronil (Ikeda et al., 2001; Zhao et al., 2014). Other convulsants considered active on these sites included pitrazepin and tetramethylene disulfotetramine (TETS) (Zhao et al., 2014). This identification of toxin/insecticide binding sites on $GABA_AR$ chloride channels was supported by strong evidence. First, genetic variants of insects that were resistant to the insecticide dieldrin allowed cloning of the "resistance to dieldrin" (*rdl*) gene, which turned out to be the insect homolog protein of the mammalian $GABA_AR$ β subunit (Ffrench-Constant, Rocheleau, Steichen, & Chalmers, 1993); the allelic residue conferring dieldrin resistance in the *rdl* gene sequence was located in the TMD-M2 helical membrane-spanning region implicated in pentameric cys-loop LGIC receptor proteins as the channel domain (Galzi & Changeux, 1994; Macdonald & Olsen, 1994). Additionally, mutagenesis of the amino acid implicated in dieldrin resistance in the $GABA_AR$ β subunit to a cysteine allowed an insecticide analogue modified with a sulfhydryl reagent moiety to bind covalently (affinity label) the channel at this residue, so-called proximity-accelerated covalent coupling (PACC; Perret et al., 1999); the attachment of the insecticide affinity label produced an irreversibly inhibited channel and block of the picrotoxin/TBPS binding sites. Some ligands, e.g., TBPS, had more potent actions on mammalian $GABA_AR$ than on insect, while others had higher potency on insects, such as fipronil. The latter showed micromolar affinity for vertebrate and nanomolar affinity for insect $GABA_AR$ (Ikeda et al., 2001). Modern techniques such as homology structural modeling and molecular dynamics simulations show that the different toxin ligands bind at overlapping but nonidentical sites in the protein channel that indeed vary with subunit subtypes (Zhao et al., 2014). Nevertheless, the sum of this work demonstrated conclusively that the picrotoxin site was on the $GABA_AR$ protein, located apparently in the channel, at a site distinct from the GABA site, and cocrystallography of the nematode GluCl LGIC protein with picrotoxinin showed picrotoxinin bound within the channel (Hibbs & Gouaux, 2011), consistent with models of the $GABA_AR$ (Olsen, 2014; Zhao et al., 2014). One could use the protein sequence to model the structure of the channel and deduce other amino acids critical to channel function, insecticide binding, conductance ion selectivity, and desensitization (Chen et al., 2006). The first X-ray crystal structure of a human $GABA_AR$ has just been published (Miller & Aricescu, 2014). This should greatly enhance our understanding of the structure and function of the protein and its modulation by drugs.

1.2.4 GABA$_A$R: Summary based on the three ligands

(a) All the binding sites are on the same protein, verified biochemically, and on cloned products.

(b) The three ligand categories help establish heterogeneity of binding populations consistent with receptor subtypes, verified by cloning and subsequent analysis of the separate protein distributions and pharmacological and physiological properties.

(c) Cloning of the GABA$_A$R was achieved by protein purification, partial sequencing, and homology screening of a cDNA library from brain (Schofield et al., 1987; Sigel & Steinmann, 2012). This resulted in a functional GABA$_A$R expressed from two subunit genes, indicating high homology with the other members of the pentameric cys-loop LGIC superfamily (Dutertre et al., 2012; Galzi & Changeux, 1994). This then led to identification of 19 homologous subunit genes, grouped into families by degree of sequence identity (α1–6, β1–3, γ1–3, δ, ε, θ, π, ρ1–3) (Barnard et al., 1998; Lüddens et al., 1995; Whiting et al., 1995). Genetic engineering approaches helped to understand subtypes and their functions.

(d) The wealth of ligands developed, especially for the BZ site, also provide tools for other methods useful in receptor analysis, especially biochemistry, namely, affinity labeling and affinity chromatography. These reagents allow purification of proteins and, coupled with site-directed mutagenesis, identification of functional domains including especially binding site locations for both high- and low-affinity ligands, including agonists, chloride channel blockers, and allosteric modulators (see below).

1.3. Identification of ligand binding sites and their three-dimensional location: Affinity labeling, mutagenesis, X-ray crystal structures, and homology modeling

1.3.1 Picrotoxin sites lead to discovery of the anesthetic sites

Ticku and Olsen (1978) found that picrotoxin binding was inhibited by barbiturates with sedative/hypnotic/anesthetic efficacy. This was consistent with evidence from electrophysiological recordings showing that pharmacologically active barbiturates and related drugs enhanced GABA currents at inhibitory GABA$_A$R synapses (Bowery & Dray, 1976), an effect that could be distinguished from that of the well-known GABA positive modulators, the BZs (Macdonald & Olsen, 1994; Study & Barker, 1981). With little knowledge of structure–activity relationships for barbiturates and related drugs, the Olsen lab (Ticku et al., 1978; Ticku & Olsen, 1978; Wong, Leeb-Lundberg, Teichberg, & Olsen, 1984) compared a series of these drugs active on binding to their ability to enhance $GABA_AR$ channels with an *in vitro* $^{36}Cl^-$ flux assay. It turned out that all sorts of CNS depressants with *in vivo* efficacy ranging from sedation to hypnosis to anesthesia were found to be PAMs for $GABA_AR$ in studies using electrophysiological recordings in neurons (Johnston, 1997; Macdonald & Olsen, 1994; Olsen, 1982; Olsen, Fischer, & Dunwiddie, 1986; Olsen & Li, 2011; Supavilai et al., 1982), and they allosterically inhibited the binding of picrotoxin/TBPS, and allosterically enhanced the binding of GABA agonists (Johnston, 1997; Olsen & Snowman, 1982; Olsen, Yang, et al., 1986) and BZs (Leeb-Lundberg, Snowman, & Olsen, 1980, 1981; Supavilai et al., 1982), and inhibited GABA antagonist (Wong, Snowman, Leeb-Lundberg, & Olsen, 1984) and BZ inverse agonist binding (Braestrup & Nielsen, 1981). This activity was found for those compounds and only those compounds, including stereoisomers, that were active pharmacologically *in vivo* and *in vitro*. The allosteric interactions *in vitro* were not limited to barbiturates but also seen for a variety of volatile and intravenous anesthetics of varying chemical structure, including long-chain alcohols and neuroactive steroids (Johnston, 1997; Olsen, 1982; Olsen, Yang, et al., 1986; Supavilai et al., 1982). Thus, the $GABA_AR$ protein family has a very large number of sites for allosteric ligands (Fig. 1). The "anesthetic" sites, first biochemically demonstrated indirectly by allosteric inhibition of the picrotoxin binding sites, are distinct from the picrotoxinin site. The inhibition of picrotoxin/TBPS binding to the $GABA_AR$ channel by barbiturates and related anesthetics appeared to be possibly competitive but was soon found to be allosteric (Olsen, Yang, et al., 1986).

The pharmacological specificity including stereoisomers of differing potency for barbiturate modulation of $GABA_AR$ provided an *in vitro* "receptor-specific" assay for such drugs that could be utilized as a tool to characterize the drug mechanism of action, screen for activity of unknown analogues, and assay the presence of the receptor (Olsen, 1982). In other words, even though all the known ligands had insufficient potency to be used for radioligand binding, they could still be quantitated and studied by their indirect action on high-affinity ligands. In the absence of tissue-like electric organs from rays and eels that expressed nicotinic acetylcholine receptors in amounts dozens of times more than mammalian muscle or brain, tissues containing the actual sites of action of drugs like anesthetics, these sites were technically difficult to study. Furthermore, even though the highest affinity (submicromolar) ligands among these PAM ligands (notably neurosteroids) (see Bianchi & Macdonald, 2003; Lambert, Belelli, Peden, Vardy, & Peters, 2003; Olsen, Cagetti, & Wallner, 2003; Zheleznova, Sedelnikova, & Weiss, 2008 and below) were made radioactive, even the world's experts in binding could not develop a specific binding assay for any of them due to their horrible nonspecific binding to tissues (e.g., Gee, Bolger, Brinton, Coirini, & McEwen, 1988; Olsen, 1982; Squires et al., 1983; Supavilai et al., 1982).

One category of PAMs for $GABA_AR$ that may be extremely important because they are active as endogenous modulators are the neurosteroids. This topic is reviewed nicely by numerous other authors, and this paragraph will be limited to a brief discussion of the use of the pharmacology as PAMs for $GABA_AR$ and for analysis of $GABA_AR$ subtypes, and identification of ligand binding sites and their mechanism of allosteric modulation. The female sex steroid hormone progesterone itself has sedative/hypnotic efficacy, and a synthetic analogue alphaxalone (5α-pregnan-3α-ol-11, 20-dione) marketed by Glaxo, was shown to have clinically useful intravenous anesthetic efficacy, along with PAM activity at $GABA_AR$ (Harrison & Simmonds, 1984; Smith, 2003; Smith, Waterhouse, & Woodward, 1987). Obviously, the anesthetic was an analogue of the nonhormone metabolite dihydroprogesterone (5α-pregnan-3α-ol-20-one (alloprenanolone, AKA 3α-5α-THP)) which was shown to be the prototype for activity on $GABA_AR$, an activity likewise demonstrated by 3α-5β-THP (pregnanolone) and 3α-5α-THDOC (3α-21-dihydroxy-5α-pregnan-20-one, AKA 3α-5α-tetrahydrodeoxycorticosterone). Note that 3β isomers are inactive.

Majewska, Harrison, Schwartz, Barker, and Paul (1986) showed that steroid hormone metabolites are endogenous potent PAMs for $GABA_AR$. Although neuron or even brain regional specific biosynthesis of the neurosteroids has not been demonstrated, the brain levels vary with the precursor hormone, e.g., cortisone and progesterone, cycling, and are increased in critical brain areas during a variety of physiological states, with a net effect of increased anxiolytic GABAergic inhibitory tone (Barbaccia, Serra, Purdy, & Biggio, 2001; Concas et al., 1998; Maguire, Stell, Rafizadeh, & Mody, 2005; Morrow et al., 1999; Paul & Purdy, 1992; Smith, 2003). Binding and neuron electrophysiological recording studies demonstrated allosteric enhancement of synaptic and extrasynaptic GABA$_A$R currents (Gee et al., 1988; Majewska et al., 1986; Smith, 2003; Turner, Ransom, Yang, & Olsen, 1989). Some evidence for more potent/efficacious actions on extrasynaptic high-affinity δ subunit-containing GABA$_A$R than on synaptic GABA$_A$R (Bianchi & Macdonald, 2003; Stell, Brickley, Tang, Farrant, & Mody, 2003; Wallner, Hanchar, & Olsen, 2003; Zheleznova et al., 2008) is consistent with the evidence that only low nanomolar levels of endogenous neurosteroids are produced under any conditions (Barbaccia et al., 2001; Concas et al., 1998; Morrow et al., 1999; Paul & Purdy, 1992; Smith, 2003). Likewise, in the hippocampal formation CA1 region, pyramidal neurons mediate tonic inhibitory GABA$_A$R currents with extrasynaptic α5 subunit-containing GABA$_A$R; general anesthetics as well as neurosteroids enhance the α5-GABA$_A$R tonic inhibition with resulting inhibition of learning and memory and production of amnesia (Wang & Orser, 2011). Also, α5-selective antagonists, like inverse agonist BZ ligands, reverse this learning and memory inhibition and improve cognitive function (Dawson et al., 2006), possibly in Alzheimer's disease.

1.3.2 GABA and BZ sites at subunit interfaces

Möhler, Battersby, and Richards (1980) discovered that [^{3}H]flunitrazepam could photoaffinity label the BZ sites in brain, and had a sufficient affinity ($K_D \sim 10$ nM) to detect binding in crude homogenates or brain sections, and yield a covalent attachment to a polypeptide on SDS–PAGE at ~50 kDa. This affinity label binding was "receptor specific" and could be employed to localize BZ binding sites (GABA$_A$R) in brain sections (Möhler et al., 1980). Others immediately showed (Sieghart & Karobath, 1980) labeling of numerous (four or more) polypeptides in brain membranes on more sensitive SDS–PAGE gels. This showed that radioactive BZ ligands bound to multiple polypeptide bands of similar but not identical molecular weights in brain, presumably (and in retrospect, correctly analyzed) as different gene products, i.e., subunits with different protein sequences. The binding site for BZs was analyzed by mutagenesis and by photoaffinity labeling. The classical agonist [^{3}H]flunitrazepam was found to bind to α1 (and also α2, α3, and α5) H102 in ECD Loop A (cow numbering; rat = 101) (Duncalfe et al., 1996; Smith & Olsen, 2000). Flunitrazepam as a classical BZ agonist did not bind to α4 or α6 (βγ2) subtypes (Barnard et al., 1998; Hevers & Lüddens, 1998). The antagonists and inverse agonists bind the α4 and α6 (βγ2) subtypes, and the binding site on all six α subunits was believed to contain the same residues as flunitrazepam in α1, confirmed by photoaffinity labeling with [^{3}H]Ro15-4513, identified (Sawyer et al., 2002) as attached covalently to Y209 (Loop C, ECD). The BZ binding site was shown to be located at the α+/γ- interface, as opposed to the GABA site at the β+/α- interface. The surprising observation was that the BZ binding sites were in the same homologous sequences (Loops A–H) as the GABA binding sites, which were the same utilized for agonist binding in all members of the cys-loop LGIC superfamily (Amin & Weiss, 1993; Duncalfe et al., 1996; Galzi & Changeux, 1994; Harrison & Lummis, 2006 Sawyer et al., 2002). This was immediately also recognized by workers on the $GABA_AR$ (Sigel & Buhr, 1997; Smith & Olsen, 1995) and indicated that the BZ sites were modified GABA agonist sites and these exogenous/synthetic PAM ligands fortuitously found a site that was almost but not quite an agonist site! This remarkable situation may be unprecedented in protein chemistry. The BZ site was compared for the various subunit subtypes by quantitative structure–activity relationships (e.g., Cook et al., 2005), and the mutagenesis *plus* affinity labeling experiments, i.e., cysteine replacement and covalent binding of a ligand analogue containing a sulfhydryl reagent (Tan et al., 2007). With the X-ray crystal structure of the invertebrate acetylcholine binding protein (Brejc et al., 2001) available as template, homology structural models of the ECD*s* of nAChRs and $GABA_ARs$ were constructed (Fig. 1), including the GABA (Sander et al., 2011) and BZ sites (Ernst, Brauchart, Boresch, & Sieghart, 2003; Ernst, Bruckner, Boresch, & Sieghart, 2005; Middendorp et al., 2014; Richter et al., 2012).

1.3.3 Benzodiazepine sites lead to discovery of the ethanol (EtOH)-sensitive benzodiazepine (BZ) sites, distinct from the low-affinity TMD sites for long-chain alcohols (anesthetics), and a candidate for the "one glass of wine" ethanol receptor site

$GABA_AR$ have long been the focus for acute alcohol actions. Evidence was reported for behaviorally relevant low millimolar alcohol actions on tonic inhibitory GABA currents (Hanchar, Dodson, Olsen, Otis, & Wallner, 2005; Liang et al., 2006; Wei, Faria, & Mody, 2004) mediated by extrasynaptic α4/6, β, δ subunit-containing $GABA_AR$. Low-dose ($\leq$30 mM) EtOH enhances recombinantly expressed δ subunit-containing subtypes of $GABA_AR$, as well as tonic inhibitory GABAR-mediated currents in cells that express the extrasynaptic δ-$GABA_AR$ (Sundstrom-Poromaa et al., 2002; Wallner et al., 2003; Wallner, Hanchar, & Olsen, 2006; Wallner & Olsen, 2008). Korpi, Kleingoor, Kettenmann, and Seeburg

(1993) found a point mutation—or allelic variant—in the $GABA_AR$ α6 subunit (R100Q) in a rat strain that exhibited *in vivo* behavioral hypersensitivity to BZs and also EtOH (Korpi et al., 1993). The BZ sensitivity was explained by the α6-R100Q mutation, which gave high-affinity BZ enhancement when expressed with β and γ2 subunits *in vitro* in oocytes. Neither wild type nor mutant receptors showed enhancement by EtOH (up to 100 mM) when expressed with β and γ2 subunits. Hanchar et al. (2005) revisited this mutation and found that α6-R100Q, when expressed with β3 and δ subunits, gave even higher sensitivity to EtOH enhancement (low mM concentrations) in oocytes than the already sensitive wild-type α6βδ subtype; further, the rats expressing this allelic α6-R100Q showed greater sensitivity (1–20 mM) to EtOH than wild-type α6-R100 (10–30 mM) for tonic inhibitory currents carried by α6βδ−$GABA_AR$ in cerebellar granule cells studied in brain slices (Hanchar et al., 2005). The amino-acid α6-R100 is located in the binding site for traditional BZ including the imidazobenzodiazepine EtOH antagonist drug Ro15-4513. Further, many behavioral effects of EtOH, and the *in vitro* enhancement of $GABA_AR$ in neurons, in culture, and in recombinant expression cells, are reversed by Ro15-4513 (Suzdak et al., 1986; Wallner et al., 2006). They demonstrated that Ro15-4513 not only blocks low mM EtOH enhancement of δ-$GABA_AR$ currents, but also this action is reversed by the close structural analogue flumazenil (Ro15-1788), the traditional BZ antagonist. A minor subset of BZ ligands, notably imidazobenzodiazepines and β-carboline compounds, are able to mimic or inhibit Ro15-4513 activity on EtOH *in vivo* and *in vitro*, and the same ligands, and only those, are able to inhibit the binding of [^{3}H]Ro15-4513 to native and recombinant δ-$GABA_AR$ (Hanchar et al., 2006). Whereas it had been believed that no BZ ligands bound to the δ-$GABA_AR$ subtypes, this new evidence suggests that δ-GABAR have a novel BZ/EtOH site with unique pharmacology with respect to BZ ligands (Santhakumar, Wallner, & Otis, 2007; Wallner et al., 2006). Thus, this work defines an EtOH receptor, when it was not clear that there is such a thing in the traditional sense (Wallner & Olsen, 2008). Perkins, Trudell, Crawford, Alkana, and Davies (2010) showed that there is an ECD region (Loop 2) in glycine LGIC receptors that affects EtOH sensitivity, and introduction of the GABAR δ subunit sequence for that region increases EtOH sensitivity, both in glycine and in $GABA_AR$ (Perkins et al., 2010). This domain is implicated in coupling agonist binding to channel gating through physical interaction between Loop 2, which is near the postulated ECD EtOH binding site, and the M2–3 linker (Kash, Jenkins, Kelley, Trudell, & Harrison, 2003), which is near the TMD anesthetic binding sites (Bertaccini, Yoluk, Lindahl, & Trudell, 2013; Olsen et al., 2014). This story needs further study to evaluate its importance for both agonist and PAM function.

Further mutagenesis of residues in the extracellular BZ binding domain (ECD) of α6βδ-GABA$_A$R revealed preliminary evidence for an EtOH-sensitive Ro15-4513 binding site at the α+/β- interface (Fig. 2, reproduced from Wallner et al., 2014). Starting with sequence loops in the major subunit for this binding site (α), including α6-R100, they predicted sequences for the minor subunit potentially situated abutting this α pocket domain, which could be either on δ or on β. Mutagenesis of residues in the sequence of the δ subunit, e.g., H68A (Meera, Olsen, Otis, & Wallner, 2010) imparts classical BZ site binding to the heteropentamer, but does not disturb the EtOH-sensitive Ro15-4513 sites in δ-GABAR. This receptor appears to carry two distinct BZ sites. These authors used BZ sensitivity as a marker for (mutated) δ subunit expression in HEK cells. On the other hand, the selectivity of β subunits of the EtOH binding (β3 > β2 ≫ β1; Wallner et al., 2003) can be identified in chimera and point mutagenesis studies to involve this loop in the β subunit domain, #62–66 in β3, Loop "D" (Fig. 2, from Wallner et al., 2014). Based on involvement of these two amino-acid residues, α6-R100 and β3-Y66, they propose a model in which δ subunit-containing (and possibly other) GABA$_A$R contain a unique EtOH site at the α+/β- ECD subunit interface. There are other δ-selective PAM ligands that might also interact here, or at a δ-interface (Jensen et al., 2013). This same interface in non-δ-GABAR has been demonstrated to show a significant affinity for a subset of BZ ligands including pyrazoloquinolines and possibly other PAM (Ramerstorfer et al., 2011).

Thus, an ECD site for EtOH modulation may resemble the traditional binding and resulting efficacy of ligands like agonists (e.g., GABA) and more well-understood modulators (e.g., BZ).

A recent review (Olsen et al., 2014) covers most of the evidence for the TMD anesthetic site with low affinity for EtOH and the ECD nonanesthetic high-affinity ethanol-sensitive BZ sites. It is our opinion, supported by considerable literature, that subtypes of the GABA$_A$R represent the brain target for EtOH that is closest to that relevant to blood and brain concentrations affecting humans (~17 mM) with the well-known intoxicating actions, and that myriad reports of EtOH acting on other neurotransmitter systems and ion channels require much higher concentrations of EtOH (>50 mM) that may be involved in anesthesia and fatal overdose for EtOH in humans (Wallner et al., 2014; Wallner & Olsen, 2008). Questions remain in the

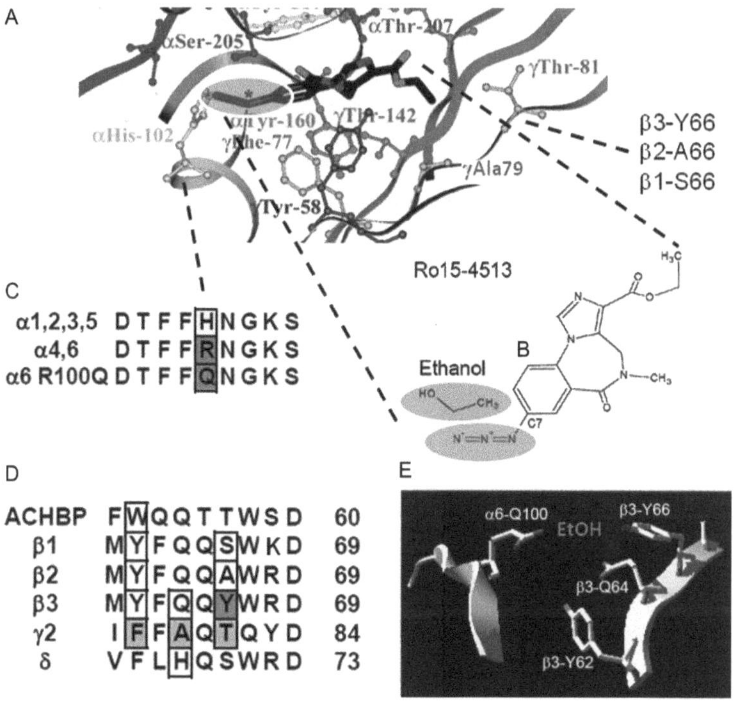

Figure 2 Structural alcohol site model. (A) Ro15-4513 docking model in the classical α(+)γ2(−) BZ binding site (adapted from fig. 10 in Sawyer et al., 2002). The docked Ro15-4513 molecule is in a stick representation with the —N=N=N$^+$ group indicated by an asterisk (structure of Ro15-4513 is enlarged in (B)). At the far left (yellow or white in the print version) in (A) is part of "Loop A" with the BZ critical H102 residue (see alignment in (C) with H (yellow or white in the print version), R (red or gray in the print version), and the α6-R100Q (green or gray in the print version) in alcohol nontolerant (ANT) rats. Panel (D) shows a protein sequence alignment of Loop D (region shown at bottom (green or white in the print version) in (A) forming a beta sheet structure with three amino-acid residues (F(Phe)77, A(Ala)79, T(Thr)81) (boxed, shaded green or light gray in the print version) pointing into and lining the benzodiazepine binding pocket). AChBP is acetylcholine binding protein, on which the structural models are built. Also boxed in this alignment are "alcohol residues" β66 (white in β1 and β2, red (dark gray in the print version) in β3), and the three critical "Loop D" γ residues F77, A79, T81, and residue δH68 (white in the print version) that confers diazepam sensitivity to α4β3δ receptors when changed to A (green or light gray in the print version), the residue present in γ2 at the homologous position (Meera et al., 2011), and a tryptophan residue in the acetylcholine/nicotine binding site (white in the print version). Panel (E) shows a structural rendering of a possible proposed EtOH binding pocket with EtOH between α6-R100Q and β3-Y66 residues at the α(+)β(−) subunit interface; this is a possibility if—the fairly long—α6-100R/Q and the β3-66Y amino-acid side chains are fully extended toward each other. *Reproduced from fig. 3 in Wallner et al. (2014) with permission.*

minds of many in the field, both as to whether low-dose EtOH effects do involve $GABA_A$Rs at all and whether the important pharmacological effects of low-dose EtOH also involve other targets. The issue is thoroughly covered in a special issue of *Alcohol* edited by Lovinger and Homanics (2007).

1.3.4 Structure of $GABA_A$R: Mutagenesis and affinity labeling to identify functional domains especially ligand binding sites such as anesthetics and steroids sites; photoaffinity labeling; modeling; and X-ray crystal structures of LGIC, GluCl, and $GABA_A$R

Prior to identification of anesthetic binding contact points on proteins such as $GABA_A$R, mutagenesis was employed to identify amino-acid residues critical for PAM action, by comparing two related LGICs of differing sensitivity to modulation. The weakness of this approach is twofold: the residues to be studied are determined by advance guessing on the part of the investigator, and the residues identified as important to drug modulation are not proven to be actual binding pocket sites, but could be due to allosteric coupling of drug action and function. Affinity labeling on the other hand will generally identify binding pocket residues. Depending on the chemistry of the affinity labeling, a couple of things can go wrong: it is possible that the photoreactive intermediate is long-lived and/or slow to react so that it can move from the actual binding pocket in the protein to a nearby but not critical residue; additionally, there may be no residues in the binding pocket that are actually good substrates for covalent binding of the affinity label ligand; thus, some theoretically active affinity labels will turn out to have no activity (Bouchet, Rendon, Wermuth, Goeldner, & Hirth, 1987; Forman & Miller, 2011). Nevertheless, various criteria for specificity generally support the conclusion that affinity labeling is specific and valuable.

Classical and ground-breaking mutagenesis studies (mutagenesis only, not affinity labeling) demonstrated residues in the TMD that are necessary for anesthetic actions (volatile agents and long-chain alcohols). That work concluded that the two residues identified were in a single binding pocket limited to intrasubunit domains (Mihic et al., 1997). These two residues M2–15′ and M3–4′, in α or β subunits, were analyzed by systematic mutagenesis and characterized for sensitivity to numerous examples of various chemical classes of general anesthetics. In some cases, substitution of larger amino acids for the native one at these two critical positions led to reduced modulation by the anesthetic, but also equal or increased opening probability of the receptor in the absence of agonist, which was interpreted as the substituted amino acid replacing the bound anesthetic as positive allosteric modulator (Jenkins et al., 2001) in the putative binding site.

This work was consistent with identification of a binding pocket for anesthetics, e.g., sensitivity to both volatile agents and long-chain fatty acids exhibited a "cutoff point," or maximum size or volume of ligand that could fit into the binding site (Krasowski et al., 2001; Wick et al., 1998). Bali and Akabas (2004) showed that sulfhydryl reagents inactivated cysteine-substituted βM3–4′ with respect to propofol enhancement of $GABA_AR$ channels, and this was protected by excess free propofol in the reaction; however, no protection was afforded to βM2–15′ (Bali & Akabas, 2004).

Further arguing against the role of M2–15′ in the binding pocket was the observation that the action of numerous chemotypes of PAM was sensitive to the nature of the residue at this position, suggesting a possible allosteric coupling site. For example, loreclezole, etomidate, and propofol were all sensitive to this residue in the β subunits (but not α) (Belelli et al., 1997; Rudolph & Antkowiak, 2004). A convincing demonstration that these residues are important for anesthetic action *in vivo* were reports establishing that genetic knockin of point mutations in the $GABA_AR$ β subunits at position M2–15′ could eliminate the anesthetic actions (Jurd et al., 2003: β3 M265N) or sedative/hypnotic actions (Reynolds et al., 2003; β2M265S) of etomidate. This shows that $GABA_AR$ are the major molecular target for etomidate and presumably other general anesthetics and that the β3 subunit-containing subtypes of $GABA_AR$ are more important for this *in vivo* action than other β subunit-containing subtypes, suggesting an anatomic correlate of anesthetic action (Jurd et al., 2003). As mentioned above, this does not prove that this residue is in the binding pocket. However, covalent attachment of sulfhydryl analogues of alcohols like propanethiol to cysteine-substituted M2–15′ leads to irreversible enhancement of channel function, as well as occlusion of the site for any additional anesthetic modulation (Mascia, Trudell, & Harris, 2000), strongly suggesting participation of this residue in PAM binding. Crystal structure evidence is needed and may be forthcoming.

Another residue at the beginning of M1 or pre-M1 in the $GABA_AR$ β2 subunit was shown by mutagenesis to alter modulation of $GABA_AR$ binding and channel function by several chemotypes of general anesthetics, again consistent more with an allosteric coupling site than a ligand binding pocket (Carlson, Engblom, Kristiansen, Schousboe, & Olsen, 2000; Chang, Olcese, & Olsen, 2003; Engblom, Carlson, Olsen, Schousboe, & Kristiansen, 2002). This area deserves further studies on its participation.

Mutagenesis suggests critical residues as possible steroid binding sites on $GABA_AR$ (Hosie et al., 2006): one residue each in αM1 and αM4 required for PAM activity, and two residues in αM1 and βM3 needed for direct gating, proposed to be part of a single intersubunit pocket. However, the possibility of an intersubunit pocket is ruled out because these two residues are not positioned near each other in the helical wheels generated by models of the intersubunit etomidate site (Li et al., 2009, 2006) generated by cysteine substitution cross-linking data (Bali, Jansen, & Akabas, 2009; Olsen & Li, 2011; Olsen et al., 2014; Stewart et al., 2013). Mouse knockins were able to identify residues in $GABA_ARs$ critical for anesthetic modulation, such as β2/3-M2–15′ for etomidate, propofol, isoflurane (partial), but not steroids (Rudolph & Antkowiak, 2004). Li et al. (2009) showed enhancement of etomidate binding by GABA-enhancing steroids, suggesting that steroid and etomidate sites are distinct (Li et al., 2009). Affinity labeling with a GABA-active neuroactive steroid derivative identified one amino-acid residue of contact in the TMD-M3 of homomeric recombinant β3 $GABA_AR$ (Chen et al., 2012). Hopefully, the identification of residues involved in functional binding in native $GABA_AR$ of the steroid PAMs will be accomplished soon.

Relatively high-affinity ligands for the anesthetic sites were utilized to make analogues with affinity labeling chemical moieties and radiolabel them to identify the binding pocket amino acids in the $GABA_AR$ proteins. A team led by Keith Miller synthesized ^{3}H-labeled azietomidate, azibarbiturate, and azipropofol (Chiara et al., 2013; Forman & Miller, 2011; Jayakar et al., 2014; Li et al., 2006; Olsen, 2014; Olsen & Li, 2011). Radiolabeled [^{3}H]azietomidate (Li et al., 2006) was used as a photoaffinity label on purified $GABA_AR$ protein from bovine cerebral cortex. Two residues were labeled, corresponding to the TMDs within M3 of the β subunit (M286 at position 4′) and M1 of the α subunit (M236 at position 11′) (Li et al., 2006). Here, we introduce a shorthand nomenclature for residues in the membrane-spanning domain (TMD). Residues are numbered 1′–20′ starting at the extracellular end and proceeding to the intracellular end of the TMD helices. The TMD helices could be aligned relative to the membrane-spanning sequences based on cysteine scanning mutagenesis and sulfhydryl cross-linking (Bali & Akabas, 2004; Bali et al., 2009), and a three-dimensional model of the TMD generated, showing that the two labeled residues could be located near each other, consistent with a single binding pocket at the intersubunit β/α interface (Li et al., 2009, 2006). As with GABA binding, there are two copies of the β/α interface etomidate binding site per pentamer (Forman & Miller, 2011; Olsen & Li, 2011).

Mutagenesis of these two residues established that they were indeed critical for anesthetic modulation of channel functions, both for etomidate and for propofol modulation (Bali & Akabas, 2004; Bali et al., 2009; Desai, Ruesch, & Forman, 2009; Krasowski et al., 2001; Stewart, Desai, Cheng, Liu, & Forman, 2008; Stewart et al., 2013); further, the etomidate enhancement of function and modulation of agonist binding could be blocked by large but not small sulfhydryl reagents attaching to a cysteine replacement mutation at αM1–11′ or βM3–4′ (Stewart et al., 2013). These results are consistent with the conclusion that the affinity-labeled residues are part of an etomidate binding pocket (Forman & Miller, 2011; Olsen & Li, 2011).

One of the residues, β(M3–4′), described by Mihic et al. (1997) corresponded with one of the residues identified by etomidate affinity labeling (Li et al., 2006). The same protection of βM3–4′ by etomidate was found after that residue was identified as binding etomidate (Stewart et al., 2013). Mutation to Trp of either of the active site Mets (Stewart et al., 2008) resulted in increased spontaneous opening and sensitivity to GABA (suggested to represent endogenous Trp replacing the exogenous anesthetic). Cysteine replacement mutants of βM2–15′ (Desai et al., 2009) reduced etomidate efficacy but not affinity, nor GABA affinity or efficacy. They suggested that this residue is not part of the anesthetic binding pocket, but could not rule out the possibility that this residue makes contact with the ligand during etomidate-enhanced GABA-gated channel opening. These workers (Forman & Miller, 2011) noted that due to the effects of mutations on basal affinity and efficacy for GABA and modulators, one needed to compare a series of agonists and modulators regarding mutant channel properties, using the Monod/Wyman/Changeux (MWC) allosteric model. Rüsch, Zhong, and Forman (2004) described well the action of etomidate modulation as an MWC co-agonist model (Rüsch et al., 2004), in which both enhancement of agonist by etomidate, at low micromolar concentrations, and direct channel gating, at high micromolar concentrations, could be explained by etomidate binding to a single class of binding sites, now identified (Li et al., 2006). On the other hand, etomidate binding to β (M3–4′) and α (M1–11′) was completely and apparently competitively inhibited by volatile anesthetics at pharmacologically appropriate concentrations, but not by alcohols up to very high doses (Li, Chiara, Cohen, & Olsen, 2010).

In fact, the subunit interface binding etomidate is the same interface where the GABA binding site is located in the ECD about 50 Å above the etomidate site in the TMD, and suggests that intersubunit interfaces are the usual site for ligand binding pockets, both allosteric and agonist (Li et al., 2006). The importance of this conclusion cannot be underestimated. Note that the GABA binding site at subunit interfaces (Amin & Weiss, 1993; Venkatachalan & Czajkowski, 2012) involves residues at positions in the subunit sequence that are homologous with nAChR, GlyR, and 5-HT3R, i.e., the "cys-loop, pentameric LGIC" superfamily (Galzi & Changeux, 1994). We mentioned that the BZ binding sites on GABAR are modified agonist sites, located at different subunit interfaces in the ECD not used by GABA sites (Smith & Olsen, 1995).

Additional important information about these TMD modulatory sites was obtained using another aziridine affinity label analogue of a barbiturate PAM on $GABA_ARs$ (Chiara et al., 2013); interestingly, residues labeled covalently by the barbiturate ligand were positioned at other intersubunit interfaces homologous to the two that bind etomidate ($\beta+/\alpha-$), namely, the $\alpha+/\beta-$ and $\gamma+/\beta-$. The etomidate sites have low affinity for barbiturates and the barbiturate sites have low affinity for etomidate, while both show mutual allosteric inhibition. Just published is a follow-up from that group (Cohen and Miller) showing that an affinity label analogue of propofol binds to all five subunit interfaces (in recombinant α/β $GABA_AR$ lacking γ or δ) employing the pocket residues identified for etomidate and barbiturates (Jayakar et al., 2014). Thus, the sensitivity to these modulators is quite dependent on the subunit composition of the $GABA_AR$ subtype and particularly important to subtype selectivity for PAMs is the heteropentameric nature of most native cys-loop, pentameric LGIC (Chiara et al., 2013; Forman & Miller, 2011; Jayakar et al., 2014; Li et al., 2006; Olsen & Li, 2011; Olsen et al., 2014). It will be interesting to see if the remaining TMD interface $\alpha+/\gamma-$ in native $\alpha/\beta/\gamma$ and $\alpha+/\delta-$ in $\alpha/\beta/\delta$ $GABA_AR$ carry any modulatory sites. On the other hand, another group, using a different propofol-based affinity label, identified a single residue at the ECD surface of TMD-M3 (Yip et al., 2013) which is distinct from those identified by Jayakar et al. (2014).

Looking more closely at Fig. 1, I have shown the location of the known ligand binding sites, both ECD and TMD, on the homology model of the heteropentameric $\alpha/\beta/\gamma$ native $GABA_AR$ using the templates for the homomeric β3 human $GABA_AR$ (Miller & Aricescu, 2014), the prokaryotic LGIC (Corringer et al., 2010), and the invertebrate nematode GluCl (Hibbs & Gouaux, 2011). The left part shows the transmembrane portion (TMD) with the binding sites for various PAMs such as general anesthetics of varying chemotype as described above. The right part shows the ECD with the binding sites for the agonist GABA as well as the two important PAMs, the BZ and high-affinity EtOH as described above.

2. CONCLUSION

Note that the native GABAR are primarily heteropentamers, and all the templates are (so far) homomers. Thus, the heteromers, unlike homomers, exhibit different sorts of subunit interfaces and ligand binding sites depending on the subunit composition, i.e., $GABA_AR$ subtypes. This model (Fig. 1) can be considered a rough first draft with new experimental evidence hopefully smoothing out the roughness in the near future.

CONFLICT OF INTEREST

R. W. O. reports no conflict of interest.

ACKNOWLEDGMENTS

Special thanks to Dr. Martin Wallner for many years of helpful discussions, and assistance with graphics and references. Thanks to Drs. Guo-Dong Li, Jing Liang, and Kerstin Lindemeyer for helpful discussions. Thanks to Professor Jean-Pierre Changeux for mentoring and encouragement. This study was supported by grants from the US National Institutes of Health.

REFERENCES

Amin, J., & Weiss, D. S. (1993). $GABA_A$ receptor needs two homologous domains of the β-subunit for activation by GABA but not by pentobarbital. *Nature, 366*, 565–569.

Arolfo, M. P., Overstreet, D. H., Yao, L., Fan, P., Lawrence, A. J., Tao, G., et al. (2009). Suppression of heavy drinking and alcohol seeking by a selective ALDH-2 inhibitor. *Alcoholism, Clinical and Experimental Research, 33*, 1935–1944.

Bali, M., & Akabas, M. H. (2004). Defining the propofol binding site location on the $GABA_A$ receptor. *Molecular Pharmacology, 65*, 68–76.

Bali, M., Jansen, M., & Akabas, M. H. (2009). GABA-induced intersubunit conformational movement in the $GABA_A$ receptor α1M1-β2M3 transmembrane subunit interface: Experimental basis for homology modeling of an intravenous anesthetic binding site. *The Journal of Neuroscience, 29*, 3083–3092.

Barbaccia, M. L., Serra, M., Purdy, R. H., & Biggio, G. (2001). Stress and neuroactive steroids. *International Review of Neurobiology, 46*, 243–272.

Barnard, E. A., Skolnick, P., Olsen, R. W., Möhler, H., Sieghart, W., Biggio, G., et al. (1998). International Union of Pharmacology. XV. Subtypes of $GABA_A$ receptors: Classification on the basis of subunit structure and receptor function. *Pharmacological Reviews, 50*, 291–313.

Beaumont, K., Chilton, W. S., Yamamura, H. I., & Enna, S. J. (1978). Muscimol binding in rat brain: Association with synaptic GABA receptors. *Brain Research, 148*, 153–162.

Belelli, D., Lambert, J. J., Peters, J. A., Wafford, K., & Whiting, P. J. (1997). The interaction of the general anesthetic etomidate with the $GABA_A$ receptor is influenced by a single amino acid. *Proceedings of the National Academy of Sciences of the United States of America, 94*, 11031–11036.

Benke, D., Fakitsas, P., Roggenmoser, C., Michel, C., Rudolph, U., & Möhler, H. (2004). Analysis of the presence and abundance of $GABA_A$ receptors containing two different types of α subunits in murine brain using point-mutated α subunits. *The Journal of Biological Chemistry, 279*, 43654–43660.

Bertaccini, E. J., Yoluk, O., Lindahl, E. R., & Trudell, J. R. (2013). Assessment of homology templates and an anesthetic binding site within the $GABA_A$ receptor. *Anesthesiology, 119*, 1087–1095.

Bianchi, M. T., & Macdonald, R. L. (2003). Neurosteroids shift partial agonist activation of $GABA_A$ receptor channels from low- to high-efficacy gating patterns. *The Journal of Neuroscience, 23*, 10934–10943.

Boehm, S. L., 2nd, Ponomarev, I., Jennings, A. W., Whiting, P. J., Rosahl, T. W., Garrett, E. M., et al. (2004). $GABA_A$ receptor subunit mutant mice: New perspectives on alcohol actions. *Biochemical Pharmacology, 68*, 1581–1602.

Bouchet, M. J., Rendon, A., Wermuth, C. G., Goeldner, M., & Hirth, C. (1987). Aryl diazo compounds and diazonium salts as potential irreversible probes of the gamma-aminobutyric acid receptor. *Journal of Medicinal Chemistry, 30*, 2222–2227.

Bowery, N. G., Bettler, B., Froestl, W., Gallagher, J. P., Marshall, F., Raiteri, M., et al. (2002). International Union of Pharmacology. XXXIII. Mammalian $GABA_B$ receptors: Structure and function. *Pharmacological Reviews, 54*, 247–264.

Bowery, N. G., Collins, J. F., & Hill, R. G. (1976). Bicyclic phosphorus esters that are potent convulsants and GABA antagonists. *Nature, 261*, 601–603.

Bowery, N. G., & Dray, A. (1976). Barbiturate reversal of amino acid antagonism produced by convulsant agents. *Nature, 264*, 276–278.

Braestrup, C., & Nielsen, M. (1981). GABA reduces binding of ^{3}H-methyl beta-carboline-3-carboxylate to brain benzodiazepine receptors. *Nature, 294*, 472–474.

Braestrup, C., Nielsen, M., Krogsgaard-Larsen, P., & Falch, E. (1979). Partial agonists for brain GABA/benzodiazepine receptor complex. *Nature, 280*, 331–333.

Braestrup, C., Nielsen, M., & Olsen, C. E. (1980). Urinary and brain beta-carboline-3-carboxylates as potent inhibitors of brain benzodiazepine receptors. *Proceedings of the National Academy of Sciences of the United States of America*, 77, 2288–2292.

Braestrup, C., Schmiechen, R., Neef, G., Nielsen, M., & Petersen, E. N. (1982). Interaction of convulsive ligands with benzodiazepine receptors. *Science, 216*, 1241–1243.

Braestrup, C., & Squires, R. F. (1977). Specific benzodiazepine receptors in rat brain characterized by high-affinity [^{3}H]diazepam binding. *Proceedings of the National Academy of Sciences of the United States of America, 74*, 3805–3809.

Brejc, K., van Dijk, W. J., Klaassen, R. V., Schuurmans, M., van Der Oost, J., Smit, A. B., et al. (2001). Crystal structure of an ACh-binding protein reveals the ligand-binding domain of nicotinic receptors. *Nature, 411*, 269–276.

Carlson, B. X., Engblom, A. C., Kristiansen, U., Schousboe, A., & Olsen, R. W. (2000). A single glycine residue at the entrance to the first membrane-spanning domain of the gamma-aminobutyric acid type A receptor beta(2) subunit affects allosteric sensitivity to GABA and anesthetics. *Molecular Pharmacology, 57*, 474–484.

Casida, J. E. (1993). Insecticide action at the GABA-gated chloride channel: Recognition, progress, and prospects. *Archives of Insect Biochemistry and Physiology, 22*, 13–23.

Chandra, D., Jia, F., Liang, J., Peng, Z., Suryanarayanan, A., Werner, D. F., et al. (2006). $GABA_A$ receptor α4 subunits mediate extrasynaptic inhibition in thalamus and dentate gyrus and the action of gaboxadol. *Proceedings of the National Academy of Sciences of the United States of America, 103*, 15230–15235.

Chang, C. S., Olcese, R., & Olsen, R. W. (2003). A single M1 residue in the β2 subunit alters channel gating of $GABA_A$ receptor in anesthetic modulation and direct activation. *The Journal of Biological Chemistry, 278*, 42821–42828.

Chen, L., Durkin, K. A., & Casida, J. E. (2006). Structural model for γ-aminobutyric acid receptor noncompetitive antagonist binding: Widely diverse structures fit the same site.

Proceedings of the National Academy of Sciences of the United States of America, 103, 5185–5190.

Chen, Z.-W., Manion, B., Townsend, R. R., Reichert, D. E., Covey, D. F., Steinbach, J. H., et al. (2012). Neurosteroid analog photolabeling of a site in the third transmembrane domain of the β3 subunit of the GABA$_A$ receptor. *Molecular Pharmacology, 82,* 408–419.

Chiara, D. C., Jayakar, S. S., Zhou, X., Zhang, X., Savechenkov, P. Y., Bruzik, K. S., et al. (2013). Specificity of intersubunit general anesthetic-binding sites in the transmembrane domain of the human α1β3γ2 GABA$_A$ receptor. *The Journal of Biological Chemistry, 288,* 19343–19357.

Christopoulos, A., Changeux, J.-P., Catterall, W. A., Febbro, D., Burns, T. P., Cidlowski, J. A., et al. (2014). International Union of Basic and Clinical Pharmacology. Multi-site pharmacology: Recommendations for the nomenclature of receptor allosterism and allosteric ligands. *Pharmacological Reviews, 66,* 918–947.

Collinson, N., Kuenzi, F. M., Jarolimek, W., Maubach, K. A., Cothliff, R., Sur, C., et al. (2002). Enhanced learning and memory and altered GABAergic synaptic transmission in mice lacking the α5 subunit of the GABA$_A$ receptor. *The Journal of Neuroscience, 22,* 5572–5580.

Concas, A., Mostallino, M. C., Porcu, P., Follesa, P., Barbaccia, M. L., Trabucchi, M., et al. (1998). Role of brain allopregnanolone in the plasticity of GABA$_A$ receptor in rat brain during pregnancy and after delivery. *Proceedings of the National Academy of Sciences of the United States of America, 95,* 13284–13289.

Cook, J. B., Foster, K. L., Eiler, W. J., 2nd, McKay, P. F., Woods, J., 2nd, Harvey, S. C., et al. (2005). Selective GABA$_A$ α5 benzodiazepine inverse agonist antagonizes the neurobehavioral actions of alcohol. *Alcoholism, Clinical and Experimental Research, 29,* 1390–1401.

Corringer, P.-J., Baaden, M., Bocquet, N., Delarue, M., Dufresne, V., Nury, H., et al. (2010). Atomic structure and dynamics of pentameric ligand-gated ion channels: New insight from bacterial homologues. *Journal of Physiology (London), 588,* 565–572.

Costa, E., & Guidotti, A. (1979). Molecular mechanisms in the receptor action of benzodiazepines. *Annual Review of Pharmacology and Toxicology, 19,* 531–545.

Curtis, D. R., & Johnston, G. A. (1974). Amino acid transmitters in the mammalian central nervous system. *Ergebnisse der Physiologie, Biologischen Chemie und Experimentellen Pharmakologie, 69,* 97–188.

Dawson, G. R., Maubach, K. A., Collinson, N., Cobain, M., Everitt, B. J., MacLeod, A. M., et al. (2006). An inverse agonist selective for α5 subunit-containing GABA$_A$ receptors enhances cognition. *The Journal of Pharmacology and Experimental Therapeutics, 316,* 1335–1345.

Desai, R., Ruesch, D., & Forman, S. A. (2009). GABA$_A$ receptor mutations at β2N265 alter etomidate efficacy while preserving basal and agonist-dependent activity. *Anesthesiology, 111,* 774–784.

Dixon, C. I., Walker, S. E., King, S. L., & Stephens, D. N. (2012). Deletion of the gabra2 gene results in hypersensitivity to the acute effects of ethanol but does not alter ethanol self administration. *PLoS One, 7,* e47135.

Duncalfe, L. L., Carpenter, M. R., Smillie, L. B., Martin, I. L., & Dunn, S. M. (1996). The major site of photoaffinity labeling of the GABA$_A$ receptor by [^{3}H]flunitrazepam is histidine 102 of the alpha subunit. *The Journal of Biological Chemistry, 271,* 9209–9214.

Dutertre, S., Becker, C. M., & Betz, H. (2012). Inhibitory glycine receptors: An update. *The Journal of Biological Chemistry, 287,* 40216–40223.

Egebjerg, J., Schousboe, A., & Krogsgaard-Larsen, P. (Eds.), (2002). *Glutamate and GABA receptors and transporters: Structure, function and pharmacology.* London: Taylor & Francis.

Engblom, A. C., Carlson, B. X., Olsen, R. W., Schousboe, A., & Kristiansen, U. (2002). Point mutation in the first transmembrane region of the beta 2 subunit of the gamma-aminobutyric acid type A receptor alters desensitization kinetics of gamma-aminobutyric acid- and anesthetic-induced channel gating. *The Journal of Biological Chemistry*, *277*, 17438–17447.

Engin, E., Bakhurin, K. I., Smith, K. S., Hines, R. M., Reynolds, L. M., Tang, W., et al. (2014). Neural basis of benzodiazepine reward: Requirement for α2 containing $GABA_A$ receptors in the nucleus accumbens. *Neuropsychopharmacology*, *39*, 1805–1815.

Enna, S. J. (Ed.), (1983). *The GABA receptors*. Totowa, NJ: Humana Press.

Enna, S. J., & Snyder, S. H. (1975). Properties of γ-aminobutyric acid (GABA) receptor binding in rat brain synaptic membrane fractions. *Brain Research*, *100*, 81–97.

Ernst, M., Brauchart, D., Boresch, S., & Sieghart, W. (2003). Comparative modeling of $GABA_A$ receptors: Limits, insights, future developments. *Neuroscience*, *119*, 933–943.

Ernst, M., Bruckner, S., Boresch, S., & Sieghart, W. (2005). Comparative models of $GABA_A$ receptor extracellular and transmembrane domains: Important insights in pharmacology and function. *Molecular Pharmacology*, *68*, 1291–1300.

Ffrench-Constant, R. H., Rocheleau, T. A., Steichen, J. C., & Chalmers, A. E. (1993). A point mutation in a Drosophila GABA receptor confers insecticide resistance. *Nature*, *363*, 449–451.

Forman, S. A., & Miller, K. W. (2011). Anesthetic sites and allosteric mechanisms of action on Cys-loop ligand-gated ion channels. *Canadian Journal of Anaesthesia*, *58*, 191–205.

Galzi, J.-L., & Changeux, J.-P. (1994). Ligand-gated ion channels as unconventional allosteric proteins. *Current Opinion in Structural Biology*, *4*, 554–565.

Gavish, M., & Snyder, S. H. (1980). Benzodiazepine recognition sites on GABA receptors. *Nature*, *287*, 651–652.

Gee, K. W., Bolger, M. B., Brinton, R. E., Coirini, H., & McEwen, B. S. (1988). Steroid modulation of the chloride ionophore in rat brain: Structure-activity requirements, regional dependence and mechanism of action. *The Journal of Pharmacology and Experimental Therapeutics*, *246*, 803–812.

Greenlee, D. V., Van Ness, P. C., & Olsen, R. W. (1978). Gamma-aminobutyric acid binding in mammalian brain: Receptor-like specificity of sodium-independent sites. *Journal of Neurochemistry*, *31*, 933–938.

Günther, U., Benson, J., Benke, D., Fritschy, J. M., Reyes, G., Knoflach, F., et al. (1995). Benzodiazepine-insensitive mice generated by targeted disruption of the γ2 subunit gene of $GABA_A$ receptors. *Proceedings of the National Academy of Sciences of the United States of America*, *92*, 7749–7753.

Haefely, W. (1982). Neurophysiology of benzodiazepines: Summary. In E. Usdin, et al. (Eds.), *Pharmacology of benzodiazepines* (pp. 509–516). Surrey: Macmillan.

Hanchar, H. J., Chutsrinopkun, P., Meera, P., Supavilai, P., Sieghart, W., Wallner, M., et al. (2006). Ethanol potently and competitively inhibits binding of the alcohol antagonist Ro15-4513 to α4/6β3δ $GABA_A$ receptors. *Proceedings of the National Academy of Sciences of the United States of America*, *103*, 8546–8550.

Hanchar, H. J., Dodson, P. D., Olsen, R. W., Otis, T. S., & Wallner, M. (2005). Alcohol-induced motor impairment caused by increased extrasynaptic $GABA_A$ receptor activity. *Nature Neuroscience*, *8*, 339–345.

Harrison, N. J., & Lummis, S. C. (2006). Locating the carboxylate group of GABA in the homomeric rho $GABA_A$ receptor ligand-binding pocket. *The Journal of Biological Chemistry*, *281*, 24455–24461.

Harrison, N. L., & Simmonds, M. A. (1984). Modulation of the GABA receptor complex by a steroid anaesthetic. *Brain Research*, *323*, 287–292.

Hevers, W., & Lüddens, H. (1998). The diversity of $GABA_A$ receptors. Pharmacological and electrophysiological properties of $GABA_A$ channel subtypes. *Molecular Neurobiology*, *18*, 35–86.

Hibbs, R. E., & Gouaux, E. (2011). Principles of activation and permeation in an anion-selective Cys-loop receptor. *Nature, 474*, 54–60.

Homanics, G. E., DeLorey, T. M., Firestone, L. L., Quinlan, J. J., Handforth, A., Harrison, N. L., et al. (1997). Mice devoid of $GABA_A$ receptor β3 subunit have epilepsy, cleft palate, and hypersensitive behavior. *Proceedings of the National Academy of Sciences of the United States of America, 94*, 4143–4148.

Hosie, A. M., Wilkins, M. E., da Silva, H. M., & Smart, T. G. (2006). Endogenous neurosteroids regulate $GABA_A$ receptors through two discrete transmembrane sites. *Nature, 444*, 486–489.

Hunkeler, W., Möhler, H., Pieri, L., Polc, P., Bonetti, E. P., Cumin, R., et al. (1981). Selective antagonists of benzodiazepines. *Nature, 290*, 514–516.

Ikeda, T., Zhao, X., Nagata, K., Kono, Y., Shono, T., Yeh, J. Z., et al. (2001). Fipronil modulation of $GABA_A$ receptors in rat dorsal root ganglion neurons. *The Journal of Pharmacology and Experimental Therapeutics, 296*, 914–921.

Jarboe, C. H., Poerter, L. A., & Buckler, R. T. (1968). Structural aspects of picrotoxinin action. *Journal of Medicinal Chemistry, 11*, 729–731.

Jayakar, S. S., Zhou, X., Chiara, D. C., Dostalova, Z., Savechenkov, P. Y., Bruzik, K. S., et al. (2014). Multiple propofol binding sites in a $GABA_A$ receptor identified using a photoreactive propofol analog. *Journal of Biological Chemistry, 289*, 27456–27468.

Jenkins, A., Greenblatt, E. P., Faulkner, H. J., Bertaccini, E., Light, A., Lin, A., et al. (2001). Evidence for a common binding cavity for three general anesthetics within the $GABA_A$ receptor. *The Journal of Neuroscience, 21*, RC136.

Jensen, M. L., Wafford, K. A., Brown, A. R., Belelli, D., Lambert, J. J., & Mirza, N. R. (2013). A study of subunit selectivity, mechanism and site of action of the δ selective compound 2 (DS2) at human recombinant and rodent native $GABA_A$ receptors. *British Journal of Pharmacology, 168*, 1118–1132.

Johnston, G. A. R. (1997). $GABA_C$ receptors: Molecular biology, pharmacology, and physiology. In S. J. Enna, & N. G. Bowery (Eds.), *The GABA receptors: Vol. 2* (pp. 297–323). Totowa: Humana Press.

Johnston, G. A. R. (2000). GABA chemistry: Analogs of GABA as therapeutic and investigational agents. In D. L. Martin, & R. W. Olsen (Eds.), *GABA in the nervous system: The view at fifty years* (pp. 65–80). Philadelphia: Lippincott Williams & Wilkins.

Johnston, G. A. R., Allan, R. D., Kennedy, S. M. E., & Twitchin, B. (1979). Systematic study of GABA analogues of restricted conformation. In H. Kofod, P. Krogsgaard-Larsen, & J. Scheel-Krüger (Eds.), *GABA-neurotransmitters* (pp. 149–164). Copenhagen: Munksgaard.

Jurd, R., Arras, M., Lambert, S., Drexler, B., Siegwart, R., Crestani, F., et al. (2003). General anesthetic actions in vivo strongly attenuated by a point mutation in the $GABA_A$ receptor β3 subunit. *FASEB Journal, 17*, 250–252.

Kahnberg, P., Lager, E., Rosenberg, C., Schougaard, J., Camet, L., Sterner, O., et al. (2002). Refinement and evaluation of a pharmacophore model for flavone derivatives binding to the benzodiazepine site of the $GABA_A$ receptor. *Journal of Medicinal Chemistry, 45*, 4188–4201.

Karobath, M., Placheta, P., Lippitsch, M., & Krogsgaard-Larsen, P. (1979). Is stimulation of benzodiazepine receptor binding mediated by a novel GABA receptor? *Nature, 278*, 748–749.

Kash, T. L., Jenkins, A., Kelley, J. C., Trudell, J. R., & Harrison, N. L. (2003). Coupling of agonist binding to channel gating in the $GABA_A$ receptor. *Nature, 421*, 272–275.

King, R. G., Nielsen, M., Stauber, G. B., & Olsen, R. W. (1987). Convulsant/barbiturate activity on the soluble gamma-aminobutyric acid-benzodiazepine receptor complex. *European Journal of Biochemistry, 169*, 555–562.

Korpi, E. R., Kleingoor, C., Kettenmann, H., & Seeburg, P. H. (1993). Benzodiazepine-induced motor impairment linked to point mutation in cerebellar $GABA_A$ receptor. *Nature, 361*, 356–359.

Krasowski, M. D., Nishikawa, K., Nikolaeva, N., Lin, A., & Harrison, N. L. (2001). Methionine 286 in transmembrane domain 3 of the $GABA_A$ receptor β subunit controls a binding cavity for propofol and other alkylphenol general anesthetics. *Neuropharmacology, 41,* 952–964.

Krogsgaard-Larsen, P., Frolund, B., Liljefors, T., & Ebert, B. (2004). $GABA_A$ agonists and partial agonists: THIP (Gaboxadol) as a non-opioid analgesic and a novel type of hypnotic. *Biochemical Pharmacology, 68,* 1573–1580.

Krogsgaard-Larsen, P., Honore, T., & Thyssen, K. (1979). GABA receptor agonists: Design and structure-activity studies. In H. Kofod, P. Krogsgaard-Larsen, & J. Scheel-Krüger (Eds.), *GABA-neurotransmitters* (pp. 201–216). Copenhagen: Munksgaard.

Krogsgaard-Larsen, P., & Johnston, G. A. (1978). Structure-activity studies on the inhibition of GABA binding to rat brain membranes by muscimol and related compounds. *Journal of Neurochemistry, 30,* 1377–1382.

Kumar, S., Porcu, P., Werner, D. F., Matthews, D. B., Diaz-Granados, J. L., Helfand, R. S., et al. (2009). The role of $GABA_A$ receptors in the acute and chronic effects of ethanol: A decade of progress. *Psychopharmacology, 205,* 529–564.

Lambert, J. J., Belelli, D., Peden, D. R., Vardy, A. W., & Peters, J. A. (2003). Neurosteroid modulation of $GABA_A$ receptors. *Progress in Neurobiology, 71,* 67–80.

Leeb-Lundberg, F., Snowman, A., & Olsen, R. W. (1980). Barbiturate receptor sites are coupled to benzodiazepine receptors. *Proceedings of the National Academy of Sciences of the United States of America,* 77, 7468–7472.

Leeb-Lundberg, F., Snowman, A., & Olsen, R. W. (1981). Interaction of anticonvulsants with the barbiturate-benzodiazepine-GABA receptor complex. *European Journal of Pharmacology, 72,* 125–129.

Li, G.-D., Chiara, D. C., Cohen, J. B., & Olsen, R. W. (2009). Neurosteroids allosterically modulate binding of the anesthetic etomidate to $GABA_A$ receptors. *The Journal of Biological Chemistry, 284,* 11771–11775.

Li, G.-D., Chiara, D. C., Cohen, J. B., & Olsen, R. W. (2010). Numerous classes of general anesthetics inhibit etomidate binding to $GABA_A$ receptors. *The Journal of Biological Chemistry, 285,* 8615–8620.

Li, G.-D., Chiara, D. C., Sawyer, G. W., Husain, S. S., Olsen, R. W., & Cohen, J. B. (2006). Identification of a $GABA_A$ receptor anesthetic binding site at subunit interfaces by photolabeling with an etomidate analog. *The Journal of Neuroscience, 26,* 11599–11605.

Liang, J., Zhang, N., Cagetti, E., Houser, C. R., Olsen, R. W., & Spigelman, I. (2006). Chronic intermittent ethanol-induced switch of ethanol actions from extrasynaptic to synaptic hippocampal $GABA_A$ receptors. *The Journal of Neuroscience, 26,* 1749–1758.

Liu, J., Yang, A. R., Kelly, T., Puche, A., Esoga, C., June, H. L., Jr., et al. (2011). Binge alcohol drinking is associated with $GABA_A$ α2-regulated Toll-like receptor 4 (TLR4) expression in the central amygdala. *Proceedings of the National Academy of Sciences of the United States of America, 108,* 4465–4470.

Lloyd, K. G., & Zivkovic, B. (1988). Specificity within the $GABA_A$ receptor supramolecular complex: A consideration of the new omega 1-receptor selective imidazopyridine hypnotic zolpidem. *Pharmacology, Biochemistry, and Behavior, 29,* 781–783.

Lo, M. M., Niehoff, D. L., Kuhar, M. J., & Snyder, S. H. (1983). Differential localization of type I and type II benzodiazepine binding sites in substantia nigra. *Nature, 306,* 57–60.

Lovinger, D. M., & Homanics, G. E. (2007). Editorial: Tonic for what ails us: High affinity $GABA_A$ receptors and alcohol. *Alcohol, 41,* 139–143.

Low, K., Crestani, F., Keist, R., Benke, D., Brunig, I., Benson, J. A., et al. (2000). Molecular and neuronal substrate for the selective attenuation of anxiety. *Science, 290,* 131–134.

Lüddens, H., Korpi, E. R., & Seeburg, P. H. (1995). GABA$_A$/benzodiazepine receptor heterogeneity: Neurophysiological implications. *Neuropharmacology, 34*, 245–254.

Lukas, S. E., Penetar, D., Su, Z., Geaghan, T., Maywalt, M., Tracy, M., et al. (2013). A standardized kudzu extract (NPI-031) reduces alcohol consumption in nontreatment-seeking male heavy drinkers. *Psychopharmacology, 226*, 65–73.

Macdonald, R. L., & Olsen, R. W. (1994). GABA$_A$ receptor channels. *Annual Review of Neuroscience, 17*, 569–602.

Maguire, J. L., Stell, B. M., Rafizadeh, M., & Mody, I. (2005). Ovarian cycle-linked changes in GABA$_A$ receptors mediating tonic inhibition alter seizure susceptibility and anxiety. *Nature Neuroscience, 8*, 797–804.

Majewska, M. D., Harrison, N. L., Schwartz, R. D., Barker, J. L., & Paul, S. M. (1986). Steroid hormone metabolites are barbiturate-like modulators of the GABA receptor. *Science, 232*, 1004–1007.

Mascia, M. P., Trudell, J. R., & Harris, R. A. (2000). Specific binding sites for alcohols and anesthetics on ligand-gated ion channels. *Proceedings of the National Academy of Sciences of the United States of America, 97*, 9305–9310.

McHugh, T. J., Jones, M. W., Quinn, J. J., Balthasar, N., Coppari, R., Elmquist, J. K., et al. (2007). Dentate gyrus NMDA receptors mediate rapid pattern separation in the hippocampal network. *Science, 317*, 94–99.

McKernan, R. M., Rosahl, T. W., Reynolds, D. S., Sur, C., Wafford, K. A., Atack, J. R., et al. (2000). Sedative but not anxiolytic properties of benzodiazepines are mediated by the GABA$_A$ receptor α1 subtype. *Nature Neuroscience, 3*, 587–592.

Meera, P., Olsen, R. W., Otis, T. S., & Wallner, M. (2010). Alcohol- and alcohol antagonist-sensitive human GABA$_A$ receptors: Tracking δ subunit incorporation into functional receptors. *Molecular Pharmacology, 78*, 918–924.

Meera, P., Wallner, M., & Otis, T. S. (2011). Molecular basis for the high THIP/gaboxadol sensitivity of extrasynaptic GABA$_A$ receptors. *Journal of Neurophysiology, 106*, 2057–2064.

Meunier, J. C., Olsen, R., Menez, A., Morgat, J. L., Fromageot, P., Ronseray, A. M., et al. (1971). Some physical properties of the acetylcholine protein receptor studied with a radioactive neurotoxin. *Comptes Rendus Hebdomadaires des Séances de l'Académie des Sciences, 273*, 595–598.

Middendorp, S. J., Hurni, E., Schonberger, M., Stein, M., Pangerl, M., Trauner, D., et al. (2014). Relative positioning of classical benzodiazepines to the gamma-subunit of GABA receptors. *ACS Chemical Biology, 15*, 1846–1853.

Mihalek, R. M., Banerjee, P. K., Korpi, E. R., Quinlan, J. J., Firestone, L. L., Mi, Z. P., et al. (1999). Attenuated sensitivity to neuroactive steroids in GABA$_A$ receptor δ subunit knockout mice. *Proceedings of the National Academy of Sciences of the United States of America, 96*, 12905–12910.

Mihic, S. J., Ye, Q., Wick, M. J., Koltchine, V. V., Krasowski, M. D., Finn, S. E., et al. (1997). Sites of alcohol and volatile anaesthetic action on GABA$_A$ and glycine receptors. *Nature, 389*, 385–389.

Miller, P. S., & Aricescu, A. R. (2014). Crystal structure of a human GABA receptor. *Nature, 512*, 270–275.

Möhler, H. (2012). The GABA system in anxiety and depression and its therapeutic potential. *Neuropharmacology, 62*, 42–53.

Möhler, H. (2014). Endogenous benzodiazepine site peptide ligands operating bidirectionally in vivo in neurogenesis and thalamic oscillations. *Neurochemical Research, 39*, 1032–1036.

Möhler, H., Battersby, M. K., & Richards, J. G. (1980). Benzodiazepine receptor protein identified and visualized in brain tissue by a photoaffinity label. *Proceedings of the National Academy of Sciences of the United States of America, 77*, 1666–1670.

Möhler, H., & Okada, T. (1977a). Benzodiazepine receptor: Demonstration in the central nervous system. *Science, 198*, 849–851.

Möhler, H., & Okada, T. (1977b). GABA receptor binding with ^{3}H (+) bicuculline-methiodide in rat CNS. *Nature, 267*, 65–67.

Morrow, A. L., Janis, G. C., VanDoren, M. J., Matthews, D. B., Samson, H. H., Janak, P. H., et al. (1999). Neurosteroids mediate pharmacological effects of ethanol: A new mechanism of ethanol action? *Alcoholism, Clinical and Experimental Research, 23*, 1933–1940.

Ng, C. K., Kim, H. L., Gavande, N., Yamamoto, I., Kumar, R. J., Mewett, K. N., et al. (2011). Medicinal chemistry of rho $GABA_C$ receptors. *Future Medicinal Chemistry, 3*, 197–209.

Nie, H., Rewal, M., Gill, T. M., Ron, D., & Janak, P. H. (2011). From the cover: Extrasynaptic δ-containing $GABA_A$ receptors in the nucleus accumbens dorsomedial shell contribute to alcohol intake. *Proceedings of the National Academy of Sciences of the United States of America, 108*, 4459–4464.

Olsen, R. W. (1982). Drug interactions at the GABA receptor-ionophore complex. *Annual Review of Pharmacology and Toxicology, 22*, 245–277.

Olsen, R. W. (2014). Analysis of γ-aminobutyric acid (GABA) type A receptor subtypes using isosteric and allosteric ligands. *Neurochemical Research, 39*(10), 1924–1941, Special issue in honor of Povl Krogsgaard-Larsen.

Olsen, R. W., Bergman, M. O., Van Ness, P. C., Lummis, S. C., Watkins, A. E., Napias, C., et al. (1981). γ-Aminobutyric acid receptor binding in mammalian brain: Heterogeneity of binding sites. *Molecular Pharmacology, 19*, 217–227.

Olsen, R. W., Bureau, M., Houser, C. R., Delgado-Escueta, A. V., Richards, J. G., & Möhler, H. (1992). GABA/benzodiazepine receptors in human focal epilepsy. *Epilepsy Research. Supplement, 8*, 383–391.

Olsen, R. W., Cagetti, E., & Wallner, M. (2003). Subunit specificity of steroid modulation of $GABA_A$ receptors. In S. S. Smith (Ed.), *Neurosteroid effects in the central nervous system: The role of the $GABA_A$ receptor* (pp. 47–61). Boca Raton: CRC Press.

Olsen, R. W., Fischer, J. B., & Dunwiddie, T. V. (1986). Barbiturate enhancement of GABA receptor binding and function as a mechanism of anesthesia. In S. Roth, & K. W. Miller (Eds.), *Molecular and cellular mechanisms of anaesthetics* (pp. 165–177). New York: Plenum Publishing Corporation.

Olsen, R. W., & Homanics, G. E. (2000). Function of $GABA_A$ receptors. Insights from mutant and knockout mice. In D. L. Martin, & R. W. Olsen (Eds.), *GABA in the nervous system: The view at fifty years* (pp. 81–96). Philadelphia, PA: Lippincott Williams & Wilkins.

Olsen, R. W., & Li, G.-D. (2011). $GABA_A$ receptors as molecular targets of general anesthetics: Identification of binding sites provides clues to allosteric modulation. *Canadian Journal of Anaesthesia, 58*, 206–215.

Olsen, R. W., Li, G.-D., Wallner, M., Trudell, J. R., Bertaccini, E. J., Lindahl, E., et al. (2014). Structural models of ligand-gated ion channels: Sites of action for anesthetics and ethanol. *Alcoholism, Clinical and Experimental Research, 38*, 595–603.

Olsen, R. W., McCabe, R. T., & Wamsley, J. K. (1990). $GABA_A$ receptor subtypes: Autoradiographic comparison of GABA, benzodiazepine, and convulsant binding sites in the rat central nervous system. *Journal of Chemical Neuroanatomy, 3*, 59–76.

Olsen, R. W., & Sieghart, W. (2008). International Union of Pharmacology. LXX. Subtypes of $GABA_A$ receptors: Classification on the basis of subunit composition, pharmacology, and function. Update. *Pharmacological Reviews, 60*, 243–260.

Olsen, R. W., Snowhill, E. W., & Wamsley, J. K. (1984). Autoradiographic localization of low affinity GABA receptors with [^{3}H]bicuculline methochloride. *European Journal of Pharmacology, 99*, 247–248.

Olsen, R. W., & Snowman, A. M. (1982). Chloride-dependent enhancement by barbiturates of GABA receptor binding. *The Journal of Neuroscience, 2*, 1812–1823.

Olsen, R. W., Yang, J., King, R. G., Dilber, A., Stauber, G. B., & Ransom, R. W. (1986). Barbiturate and benzodiazepine modulation of GABA receptor binding and function. *Life Sciences*, *39*, 1969–1976.

Papadopoulos, V., Berkovich, A., Krueger, K. E., Costa, E., & Guidotti, A. (1991). Diazepam binding inhibitor and its processing products stimulate mitochondrial steroid biosynthesis via an interaction with mitochondrial benzodiazepine receptors. *Endocrinology*, *129*, 1481–1488.

Paul, S. M., & Purdy, R. H. (1992). Neuroactive steroids. *FASEB Journal*, *6*, 2311–2322.

Perkins, D. I., Trudell, J. R., Crawford, D. K., Alkana, R. L., & Davies, D. L. (2010). Molecular targets and mechanisms for ethanol action in glycine receptors. *Pharmacology & Therapeutics*, *127*, 53–65.

Perret, P., Sarda, X., Wolff, M., Wu, T. T., Bushey, D., & Goeldner, M. (1999). Interaction of non-competitive blockers within the $GABA_A$ chloride channel using chemically reactive probes as chemical sensors for cysteine mutants. *The Journal of Biological Chemistry*, *274*, 25350–25354.

Pert, C. B., & Snyder, S. H. (1973). Opiate receptor: Demonstration in nervous tissue. *Science*, *179*, 1011–1014.

Ramerstorfer, J., Furtmuller, R., Sarto-Jackson, I., Varagic, Z., Sieghart, W., & Ernst, M. (2011). The $GABA_A$ receptor α+β- interface: A novel target for subtype selective drugs. *The Journal of Neuroscience*, *31*, 870–877.

Rewal, M., Jurd, R., Gill, T. M., He, D. Y., Ron, D., & Janak, P. H. (2009). Alpha4-containing $GABA_A$ receptors in the nucleus accumbens mediate moderate intake of alcohol. *The Journal of Neuroscience*, *29*, 543–549.

Reynolds, D. S., Rosahl, T. W., Cirone, J., O'Meara, G. F., Haythornthwaite, A., Newman, R. J., et al. (2003). Sedation and anesthesia mediated by distinct $GABA_A$ receptor isoforms. *The Journal of Neuroscience*, *23*, 8608–8617.

Richter, L., de Graaf, C., Sieghart, W., Varagic, Z., Morzinger, M., de Esch, I. J., et al. (2012). Diazepam-bound $GABA_A$ receptor models identify new benzodiazepine binding-site ligands. *Nature Chemical Biology*, *8*, 455–464.

Roberts, E., Chase, T. N., & Tower, D. (Eds.), (1975). *GABA in nervous system function*. New York: Raven Press.

Rudolph, U., & Antkowiak, B. (2004). Molecular and neuronal substrates for general anaesthetics. *Nature Reviews. Neuroscience*, *5*, 709–720.

Rudolph, U., Crestani, F., Benke, D., Brunig, I., Benson, J. A., Fritschy, J. M., et al. (1999). Benzodiazepine actions mediated by specific $GABA_A$ receptor subtypes. *Nature*, *401*, 796–800.

Rudolph, U., & Möhler, H. (2004). Analysis of $GABA_A$ receptor function and dissection of the pharmacology of benzodiazepines and general anesthetics through mouse genetics. *Annual Review of Pharmacology and Toxicology*, *44*, 475–498.

Rudolph, U., & Möhler, H. (2006). GABA-based therapeutic approaches: $GABA_A$ receptor subtype functions. *Current Opinion in Pharmacology*, *6*, 18–23.

Rudolph, U., & Möhler, H. (2014). $GABA_A$ receptor subtypes: Therapeutic potential in Down syndrome, affective disorders, schizophrenia, and autism. *Annual Review of Pharmacology and Toxicology*, *54*, 483–507.

Rüsch, D., Zhong, H., & Forman, S. A. (2004). Gating allosterism at a single class of etomidate sites on $\alpha 1\beta 2\gamma 2L$ $GABA_A$ receptors accounts for both direct activation and agonist modulation. *The Journal of Biological Chemistry*, *279*, 20982–20992.

Sander, T., Frolund, B., Bruun, A. T., Ivanov, I., McCammon, J. A., & Balle, T. (2011). New insights into the $GABA_A$ receptor structure and orthosteric ligand binding: Receptor modeling guided by experimental data. *Proteins*, *79*, 1458–1477.

Santhakumar, V., Wallner, M., & Otis, T. (2007). Ethanol acts directly on extrasynaptic subtypes of $GABA_A$ receptors to increase tonic inhibition. *Alcohol*, *41*, 211–221.

Sawyer, G. W., Chiara, D. C., Olsen, R. W., & Cohen, J. B. (2002). Identification of the bovine $GABA_A$ receptor α subunit residues photolabeled by the imidazobenzodiazepine [^{3}H]Ro15-4513. *The Journal of Biological Chemistry, 277*, 50036–50045.

Schofield, P. R., Darlison, M. G., Fujita, N., Burt, D. R., Stephenson, F. A., Rodriguez, H., et al. (1987). Sequence and functional expression of the $GABA_A$ receptor shows a ligand-gated receptor super-family. *Nature, 328*, 221–227.

Shen, Y., Lindemeyer, A. K., Gonzalez, C., Shao, X. M., Spigelman, I., Olsen, R. W., et al. (2012). Dihydromyricetin as a novel anti-alcohol intoxication medication. *The Journal of Neuroscience, 32*, 390–401.

Sieghart, W., & Karobath, M. (1980). Molecular heterogeneity of benzodiazepine receptors. *Nature, 286*, 285–287.

Sigel, E., & Barnard, E. A. (1984). A γ-aminobutyric acid/benzodiazepine receptor complex from bovine cerebral cortex. Improved purification with preservation of regulatory sites and their interactions. *The Journal of Biological Chemistry, 259*, 7219–7223.

Sigel, E., & Buhr, A. (1997). The benzodiazepine binding site of $GABA_A$ receptors. *Trends in Pharmacological Sciences, 18*, 425–429.

Sigel, E., & Steinmann, M. E. (2012). Structure, function, and modulation of $GABA_A$ receptors. *The Journal of Biological Chemistry, 287*, 40224–40231.

Smart, T. G., & Paoletti, P. (2012). Synaptic neurotransmitter-gated receptors. *Cold Spring Harbor Perspectives in Biology, 4*.

Smith, S. S. (Ed.), (2003). *Neurosteroid effects in the central nervous system: The role of the $GABA_A$ receptor.* Boca Raton, FL: CRC Press.

Smith, G. B., & Olsen, R. W. (1994). Identification of a [^{3}H]muscimol photoaffinity substrate in the bovine $GABA_A$ receptor alpha subunit. *The Journal of Biological Chemistry, 269*, 20380–20387.

Smith, G. B., & Olsen, R. W. (1995). Functional domains of $GABA_A$ receptors. *Trends in Pharmacological Sciences, 16*, 162–168.

Smith, G. B., & Olsen, R. W. (2000). Deduction of amino acid residues in the $GABA_A$ receptor alpha subunits photoaffinity labeled with the benzodiazepine flunitrazepam. *Neuropharmacology, 39*, 55–64.

Smith, S. S., Waterhouse, B. D., & Woodward, D. J. (1987). Locally applied progesterone metabolites alter neuronal responsiveness in the cerebellum. *Brain Research Bulletin, 18*, 739–747.

Squires, R. F., Benson, D. I., Braestrup, C., Coupet, J., Klepner, C. A., Myers, V., et al. (1979). Some properties of brain specific benzodiazepine receptors: New evidence for multiple receptors. *Pharmacology, Biochemistry, and Behavior, 10*, 825–830.

Squires, R. F., Casida, J. E., Richardson, M., & Saederup, E. (1983). [^{35}S]TBPS binds with high affinity to brain-specific sites coupled to $GABA_A$ and ion recognition sites. *Molecular Pharmacology, 23*, 326–336.

Stell, B. M., Brickley, S. G., Tang, C. Y., Farrant, M., & Mody, I. (2003). Neuroactive steroids reduce neuronal excitability by selectively enhancing tonic inhibition mediated by δ subunit-containing $GABA_A$ receptors. *Proceedings of the National Academy of Sciences of the United States of America, 100*, 14439–14444.

Stewart, D., Desai, R., Cheng, Q., Liu, A., & Forman, S. A. (2008). Tryptophan mutations at azi-etomidate photo-incorporation sites on α1 or β2 subunits enhance $GABA_A$ receptor gating and reduce etomidate modulation. *Molecular Pharmacology, 74*, 1687–1695.

Stewart, D. S., Hotta, M., Li, G. D., Desai, R., Chiara, D. C., Olsen, R. W., et al. (2013). Cysteine substitutions define etomidate binding and gating linkages in the α-M1 domain of $GABA_A$ receptors. *The Journal of Biological Chemistry, 288*, 30373–30386.

Study, R. E., & Barker, J. L. (1981). Diazepam and (-)-pentobarbital: Fluctuation analysis reveals different mechanisms for potentiation of γ-aminobutyric acid responses in cultured central neurons. *Proceedings of the National Academy of Sciences of the United States of America*, *78*, 7180–7184.

Sundstrom-Poromaa, I., Smith, D. H., Gong, Q. H., Sabado, T. N., Li, X., Light, A., et al. (2002). Hormonally regulated α4β2δ $GABA_A$ receptors are a target for alcohol. *Nature Neuroscience*, *5*, 721–722.

Supavilai, P., & Karobath, M. (1984). [^{35}S]-t-butylbicyclophosphorothionate binding sites are constituents of the γ-aminobutyric acid benzodiazepine receptor complex. *The Journal of Neuroscience*, *4*, 1193–1200.

Supavilai, P., Mannonen, A., & Karobath, M. (1982). Modulation of GABA binding sites by CNS depressants and CNS convulsants. *Neurochemistry International*, *4*, 259–268.

Suzdak, P. D., Glowa, J. R., Crawley, J. N., Schwartz, R. D., Skolnick, P., & Paul, S. M. (1986). A selective imidazobenzodiazepine antagonist of ethanol in the rat. *Science*, *234*, 1243–1247.

Tallman, J. F., Paul, S. M., Skolnick, P., & Gallager, D. W. (1980). Receptors for the age of anxiety: Pharmacology of the benzodiazepines. *Science*, *207*, 274–281.

Tan, K. R., Gonthier, A., Baur, R., Ernst, M., Goeldner, M., & Sigel, E. (2007). Proximity-accelerated chemical coupling reaction in the benzodiazepine-binding site of $GABA_A$ Receptors: Superposition of different allosteric modulators. *The Journal of Biological Chemistry*, *282*, 26316–26325.

Ticku, M. K., Ban, M., & Olsen, R. W. (1978). Binding of [^{3}H]α-dihydropicrotoxinin, a GABA synaptic antagonist, to rat brain membranes. *Molecular Pharmacology*, *14*, 391–402.

Ticku, M. K., & Olsen, R. W. (1978). Interaction of barbiturates with dihydropicrotoxinin binding sites related to the GABA receptor-ionophore system. *Life Sciences*, *22*, 1643–1651.

Ticku, M. K., & Olsen, R. W. (1979). Cage convulsants inhibit picrotoxinin binding. *Neuropharmacology*, *18*, 315–318.

Turner, D. M., Ransom, R. W., Yang, J. S., & Olsen, R. W. (1989). Steroid anesthetics and naturally-occurring analogs modulate the GABA receptor complex at a site distinct from barbiturates. *The Journal of Pharmacology and Experimental Therapeutics*, *248*, 960–966.

Venkatachalan, S. P., & Czajkowski, C. (2012). Structural link between $GABA_A$ receptor agonist binding site and inner β-sheet governs channel activation and allosteric drug modulation. *The Journal of Biological Chemistry*, *287*, 6714–6724.

Wallner, M., Hanchar, H. J., & Olsen, R. W. (2003). Ethanol enhances α4β3δ and α6β3δ $GABA_A$ receptors at low concentrations known to affect humans. *Proceedings of the National Academy of Sciences of the United States of America*, *100*, 15218–15223.

Wallner, M., Hanchar, H. J., & Olsen, R. W. (2006). Low dose alcohol actions on α4β3δ $GABA_A$ receptors are reversed by the behavioral alcohol antagonist Ro15-4513. *Proceedings of the National Academy of Sciences of the United States of America*, *103*, 8540–8545.

Wallner, M., Hanchar, H. J., & Olsen, R. W. (2014). Alcohol selectivity of β3-containing $GABA_A$ receptors: Evidence for a unique extracellular alcohol/imidazobenzodiazepine Ro15-4513 binding site at the α+β- subunit interface in αβ3δ $GABA_A$ receptors. *Neurochemical Research*, *39*, 1118–1126.

Wallner, M., & Olsen, R. W. (2008). Physiology and pharmacology of alcohol: The imidazobenzodiazepine alcohol antagonist site on subtypes of $GABA_A$ receptors as an opportunity for drug development? *British Journal of Pharmacology*, *154*, 288–298.

Wang, D. S., & Orser, B. A. (2011). Inhibition of learning and memory by general anesthetics. *Canadian Journal of Anaesthesia*, *58*, 167–177.

Wei, W., Faria, L. C., & Mody, I. (2004). Low ethanol concentrations selectively augment the tonic inhibition mediated by δ subunit-containing $GABA_A$ receptors in hippocampal neurons. *The Journal of Neuroscience*, *24*, 8379–8382.

Whiting, P. J. (2003). $GABA_A$ receptor subtypes in the brain: A paradigm for CNS drug discovery? *Drug Discovery Today*, *8*, 445–450.

Whiting, P. J., McKernan, R. M., & Wafford, K. A. (1995). Structure and pharmacology of vertebrate $GABA_A$ receptor subtypes. *International Review of Neurobiology, 38*, 95–138.

Wick, M. J., Mihic, S. J., Ueno, S., Mascia, M. P., Trudell, J. R., Brozowski, S. J., et al. (1998). Mutations of gamma-aminobutyric acid and glycine receptors change alcohol cutoff: Evidence for an alcohol receptor? *Proceedings of the National Academy of Sciences of the United States of America, 95*, 6504–6509.

Wong, E. H., Leeb-Lundberg, L. M., Teichberg, V. I., & Olsen, R. W. (1984). γ-Aminobutyric acid activation of $^{36}Cl^-$ flux in rat hippocampal slices and its potentiation by barbiturates. *Brain Research, 303*, 267–275.

Wong, E. H., Snowman, A. M., Leeb-Lundberg, L. M., & Olsen, R. W. (1984). Barbiturates allosterically inhibit GABA antagonist and benzodiazepine inverse agonist binding. *European Journal of Pharmacology, 102*, 205–212.

Yamamura, H. I., Enna, S. J., & Kuhar, M. J. (Eds.), (1978). *Neurotransmitter receptor binding*. New York: Raven.

Yip, G. M., Chen, Z. W., Edge, C. J., Smith, E. H., Dickinson, R., Hohenester, E., et al. (2013). A propofol binding site on mammalian $GABA_A$ receptors identified by photolabeling. *Nature Chemical Biology, 9*, 715–720.

Zhao, C., Hwang, S. H., Buchholz, B. A., Carpenter, T. S., Lightstone, F., Yang, J., et al. (2014). $GABA_A$ receptor target of tetramethylenedisulfotetramine. *Proceedings of the National Academy of Sciences of the United States of America, 111*, 8607–8612.

Zheleznova, N., Sedelnikova, A., & Weiss, D. S. (2008). $\alpha1\beta2\delta$, a silent $GABA_A$ receptor: Recruitment by tracazolate and neurosteroids. *British Journal of Pharmacology, 153*, 1062–1071.

Zukin, S. R., Young, A. B., & Snyder, S. H. (1974). Gamma-aminobutyric acid binding to receptor sites in the rat central nervous system. *Proceedings of the National Academy of Sciences of the United States of America, 71*, 4802–4807.

CHAPTER EIGHT

Diversity in GABAergic Signaling

Kaspar Vogt[1]
International Institute for Integrative Sleep Medicine (WPI-IIIS), University of Tsukuba, Tsukuba, Ibaraki, Japan
[1]Corresponding author: e-mail address: vogt.kaspar.fu@u.tsukuba.ac.jp

Contents

Abstract

$GABA_A$ receptor-mediated synaptic transmission is responsible for inhibitory control of neural function in the brain. Recent progress has shown that $GABA_A$ receptors also provide a wide range of additional functions beyond simple inhibition. This diversity of functions is mediated by a large variety of different interneuron classes acting on a diverse population of receptor subtypes. Here, I will focus on an additional source of GABAergic signaling diversity, caused by the highly variable ion signaling mechanism of $GABA_A$ receptors. In concert with the other two sources of GABAergic heterogeneity, this variability in signaling allows for a wide array of GABAergic effects that are crucial for the development of the brain and its function.

ABBREVIATIONS

$5HT_3$ 5-hydroxytryptamine (serotonin) type 3
CA1 cornu ammonis area 1
CFP cyan fluorescent protein
CNS central nervous system

Advances in Pharmacology, Volume 73
ISSN 1054-3589
http://dx.doi.org/10.1016/bs.apha.2014.11.009

Cys cysteine
FRET Förster resonance energy transfer
GABA gamma-aminobutyric acid
KCC2 K-Cl cotransporter 2
NKCC1 Na-K-Cl cotransporter 1
YFP yellow fluorescent protein

1. INTRODUCTION

GABAergic interneurons represent roughly one-fifth of the neuronal population in the brain. They show remarkable diversity in their morphology, firing patterns, and cytochemical marker expression. A truly comprehensive classification is still not possible, but significant progress has been made to understand the role of different interneuron subtypes in brain function (Ascoli et al., 2008; DeFelipe et al., 2013).

One of the key targets of released gamma-aminobutyric acid (GABA) is a family of heteropentameric, anion selective channels, so-called $GABA_A$ receptors (Hevers & Luddens, 1998). Together with nicotinic acetylcholine-, serotonin $5HT_3$ and glycine receptors, $GABA_A$ receptors belong to the cysteine (Cys) loop ligand-gated ion channel superfamily (Betz, 1990). A large number of $GABA_A$ receptor subunit isoforms have been found and even though only a fraction of the possible combinations are assembled into functional receptors, a large number of molecularly distinct $GABA_A$ receptor subtypes have been described (Hevers & Luddens, 1998; Rudolph & Mohler, 2004).

On the signaling side, the different receptor subtypes show some differences in GABA affinity, single-channel conductance, and kinetics (Hevers & Luddens, 1998). However, their basic signaling mechanism is the same: Binding of GABA to $GABA_A$ receptors increases the open probability of their central anion selective pore with a dominant (80%) chloride and a smaller (20%) bicarbonate permeability (Bormann, Hamill, & Sakmann, 1987; Kaila, 1994; Kaila & Voipio, 1987). Bicarbonate permeability has not been assessed for many subtypes of $GABA_A$ receptor and may vary according to subunit composition. Channel opening therefore causes the flow of chloride and bicarbonate anions across the cell membrane. An increasing body of literature shows that this can cause either excitatory or inhibitory effects depending on a large number of circumstances (Deidda, Bozarth, & Cancedda, 2014; Doyon et al., 2011). Excitatory in this context means promoting action potential generation and inhibition means a decrease in the likelihood of action potential generation in the target neurons and neural networks. How can such a straightforward signaling mechanism be so variable? The opening of $GABA_A$ receptor channels can influence neurons in several ways. The increased membrane conductance shunts other transmembrane currents and can thus dampen changes in the membrane potential that would otherwise occur. The net effect is inhibitory and shunting inhibition is a significant effect of $GABA_A$ receptor activation. Shunting inhibition can occur in the complete absence of any membrane potential changes and its importance is thus sometimes underestimated.

If the reversal potential for the flowing anions is not at the momentary membrane potential of the neuron, a net charge will flow across the cell membrane and change the membrane potential according to the net flow of ions. The equilibrium potential for bicarbonate ions (−10 mV) (Kaila & Voipio, 1987; Rivera, Voipio, & Kaila, 2005) is significantly depolarized from the resting membrane potential and above the sodium action potential threshold and their outward flow at resting membrane potentials will tend to depolarize neurons. Bicarbonate ion gradients are largely fixed by the neuronal pH homeostasis and thus not variable. The reversal potential for chloride ions, however, is highly variable and can give rise to de- and hyperpolarizing responses in neurons. This variability in the chloride reversal potential is at the heart of the large variability in $GABA_A$ receptor signaling. $GABA_A$ receptor-mediated responses vary from de- to hyperpolarizing responses over a large temporal and spatial domain, ranging from development timescales to fractions of a second and from gender differences to brain regions all the way to subcellular compartments.

For the remainder of the review, we will therefore focus on the effect of chloride handling on $GABA_A$ receptor signaling and its influence on neurons and neuronal compartment. It is crucial to note that the terms "excitatory" and "depolarizing" or "inhibitory" and "hyperpolarizing" are not synonyms in this context (Kaila, 1994; Morita, Tsumoto, & Aihara, 2006). The net effect of $GABA_A$ receptor activation depends on the shunting effect and on the size and velocity of a possible change in membrane potential as well as on the reaction of the voltage-gated ion channels in the affected compartments. For example, a slow depolarization of a compartment containing voltage-activated channels can be inhibitory both because of shunting and because of channel inactivation. Conversely, a hyperpolarization may cause a deinactivation of a voltage-gated channel and cause subsequent rebound firing, when the $GABA_A$ receptors close and shunting stops.

2. FACTORS SHAPING THE NEURONAL TRANSMEMBRANE CHLORIDE GRADIENT

Many factors affect the distribution of chloride ions across cell membranes (Blaesse, Airaksinen, Rivera, & Kaila, 2009; Kaila, 1994). This review is focused on the ones most relevant for mammalian neurons under physiological conditions. For this review, a fixed extracellular concentration of chloride ions at 110 mM will be assumed; thus, the factors considered will affect the transmembrane gradient by affecting the intracellular concentration of chloride. The main factors controlling intracellular chloride concentration are membrane potential, chloride transporters, and Gibbs–Donnan effects of impermeable anions.

2.1. Resting membrane potential

The resting membrane potential exerts an electrical force on the negatively charged chloride ions. If the membrane potential were the only force acting and provided a sufficient chloride conductance, the steady-state condition would produce a chloride gradient that follows Nernst's law at the resting membrane potential and thereby fix the chloride reversal potential at the resting membrane potential. Countless experiments have shown that the chloride equilibrium potential can be more positive or more negative than the resting membrane potential; thus in addition to the passive equilibrium condition, active transport is shaping the chloride distribution across the cell membrane. Two members of the cation-chloride cotransporter family (Gagnon & Delpire, 2013) represent the most important chloride pumps for neurons. One member of the sodium potassium chloride cotransporter family: Na-K-Cl cotransporter 1 (NKCC1) with a pumping stochiometry of 1Na, 1K, 2Cl, and one member of the potassium chloride cotransporter family: K-Cl cotransporter 2 (KCC2) with a 1K, 1Cl stochiometry (Blaesse et al., 2009; Payne, Rivera, Voipio, & Kaila, 2003). Both pumps use the energy of the gradients established by the Na/K ATPase to pump chloride against a potential electrochemical gradient.

2.2. K-Cl cotransporter 2

KCC2 is exclusively found in neurons (Payne, Stevenson, & Donaldson, 1996). Equilibrium conditions for KCC2-mediated transport will establish a chloride gradient that is basically the opposite of the transmembrane potassium gradient and a chloride equilibrium potential that is identical to the potassium reversal potential. This is below the resting membrane potential for neurons, since the resting membrane is dominated, but not completely determined by the potassium equilibrium potential. Any changes in the concentrations of any of the pumped ion species will therefore cause a net flow of the other species. Increases in the intracellular Cl concentration will produce K efflux, while extracellular accumulation of K will cause an intracellular Cl accumulation (Blaesse et al., 2009).

KCC2 knockout mice are not viable after birth due to a failure to generate rhythmic activity in the pre-Bötzinger complex and thus failing to breathe (Hubner et al., 2001). Mice in which KCC2 expression is reduced to 5–10% due to knockout of the main KCC2b isoform show hyperexcitability and epileptic seizures (Woo et al., 2002). Ablation of KCC2 in adult mice results in a fatal hyperexcitability shortly after the loss of KCC2 expression (Gagnon & Delpire, 2013). The interpretation of KCC2 expression data is complicated by the fact that KCC2 has functions in spine maturation, independent of its transport function (Blaesse & Schmidt, 2014; Fiumelli et al., 2013; Li et al., 2007; Puskarjov et al., 2014).

Available data indicate that KCC2 transcription, surface expression, and function are all dynamically regulated (Blaesse et al., 2009; Medina et al., 2014).

2.3. Na-K-Cl cotransporter 1

NKCC1 is widely expressed in the central nervous system (CNS) both in glia and in neurons (Payne et al., 2003; Yan, Dempsey, & Sun, 2001). The transport stochiometry for NKCC1 favors a chloride accumulation above the value determined by the resting membrane potential and thus a depolarized chloride reversal potential. Knockout of the NKCC1 gene produces defects in inner ear function due to faulty endolymph production, but no overt neurological phenotype (Payne et al., 2003). Conflicting results were obtained by investigations of the influence of NKCC1 knockout on general excitability, with one group finding decreased excitability in NKCC1 knockout mice (Dzhala et al., 2005), while others have found that NKCC1 prevents hyperexcitability (Zhu, Polley, Mathews, & Delpire, 2008). Equilibrium conditions would be reached at an intracellular chloride concentration of roughly 60 mM; however, NKCC1 is not at equilibrium under physiological conditions (Russell, 2000). NKCC1 activity is controlled by chloride concentration-dependent phosphorylation (Blaesse et al., 2009; Russell, 2000).

2.4. Impermeable anions

The intracellular milieu contains a large number of anions in the form of negative charges on proteins and nucleic acids, which exhibit low mobility and are not membrane permeable. Similarly, the extracellular matrix also contains a significant number of fixed negative charges. These impermeable anions give rise to Gibbs–Donnan effects which may influence the transmembrane chloride gradient. For a Gibbs–Donnan equilibrium two conditions have to be met. First electroneutrality: the sum of positive and negative charges in each compartment has to be the same. Second: the product of the concentration of diffusible ions (potassium and chloride in this case) has to be equal inside and outside. These effects are purely passive and cannot produce distributions that allow a net flow of chloride ions upon $GABA_A$ receptor opening; that is, only shunting inhibition could occur under these conditions. Recently, Gibbs–Donnan effects, and not chloride pumps, were proposed to play the determining role in setting the conditions for chloride distribution across neural cell membranes in the cortex (Glykys et al., 2014; but see Luhmann, Kirischuk, & Kilb, 2014; Voipio et al., 2014). Given that $GABA_A$ receptor-mediated hyper- and depolarizing responses are routinely observed, active chloride transport necessarily plays a crucial role in setting the neuronal chloride concentration (Kaila, Price, Payne, Puskarjov, & Voipio, 2014). Unevenly distributed immobile anions do not affect the transmembrane driving force for chloride ions locally. The same effect that lowers the chloride concentration in the vicinity of the fixed anion intracellularly also impedes the inflow of chloride from the extracellular space. Therefore, this mechanism cannot account for different driving forces between different subcellular compartments—such differences also must be maintained by active, energy-dependent mechanisms (Kaila, Price, et al., 2014).

3. EXPERIMENTAL TECHNIQUES TO STUDY CHLORIDE HOMEOSTASIS AND E-$GABA_A$

Experimental investigation of $GABA_A$ receptor signaling regularly produces conflicting results. On the one hand, this is a result of the considerable technical difficulty in maintaining proper experimental conditions. On the other hand, the unstable nature of the chloride homeostasis itself may produce varying, seemingly conflicting results, if important parameters such as species, age, sex, cell type, or activity levels are not taken into account.

3.1. Classical electrophysiology

Conventional whole-cell recordings dialyze the highly mobile chloride ions. Intracellular chloride is then dependent on intra-pipette chloride concentrations and endogenous chloride handling is compromised. Chloride concentrations do not equilibrate immediately and uniformly, in particular in geometrically complex structures, such as neurons. Deliberate chloride loading can provide useful insight into the chloride handling capacity of neurons (Khirug et al., 2005). However, $GABA_A$ receptor-mediated responses are likely no longer the same as before the whole-cell recording was established.

One widely used method to avoid perturbing the chloride homeostasis is to use the perforated patch recording technique (Horn & Marty, 1988) in which a pore-forming agent is added to the patch solution. The topical antibiotic gramicidin, isolated from the soil bacterium *Bacillus brevis*, is commonly used to study GABAergic signals (Kyrozis & Reichling, 1995) because of its selective permeability to monovalent cations (Myers & Haydon, 1972). Care has to be taken to avoid spurious membrane rupture during the recording. Addition of a fluorescent tracer, which cannot penetrate the gramicidin-formed pores, to the patch solution makes it possible to continuously monitor membrane integrity. Addition of readily gramicidin-permeable cesium ions to the patch solution may improve space clamp, but will disrupt chloride homeostasis by interfering with cation-chloride cotransporter function (van Brederode, Takigawa, & Alzheimer, 2001).

Micro- or unitary field recordings have recently gained interest for the study of GABAergic signals, since they offer the opportunity to study synaptic signals without perturbing the intracellular milieu (Bazelot, Dinocourt, Cohen, & Miles, 2010; Glickfeld, Roberts, Somogyi, & Scanziani, 2009; Sauer, Struber, & Bartos, 2012). Since transmembrane chloride gradients depend on both the intra- and extracellular milieu, care has to be taken that the latter also remains unperturbed during recordings with field electrodes.

3.2. Imaging

As further detailed below, there are numerous findings that indicate compartmentalized chloride handling and consequently $GABA_A$ receptor-mediated signaling in neurons (Alger & Nicoll, 1979; Gulledge & Stuart, 2003; Szabadics et al., 2006). Over the last few years, several imaging methods have been used to study chloride handling in neurons. Two general approaches have been taken. Chloride ion-sensitive fluorophores have been developed, both as organic dyes and as genetically encoded sensors. Voltage-sensitive dye imaging allows the recording of $GABA_A$ receptor-mediated transmembrane voltage changes at subcellular resolution.

3.2.1 Chloride-sensitive dyes

Certain quinoline-based fluorescent dyes exhibit strong quenching by chloride ions. While these dyes have been available for a while, their widespread use is hampered by several factors. They are plagued by significant toxicity, relatively high rate of diffusion across cell membranes and a need for activation by UV light with a resulting high background autofluorescence and strong bleaching (Bregestovski, Waseem, & Mukhtarov, 2009). The membrane permeable dye [*N*-(6-methoxyquinolyl)acetoethyl ester] (MQAE) has a relatively high sensitivity for chloride with a K_d of 40 mM and endogenous chloride transients can be measured using two-photon excitation, which avoids some of the issues of UV illumination (Marandi, Konnerth, & Garaschuk, 2002). MQAE is not ratiometric and thus not suitable for determining baseline chloride levels (Marandi et al., 2002).

3.2.2 Genetically encoded chloride sensors

Genetically encoded chloride sensors use the halide sensitivity of yellow fluorescent protein (YFP) to measure intracellular chloride levels (Jayaraman, Haggie, Wachter, Remington, & Verkman, 2000). By linking a chloride-sensitive YFP-based sensor to cyan fluorescent protein (CFP), the ratiometric, FRET-based chloride sensor Clomeleon was developed (Kuner & Augustine, 2000). Clomeleon can be expressed in select cell types and can be used to measure chloride transients as well as resting chloride levels (Berglund et al., 2006). Clomeleon is stable over long time periods and no cytotoxicity has been observed. It has a low affinity for chloride with an EC_{50} of about 160 mM (Kuner & Augustine, 2000). Imaging systems with particularly low noise levels are required when using Clomeleon with the aim to reliably detect low chloride levels: at 5 mM intracellular chloride only 3% of the dye will be chloride-bound and an increase to 10 mM will roughly double that to 6%. Measurements of low chloride concentrations are also sensitive to the possibility of uneven bleaching of the YFP and CFP components of the sensor (Bregestovski et al., 2009). Cl-sensor is another ratiometric YFP-based chloride indicator with higher chloride sensitivity with an EC_{50} of around 30 mM (Markova, Mukhtarov, Real, Jacob, & Bregestovski, 2008), which should improve the reliability of measurements of low intracellular chloride levels.

Chloride imaging cannot detect either shunting inhibition or bicarbonate flux and therefore only allows indirect statements on the impact of a GABAergic signal on the excitability of the postsynaptic neuron.

3.2.3 Voltage-sensitive dye imaging

Another technique to study GABAergic signaling with high spatial resolution is transmembrane voltage imaging. While genetically encoded voltage sensors have made steady progress over the last few years (Han et al., 2013), currently the superior kinetic properties and better signal-to-noise ratio of voltage-sensitive dyes favor the latter for this application. Using stable, high-intensity light sources, changes in the transmembrane voltage of a few millivolts can be resolved at speeds sufficient to record synaptic responses at subcellular resolution (Canepari, Willadt, Zecevic, & Vogt, 2010). An advantage of this method over chloride imaging lies in the direct observation of both excitatory and inhibitory signals and their interaction (Willadt, Nenniger, & Vogt, 2013). Substantial bleaching and potential phototoxicity limit the method to acute experiments and the study of evoked responses.

4. VARIABILITY OF GABAergic SIGNALING

The highly dynamic nature of intracellular chloride handling or "ionic plasticity" (Kaila, Ruusuvuori, Seja, Voipio, & Puskarjov, 2014) is directly reflected in the variability of GABAergic signals. I will highlight a few examples of either temporal or spatial variability of $GABA_A$ receptor-mediated responses (Table 1).

4.1. Variability in the temporal domain

Chloride concentration gradients show large variability over time frames that vary from days to seconds.

4.1.1 Developmental timeframe

Changes in the response to $GABA_A$ receptor activation during the maturation of the brain have been widely reported, covering late embryonic to juvenile preparations. One prevailing view is that $GABA_A$ receptor activation is excitatory in late embryonic stages and early postnatally, transitioning to inhibitory effects during the first 2 weeks of postnatal development (Ben-Ari, Gaiarsa, Tyzio, & Khazipov, 2007); however, the generality of this view has been questioned (Bregestovski & Bernard, 2012). Two main findings provide evidence against an overall excitatory effect of $GABA_A$ receptor

Table 1 Variability of intracellular chloride homeostasis at different timeframes and their main causes

Intracellular chloride dynamics and their main causes and consequences

Timeframe	Effect	Causes	Consequence
Seconds	Activity-dependent short-term plasticity	Activity-dependent chloride accumulation and depolarizing shift in the reversal potential	High GABAergic activity causes rapid loss of hyperpolarizing and possible switch to depolarizing $GABA_A$ receptor-mediated signals
Hours	Spike-timing-dependent plasticity of GABAergic inputs	Calcium-dependent downregulation of KCC2 and depolarizing shift in the reversal potential	Paired pre- and postsynaptic activation reduces GABAergic input by reducing chloride driving force
	Circadian shift in GABA effects in suprachiasmatic nucleus	Circadian variation of NKCC1 expression	Increased NKCC1 expression produces depolarizing shift of $GABA_A$ receptor reversal potential in dark (active) phase for rodents
Days	Change of $GABA_A$ receptor driving force around birth	Oxytocin-dependent downregulation of NKCC1 expression	Rapid hyperpolarizing shift of $GABA_A$ receptor reversal potential around birth
Weeks	Developmental shift in $GABA_A$ receptor driving force	Developmental decrease in NKCC1 expression and increase in KCC2 expression	Relatively depolarized $GABA_A$ receptor reversal potential in immature neurons
	Seasonal changes in $GABA_A$ receptor reversal potential in the suprachiasmatic nucleus	Day length-dependent shift in NKCC1 function	Short day length produces depolarizing shifts and long-day length produces hyperpolarizing shift in $GABA_A$ receptor reversal potential in dark-active rodents

See also Kaila, Ruusuvuori, et al. (2014).

activation even at very young ages. First, the antagonist bicuculline induces hyperactivity and seizures in rats as young as postnatal day 3 (Baram & Snead, 1990)—the youngest age tested. Second, ablation of KCC2 produces a hyper excitable phenotype at late embryonic stages compared to wild-type mice (Hubner et al., 2001). Nevertheless, a number of studies have shown a clear shift in the $GABA_A$ receptor reversal potential during development for many areas of the brain (Ben-Ari, 2002; Blaesse et al., 2009; Owens, Boyce, Davis, & Kriegstein, 1996; Yamada et al., 2004), accompanied by a reduction in NKCC1 expression and an increase in KCC2 expression in these regions or neurons. It is worth repeating that depolarizing $GABA_A$ receptor-mediated responses need not be excitatory, but potentially can be. Depolarizing $GABA_A$ receptor activation can cause postsynaptic calcium influx without the need to trigger classical sodium action potentials (Yamada et al., 2004), and depolarizing GABAergic signals can play an important role in the early maturation of CNS circuits (Ben-Ari, 2002; Owens & Kriegstein, 2002).

GABAergic signaling has been shown to play an important role in the maturation and synaptic integration of newborn neurons during adult neurogenesis in the dentate gyrus (Ge et al., 2006). These neurons first receive tonic $GABA_A$ receptor-mediated signals, followed by $GABA_A$ receptor-mediated fast synaptic input and finally glutamatergic signals. Disruption of the initially significantly depolarized chloride gradient by NKCC1 inhibition produced developmental deficits resulting in reduced synaptic input and dendritic hypomorphism (Ge et al., 2006).

A recent study (Glykys et al., 2014) has challenged a number of findings on neuronal chloride handling, but see Luhmann et al. (2014) and Voipio et al. (2014). Using Clomeleon imaging, the authors describe a wide range of intracellular chloride concentrations with a mean around 14 mM in cornu ammonis area 1 (CA1) hippocampal pyramidal cells that did not significantly change from postnatal days 8–9 to postnatal days 32–44. The authors postulate that impermeable anions play a much larger role in the control of intracellular chloride than previously assumed. However, Clomeleon is not ideal for the study of low intracellular chloride concentrations, and as mentioned before, immobile anion distributions cannot sustain non-equilibrium driving forces for chloride (Kaila, Price, et al., 2014).

4.1.2 Seasonal timeframe

Results from a recent study have shown that neurons in the suprachiasmatic nucleus exhibit day length-dependent chloride handling

(Farajnia, van Westering, Meijer, & Michel, 2014). Long-day length was accompanied by a relatively depolarized $GABA_A$ receptor reversal potential of −40 mV, which shifted to −50 mV in the short-day condition. Exogenous GABA application predominantly increased calcium concentrations under long-day conditions, while it reduced calcium signals in a majority of neurons in short-day conditions.

4.1.3 Day-to-day variation and circadian rhythms

The peptide hormone oxytocin has been shown to produce a brief shift in the chloride reversal potential in hippocampal, neocortical, and trigeminal neurons of pups around birth (Mazzuca et al., 2011; Tyzio et al., 2006). Oxytocin produced a transient downregulation of NKCC1 activity and a shift of the chloride reversal potential to the resting membrane potential for roughly 1 day. This produces neonate analgesia (Mazzuca et al., 2011) and may be important for the later CNS development (Tyzio et al., 2014).

The main circadian clock in the suprachiasmatic nucleus consists of predominantly GABAergic neurons. In addition to the seasonal dependence of chloride handling, circadian rhythms in chloride reversal and $GABA_A$ receptor signaling have been observed. Initial reports indicated depolarized $GABA_A$ receptor reversal potentials during the day and a shift toward hyperpolarizing potentials at night (Bos & Mirmiran, 1993; Liou & Albers, 1990). Later studies have still confirmed circadian variations, but found relatively hyperpolarized chloride gradients during the day (light phase) and depolarized gradients during the night (dark phase) (Choi et al., 2008; De Jeu & Pennartz, 2002); thus, relatively depolarizing $GABA_A$ receptor responses coincided with the active phase of rat or mouse behavior. The circadian effects were accompanied by higher sensitivity of the $GABA_A$ receptor reversal potential to the NKCC1 antagonist bumetanide and higher NKCC1 expression at night. These effects were most prominent in the dorsal part of the suprachiasmatic nucleus and individual neurons varied considerably in their responses (Choi et al., 2008), demonstrating that a significant number of neurons may have to be studied to detect such effects.

4.1.4 Chloride homeostasis-dependent long-term plasticity

Transient increases in intracellular chloride after intense GABAergic activity generally dissipate through passive equilibration and KCC2-mediated chloride efflux in a matter of seconds. Repeated coincident activation of GABAergic inputs and their hippocampal pyramidal neuron targets induced prolonged shifts in the $GABA_A$ receptor reversal potential (Woodin, Ganguly, & Poo, 2003). The authors showed a calcium-dependent downregulation of KCC2 activity accompanied by an increase in postsynaptic intracellular chloride concentration. These and other experiments clearly show that specific patterns of neural activity can influence GABAergic synaptic transmission over prolonged periods and with important consequences for network activity (Vogels et al., 2013).

4.1.5 Short-term activity- dependent chloride regulation

Some of the earliest studies on $GABA_A$ receptor signaling (Alger & Nicoll, 1979) have noted biphasic hyper- and depolarizing responses following exogenous GABA application or strong stimulation of inhibitory inputs to hippocampal pyramidal cells (Staley, Soldo, & Proctor, 1995). Prolonged opening of $GABA_A$ receptors will lead to a shift in intracellular chloride (Thompson & Gahwiler, 1989) toward an equilibrium in which its electrochemical gradient matches the membrane potential at the time of receptor opening. If the neuron is significantly depolarized during that time, subsequent $GABA_A$ receptor activation will yield depolarizing responses from the resting membrane potential. Bicarbonate permeability of $GABA_A$ receptors results in a compound reversal potential more depolarized than the chloride reversal potential (Staley et al., 1995); if open $GABA_A$ receptors represent the dominant membrane conductance, intracellular chloride can accumulate toward a level set by the bicarbonate equilibrium (Kaila, 1994) even in the absence of an additional depolarizing force. Chloride extrusion mechanisms have been shown to efficiently restore hyperpolarizing chloride gradients under physiological conditions; however, transiently reduced driving forces (Thompson & Gahwiler, 1989) and even depolarizing $GABA_A$ responses can be observed (Lee et al., 2012).

In conclusion, the reversal potential for $GABA_A$ receptors is highly dynamic and critically depends on the state of activity of a neuron (Raimondo, Markram, & Akerman, 2012). This activity state is likely to differ significantly between relatively active *in vivo* preparations and relatively silent acute brain slice preparations.

4.2. Variability in the spatial domain

For this parameter, the variability extends from species and gender differences all the way to subcellular domains.

4.2.1 Interspecies variability

Different species of animals, which are frequently used for electrophysiological studies, differ in their CNS maturation time frames. Rats and mice are born relatively immature, while guinea pigs show higher levels of maturation at birth. Comparisons of $GABA_A$ receptor reversal potentials during development time have to take these differences into account (Blaesse et al., 2009).

4.2.2 Gender differences

In an elegant study, a striking gender difference in the developmental shift of $GABA_A$ receptor reversal potentials in CA1 pyramidal neurons of rats was shown (Galanopoulou, 2008a, 2008b). Female rats exhibited a much smaller negative shift in the $GABA_A$ receptor reversal potential. They started from a significantly hyperpolarized value compared to neurons from male rats, which exhibited a more depolarized $GABA_A$ reversal potential and a more dramatic hyperpolarization over their 2 weeks postnatal development. For a review on sexually dimorphic GABAergic signaling, see Galanopoulou (2008a, 2008b).

4.2.3 Region-specific chloride homeostasis

With regard to their chloride homeostasis, neurons in the spinal cord and the brainstem mature before neurons in the thalamus, which in turn mature before neurons in the cortex (Blaesse et al., 2009). While cortical circuits may still be maturing and forming synaptic contacts, circuits in the spinal cord and brainstem need to be fully functioning for survival (Hubner et al., 2001).

4.2.4 Cell type-specific chloride homeostasis

Both in cortex and in amygdala fast-spiking neurons were found to exhibit significantly depolarized $GABA_A$ receptor reversal potentials compared to regular spiking neurons (Martina, Royer, & Pare, 2001). These spiking parameters roughly characterize interneurons and principal cells, respectively. Thus, these two different cell types show clear differences in $GABA_A$ receptor-mediated signaling. Even more specific effects were found in the hippocampus: Perisomatic-inhibiting interneurons caused depolarizing postsynaptic signals in dentate gyrus granule cells, but hyperpolarized CA1 pyramidal neurons (Sauer et al., 2012). Interestingly, this difference was not due to an absolute difference in the $GABA_A$ receptor reversal potential, but caused by a lower resting membrane potential in dentate gyrus granule cells. Measurements in the cerebellum revealed $GABA_A$ receptor reversal potentials of around -85 mV for Purkinje cells and around -58 mV for stellate and basket cells (Chavas & Marty, 2003). Thus in many areas of the brain, the driving force for $GABA_A$ receptors can vary considerably in a cell-type-specific manner.

4.2.5 Subcellular variation in chloride handling and GABAergic signaling

Subcellular domains with differential $GABA_A$ receptor-mediated signaling were observed early in the investigation of GABAergic synaptic transmission (Alger & Nicoll, 1979). Since then, a number of experimental and theoretical approaches have investigated the role of subcellular domains in chloride handling and in generating different $GABA_A$ receptor-mediated responses. These effects are of particular interest for our understanding of interneuron function, because one of the striking precision with which different interneuron subtypes target specific subcellular locations (Ascoli et al., 2008; DeFelipe et al., 2013). Depolarizing dendritic, but not somatic $GABA_A$ receptor-mediated responses were described in neocortex (Gulledge & Stuart, 2003). Of particular importance is the author's observation that the shunting (inhibitory) component and the membrane potential fluctuation (excitatory or inhibitory) have different spatial and temporal profiles. Shunting is restricted to the place and time of receptor activation, while membrane potential changes can spread spatially and outlast channel opening (Gulledge & Stuart, 2003). Somato-dendritic $GABA_A$ receptor-mediated inhibition, but axon initial segment excitation has been described in pyramidal neurons (Szabadics et al., 2006), but this finding is not universal (Bazelot et al., 2010; Glickfeld et al., 2009). The specifically elevated intracellular chloride concentration and depolarized $GABA_A$ receptor reversal potential at the axon initial segment were found together with a lack of KCC2 staining and an increased presence of NKCC1 at that site (Khirug et al., 2008). Indicating that differentially located pump proteins can give rise to intracellular chloride gradients, which would have to be actively maintained. Imaging has directly revealed chloride concentration gradients in retinal on-type bipolar cells (Duebel et al., 2006). The conditions and consequences of subcellular specialization in GABAergic signaling have been carefully modeled (Doyon et al., 2011). One of the main findings of this study indicates that subcellular chloride gradients can be actively maintained under realistic assumptions about chloride homeostasis.

5. CONCLUSION

A global decrease or even loss of $GABA_A$ receptor-mediated signaling produces hyperexcitability, epileptic seizures, and eventually death. As such, the label "inhibitory transmitter" for GABA is justified. However, as soon as specific circuits and neurons are investigated, the large repertoire of possible

$GABA_A$ receptor-mediated effects and their highly volatile nature have to be considered. Many of the relevant variables, in particular prior or concomitant synaptic activity, sex, or circadian phase are only marginally explored, often not controlled for and thus not reported.

Changes in chloride homeostasis and concomitant shifts in $GABA_A$ receptor reversal potential affect the excitability of subcellular neural compartments, neurons, and neural networks. A shift in the reversal potential toward more positive values is with some exceptions going to render the system more excitable even if the reversal potential per se does not become depolarizing or excitatory. Thus, the highly dynamic chloride homeostasis is an important regulatory element in setting the excitation–inhibition balance in the CNS.

Every technique to study $GABA_A$ receptor-mediated signals in use today has considerable drawbacks and the potential to produce significant artifacts. Ideally, multiple techniques with nonoverlapping weaknesses are used to study the rich variety of ionotropic GABAergic signals in the brain.

A better understanding of the specific dynamics of $GABA_A$ receptor signaling will hopefully soon allow us to incorporate them into the bigger picture of the molecular and morphological diversity of the GABAergic system.

CONFLICT OF INTEREST

The author declares no conflicts of interest regarding this chapter.

REFERENCES

Alger, B. E., & Nicoll, R. A. (1979). GABA-mediated biphasic inhibitory responses in hippocampus. *Nature, 281*(5729), 315–317.

Ascoli, G. A., Alonso-Nanclares, L., Anderson, S. A., Barrionuevo, G., Benavides-Piccione, R., Burkhalter, A., et al. (2008). Petilla terminology: Nomenclature of features of GABAergic interneurons of the cerebral cortex. *Nature Reviews. Neuroscience, 9*(7), 557–568.

Baram, T. Z., & Snead, O. C., 3rd. (1990). Bicuculline induced seizures in infant rats: Ontogeny of behavioral and electrocortical phenomena. *Brain Research. Developmental Brain Research, 57*(2), 291–295.

Bazelot, M., Dinocourt, C., Cohen, I., & Miles, R. (2010). Unitary inhibitory field potentials in the CA3 region of rat hippocampus. *The Journal of Physiology, 588*(Pt 12), 2077–2090.

Ben-Ari, Y. (2002). Excitatory actions of gaba during development: The nature of the nurture. *Nature Reviews. Neuroscience, 3*(9), 728–739.

Ben-Ari, Y., Gaiarsa, J. L., Tyzio, R., & Khazipov, R. (2007). GABA: A pioneer transmitter that excites immature neurons and generates primitive oscillations. *Physiological Reviews, 87*(4), 1215–1284.

Berglund, K., Schleich, W., Krieger, P., Loo, L. S., Wang, D., Cant, N. B., et al. (2006). Imaging synaptic inhibition in transgenic mice expressing the chloride indicator, Clomeleon. *Brain Cell Biology, 35*(4–6), 207–228.

Betz, H. (1990). Ligand-gated ion channels in the brain: The amino acid receptor superfamily. *Neuron, 5*(4), 383–392.

Blaesse, P., Airaksinen, M. S., Rivera, C., & Kaila, K. (2009). Cation-chloride cotransporters and neuronal function. *Neuron*, *61*(6), 820–838.

Blaesse, P., & Schmidt, T. (2014). K-Cl cotransporter KCC2-a moonlighting protein in excitatory and inhibitory synapse development and function. *Pflügers Archiv*. http://dx.doi.org/10.1007/s00424-014-1547-6.

Bormann, J., Hamill, O. P., & Sakmann, B. (1987). Mechanism of anion permeation through channels gated by glycine and gamma-aminobutyric acid in mouse cultured spinal neurones. *The Journal of Physiology*, *385*, 243–286.

Bos, N. P., & Mirmiran, M. (1993). Effects of excitatory and inhibitory amino acids on neuronal discharges in the cultured suprachiasmatic nucleus. *Brain Research Bulletin*, *31*(1–2), 67–72.

Bregestovski, P., & Bernard, C. (2012). Excitatory GABA: How a correct observation may turn out to be an experimental artifact. *Frontiers in Pharmacology*, *3*, 65.

Bregestovski, P., Waseem, T., & Mukhtarov, M. (2009). Genetically encoded optical sensors for monitoring of intracellular chloride and chloride-selective channel activity. *Frontiers in Molecular Neuroscience*, *2*, 15.

Canepari, M., Willadt, S., Zecevic, D., & Vogt, K. E. (2010). Imaging inhibitory synaptic potentials using voltage sensitive dyes. *Biophysical Journal*, *98*(9), 2032–2040.

Chavas, J., & Marty, A. (2003). Coexistence of excitatory and inhibitory GABA synapses in the cerebellar interneuron network. *The Journal of Neuroscience*, *23*(6), 2019–2031.

Choi, H. J., Lee, C. J., Schroeder, A., Kim, Y. S., Jung, S. H., Kim, J. S., et al. (2008). Excitatory actions of GABA in the suprachiasmatic nucleus. *The Journal of Neuroscience*, *28*(21), 5450–5459.

DeFelipe, J., Lopez-Cruz, P. L., Benavides-Piccione, R., Bielza, C., Larranaga, P., Anderson, S., et al. (2013). New insights into the classification and nomenclature of cortical GABAergic interneurons. *Nature Reviews. Neuroscience*, *14*(3), 202–216.

Deidda, G., Bozarth, I. F., & Cancedda, L. (2014). Modulation of GABAergic transmission in development and neurodevelopmental disorders: Investigating physiology and pathology to gain therapeutic perspectives. *Frontiers in Cellular Neuroscience*, *8*, 119.

De Jeu, M., & Pennartz, C. (2002). Circadian modulation of GABA function in the rat suprachiasmatic nucleus: Excitatory effects during the night phase. *Journal of Neurophysiology*, *87*(2), 834–844.

Doyon, N., Prescott, S. A., Castonguay, A., Godin, A. G., Kroger, H., & De Koninck, Y. (2011). Efficacy of synaptic inhibition depends on multiple, dynamically interacting mechanisms implicated in chloride homeostasis. *PLoS Computational Biology*, 7(9), e1002149.

Duebel, J., Haverkamp, S., Schleich, W., Feng, G., Augustine, G. J., Kuner, T., et al. (2006). Two-photon imaging reveals somatodendritic chloride gradient in retinal ON-type bipolar cells expressing the biosensor Clomeleon. *Neuron*, *49*(1), 81–94.

Dzhala, V. I., Talos, D. M., Sdrulla, D. A., Brumback, A. C., Mathews, G. C., Benke, T. A., et al. (2005). NKCC1 transporter facilitates seizures in the developing brain. *Nature Medicine*, *11*(11), 1205–1213.

Farajnia, S., van Westering, T. L., Meijer, J. H., & Michel, S. (2014). Seasonal induction of GABAergic excitation in the central mammalian clock. *Proceedings of the National Academy of Sciences of the United States of America*, *111*(26), 9627–9632.

Fiumelli, H., Briner, A., Puskarjov, M., Blaesse, P., Belem, B. J., Dayer, A. G., et al. (2013). An ion transport-independent role for the cation-chloride cotransporter KCC2 in dendritic spinogenesis in vivo. *Cerebral Cortex*, *23*(2), 378–388.

Gagnon, K. B., & Delpire, E. (2013). Physiology of SLC12 transporters: Lessons from inherited human genetic mutations and genetically engineered mouse knockouts. *American Journal of Physiology. Cell Physiology*, *304*(8), C693–C714.

Galanopoulou, A. S. (2008a). Dissociated gender-specific effects of recurrent seizures on GABA signaling in CA1 pyramidal neurons: Role of $GABA_A$ receptors. *The Journal of Neuroscience*, *28*(7), 1557–1567.

Galanopoulou, A. S. (2008b). Sexually dimorphic expression of KCC2 and GABA function. *Epilepsy Research*, *80*(2–3), 99–113.

Ge, S., Goh, E. L., Sailor, K. A., Kitabatake, Y., Ming, G. L., & Song, H. (2006). GABA regulates synaptic integration of newly generated neurons in the adult brain. *Nature*, *439*(7076), 589–593.

Glickfeld, L. L., Roberts, J. D., Somogyi, P., & Scanziani, M. (2009). Interneurons hyperpolarize pyramidal cells along their entire somatodendritic axis. *Nature Neuroscience*, *12*(1), 21–23.

Glykys, J., Dzhala, V., Egawa, K., Balena, T., Saponjian, Y., Kuchibhotla, K. V., et al. (2014). Local impermeant anions establish the neuronal chloride concentration. *Science*, *343*(6171), 670–675.

Gulledge, A. T., & Stuart, G. J. (2003). Excitatory actions of GABA in the cortex. *Neuron*, *37*(2), 299–309.

Han, Z., Jin, L., Platisa, J., Cohen, L. B., Baker, B. J., & Pieribone, V. A. (2013). Fluorescent protein voltage probes derived from ArcLight that respond to membrane voltage changes with fast kinetics. *PLoS One*, *8*(11), e81295.

Hevers, W., & Luddens, H. (1998). The diversity of $GABA_A$ receptors. Pharmacological and electrophysiological properties of $GABA_A$ channel subtypes. *Molecular Neurobiology*, *18*(1), 35–86.

Horn, R., & Marty, A. (1988). Muscarinic activation of ionic currents measured by a new whole-cell recording method. *The Journal of General Physiology*, *92*(2), 145–159.

Hubner, C. A., Stein, V., Hermans-Borgmeyer, I., Meyer, T., Ballanyi, K., & Jentsch, T. J. (2001). Disruption of KCC2 reveals an essential role of K-Cl cotransport already in early synaptic inhibition. *Neuron*, *30*(2), 515–524.

Jayaraman, S., Haggie, P., Wachter, R. M., Remington, S. J., & Verkman, A. S. (2000). Mechanism and cellular applications of a green fluorescent protein-based halide sensor. *The Journal of Biological Chemistry*, *275*(9), 6047–6050.

Kaila, K. (1994). Ionic basis of $GABA_A$ receptor channel function in the nervous system. *Progress in Neurobiology*, *42*(4), 489–537.

Kaila, K., Price, T. J., Payne, J. A., Puskarjov, M., & Voipio, J. (2014). Cation-chloride cotransporters in neuronal development, plasticity and disease. *Nature Reviews. Neuroscience*, *15*(10), 637–654.

Kaila, K., Ruusuvuori, E., Seja, P., Voipio, J., & Puskarjov, M. (2014). GABA actions and ionic plasticity in epilepsy. *Current Opinion in Neurobiology*, *26*, 34–41.

Kaila, K., & Voipio, J. (1987). Postsynaptic fall in intracellular pH induced by GABA-activated bicarbonate conductance. *Nature*, *330*(6144), 163–165.

Khirug, S., Huttu, K., Ludwig, A., Smirnov, S., Voipio, J., Rivera, C., et al. (2005). Distinct properties of functional KCC2 expression in immature mouse hippocampal neurons in culture and in acute slices. *The European Journal of Neuroscience*, *21*(4), 899–904.

Khirug, S., Yamada, J., Afzalov, R., Voipio, J., Khiroug, L., & Kaila, K. (2008). GABAergic depolarization of the axon initial segment in cortical principal neurons is caused by the Na-K-2Cl cotransporter NKCC1. *The Journal of Neuroscience: The Official Journal of the Society for Neuroscience*, *28*(18), 4635–4639.

Kuner, T., & Augustine, G. J. (2000). A genetically encoded ratiometric indicator for chloride: Capturing chloride transients in cultured hippocampal neurons. *Neuron*, *27*(3), 447–459.

Kyrozis, A., & Reichling, D. B. (1995). Perforated-patch recording with gramicidin avoids artifactual changes in intracellular chloride concentration. *Journal of Neuroscience Methods*, *57*(1), 27–35.

Lee, J., Woo, J., Favorov, O. V., Tommerdahl, M., Lee, C. J., & Whitsel, B. L. (2012). Columnar distribution of activity dependent gabaergic depolarization in sensorimotor cortical neurons. *Molecular Brain*, *5*, 33.

Li, H., Khirug, S., Cai, C., Ludwig, A., Blaesse, P., Kolikova, J., et al. (2007). KCC2 interacts with the dendritic cytoskeleton to promote spine development. *Neuron*, *56*(6), 1019–1033.

Liou, S. Y., & Albers, H. E. (1990). Single unit response of neurons within the hamster suprachiasmatic nucleus to GABA and low chloride perfusate during the day and night. *Brain Research Bulletin*, *25*(1), 93–98.

Luhmann, H. J., Kirischuk, S., & Kilb, W. (2014). Comment on "local impermeant anions establish the neuronal chloride concentration". *Science*, *345*(6201), 1130.

Marandi, N., Konnerth, A., & Garaschuk, O. (2002). Two-photon chloride imaging in neurons of brain slices. *Pflügers Archiv*, *445*(3), 357–365.

Markova, O., Mukhtarov, M., Real, E., Jacob, Y., & Bregestovski, P. (2008). Genetically encoded chloride indicator with improved sensitivity. *Journal of Neuroscience Methods*, *170*(1), 67–76.

Martina, M., Royer, S., & Pare, D. (2001). Cell-type-specific GABA responses and chloride homeostasis in the cortex and amygdala. *Journal of Neurophysiology*, *86*(6), 2887–2895.

Mazzuca, M., Minlebaev, M., Shakirzyanova, A., Tyzio, R., Taccola, G., Janackova, S., et al. (2011). Newborn analgesia mediated by oxytocin during delivery. *Frontiers in Cellular Neuroscience*, *5*, 3.

Medina, I., Friedel, P., Rivera, C., Kahle, K. T., Kourdougli, N., Uvarov, P., et al. (2014). Current view on the functional regulation of the neuronal K(+)-Cl(−) cotransporter KCC2. *Frontiers in Cellular Neuroscience*, *8*, 27.

Morita, K., Tsumoto, K., & Aihara, K. (2006). Bidirectional modulation of neuronal responses by depolarizing GABAergic inputs. *Biophysical Journal*, *90*(6), 1925–1938.

Myers, V. B., & Haydon, D. A. (1972). Ion transfer across lipid membranes in the presence of gramicidin A. II. The ion selectivity. *Biochimica et Biophysica Acta*, *274*(2), 313–322.

Owens, D. F., Boyce, L. H., Davis, M. B., & Kriegstein, A. R. (1996). Excitatory GABA responses in embryonic and neonatal cortical slices demonstrated by gramicidin perforated-patch recordings and calcium imaging. *The Journal of Neuroscience*, *16*(20), 6414–6423.

Owens, D. F., & Kriegstein, A. R. (2002). Is there more to GABA than synaptic inhibition? *Nature Reviews. Neuroscience*, *3*(9), 715–727.

Payne, J. A., Rivera, C., Voipio, J., & Kaila, K. (2003). Cation-chloride co-transporters in neuronal communication, development and trauma. *Trends in Neurosciences*, *26*(4), 199–206.

Payne, J. A., Stevenson, T. J., & Donaldson, L. F. (1996). Molecular characterization of a putative K Cl cotransporter in rat brain. A neuronal-specific isoform. *The Journal of Biological Chemistry*, *271*(27), 16245–16252.

Puskarjov, M., Seja, P., Heron, S. E., Williams, T. C., Ahmad, F., Iona, X., et al. (2014). A variant of KCC2 from patients with febrile seizures impairs neuronal Cl- extrusion and dendritic spine formation. *EMBO Reports*, *15*(6), 723–729.

Raimondo, J. V., Markram, H., & Akerman, C. J. (2012). Short-term ionic plasticity at GABAergic synapses. *Frontiers in Synaptic Neuroscience*, *4*, 5.

Rivera, C., Voipio, J., & Kaila, K. (2005). Two developmental switches in GABAergic signalling: The K+−Cl- cotransporter KCC2 and carbonic anhydrase CAVII. *The Journal of Physiology*, *562*(Pt 1), 27–36.

Rudolph, U., & Mohler, H. (2004). Analysis of $GABA_A$ receptor function and dissection of the pharmacology of benzodiazepines and general anesthetics through mouse genetics. *Annual Review of Pharmacology and Toxicology*, *44*, 475–498.

Russell, J. M. (2000). Sodium-potassium-chloride cotransport. *Physiological Reviews*, *80*(1), 211–276.

Sauer, J. F., Struber, M., & Bartos, M. (2012). Interneurons provide circuit-specific depolarization and hyperpolarization. *The Journal of Neuroscience*, *32*(12), 4224–4229.

Staley, K. J., Soldo, B. L., & Proctor, W. R. (1995). Ionic mechanisms of neuronal excitation by inhibitory $GABA_A$ receptors. *Science*, *269*(5226), 977–981.

Szabadics, J., Varga, C., Molnar, G., Olah, S., Barzo, P., & Tamas, G. (2006). Excitatory effect of GABAergic axo-axonic cells in cortical microcircuits. *Science*, *311*(5758), 233–235.

Thompson, S. M., & Gahwiler, B. H. (1989). Activity-dependent disinhibition. II. Effects of extracellular potassium, furosemide, and membrane potential on ECl- in hippocampal CA3 neurons. *Journal of Neurophysiology*, *61*(3), 512–523.

Tyzio, R., Cossart, R., Khalilov, I., Minlebaev, M., Hubner, C. A., Represa, A., et al. (2006). Maternal oxytocin triggers a transient inhibitory switch in GABA signaling in the fetal brain during delivery. *Science*, *314*(5806), 1788–1792.

Tyzio, R., Nardou, R., Ferrari, D. C., Tsintsadze, T., Shahrokhi, A., Eftekhari, S., et al. (2014). Oxytocin-mediated GABA inhibition during delivery attenuates autism pathogenesis in rodent offspring. *Science*, *343*(6171), 675–679.

van Brederode, J. F., Takigawa, T., & Alzheimer, C. (2001). GABA-evoked chloride currents do not differ between dendrites and somata of rat neocortical neurons. *The Journal of Physiology*, *533*(Pt 3), 711–716.

Vogels, T. P., Froemke, R. C., Doyon, N., Gilson, M., Haas, J. S., Liu, R., et al. (2013). Inhibitory synaptic plasticity: Spike timing-dependence and putative network function. *Frontiers in Neural Circuits*, 7, 119.

Voipio, J., Boron, W. F., Jones, S. W., Hopfer, U., Payne, J. A., & Kaila, K. (2014). Comment on "local impermeant anions establish the neuronal chloride concentration". *Science*, *345*(6201), 1130.

Willadt, S., Nenniger, M., & Vogt, K. E. (2013). Hippocampal feedforward inhibition focuses excitatory synaptic signals into distinct dendritic compartments. *PLoS One*, *8*(11), e80984.

Woo, N. S., Lu, J., England, R., McClellan, R., Dufour, S., Mount, D. B., et al. (2002). Hyperexcitability and epilepsy associated with disruption of the mouse neuronal-specific K-Cl cotransporter gene. *Hippocampus*, *12*(2), 258–268.

Woodin, M. A., Ganguly, K., & Poo, M. M. (2003). Coincident pre- and postsynaptic activity modifies GABAergic synapses by postsynaptic changes in Cl- transporter activity. *Neuron*, *39*(5), 807–820.

Yamada, J., Okabe, A., Toyoda, H., Kilb, W., Luhmann, H. J., & Fukuda, A. (2004). Cl- uptake promoting depolarizing GABA actions in immature rat neocortical neurones is mediated by NKCC1. *The Journal of Physiology*, *557*(Pt 3), 829–841.

Yan, Y., Dempsey, R. J., & Sun, D. (2001). Expression of Na(+)-K(+)-Cl(−) cotransporter in rat brain during development and its localization in mature astrocytes. *Brain Research*, *911*(1), 43–55.

Zhu, L., Polley, N., Mathews, G. C., & Delpire, E. (2008). NKCC1 and KCC2 prevent hyperexcitability in the mouse hippocampus. *Epilepsy Research*, *79*(2–3), 201–212.

CHAPTER NINE

The Diversity of $GABA_A$ Receptor Subunit Distribution in the Normal and Huntington's Disease Human Brain[1]

H.J. Waldvogel*,†, R.L.M. Faull*,†,1

*Centre for Brain Research, Faculty of Medical and Health Sciences, University of Auckland, Auckland, New Zealand

†Department of Anatomy with Radiology, Faculty of Medical and Health Sciences, University of Auckland, Auckland, New Zealand

[1]Corresponding author: e-mail address: rlm.faull@auckland.ac.nz

Contents

[1]Tribute to Professor Hanns Mohler

Advances in Pharmacology, Volume 73
ISSN 1054-3589
http://dx.doi.org/10.1016/bs.apha.2014.11.010

Abstract

GABA$_A$ receptors are assembled into pentameric receptor complexes from a total of 19 different subunits derived from a variety of different subunit classes (α_{1-6}, β_{1-3}, γ_{1-3}, δ, ε, θ, and π) which surround a central chloride ion channel. GABA$_A$ receptor complexes are distributed heterogeneously throughout the brain and spinal cord and are activated by the extensive GABAergic inhibitory system. In this chapter, we describe the heterogeneous distribution of six of the most widely distributed subunits (α_1, α_2, α_3, $\beta_{2,3}$, and γ_2) throughout the human basal ganglia. This review describes the studies we have carried out on the normal and Huntington's disease human basal ganglia using autoradiographic labeling and immunohistochemistry in the human basal ganglia. GABA$_A$ receptors are known to react to changing conditions in the brain in neurological disorders, especially in Huntington's disease and display a high degree of plasticity which is thought to compensate for loss of function caused by disease. In Huntington's disease, the variable loss of GABAergic medium spiny striatopallidal projection neurons is associated with a loss of GABA$_A$ receptor subunits in the striosome and/or the matrix compartments of the striatum. By contrast in the globus pallidus, a loss of the GABAergic striatal projection neurons results in a dramatic upregulation of subunits on the large postsynaptic pallidal neurons; this is thought to be a compensatory plastic mechanism resulting from the loss of striatal GABAergic input. Most interestingly, our studies have revealed that the subventricular zone overlying the caudate nucleus contains a variety of proliferating progenitor stem cells that possess a heterogeneity of GABA$_A$ receptor subunits which may play a role in human brain repair mechanisms.

1. INTRODUCTION

1.1. GABA$_A$ receptors

Gamma-aminobutyric acid (GABA) is the most widespread inhibitory neurotransmitter in the central nervous system. It acts via GABA$_A$, GABA$_B$, and GABA$_C$ receptors distributed throughout the central nervous system. There is an extensive literature showing the widespread distribution of GABA$_A$ receptors (GABA$_A$Rs) throughout the human brain as evidenced by autoradiographic binding studies using GABA$_A$R-specific ligands, such as the GABA agonist [^{3}H]muscimol, and benzodiazepines, such as [^{3}H] flunitrazepam, [^{3}H]Ro15-1788 and antibodies to specific receptor subunits (Faull & Villiger, 1988; Faull, Waldvogel, Nicholson, & Synek, 1993; Faull, Waldvogel, Nicholson, Williams, & Dragunow, 1994; Loup, Wieser, Yonekawa, Aguzzi, & Fritschy, 2000; Waldvogel et al., 1990; Whitehouse et al., 1985; Zezula, Cortes, Probst, & Palacios, 1988). GABA$_A$Rs are known to be comprised of pentameric subunit complexes surrounding an inhibitory chloride ion channel. These receptor complexes are comprised of configurations of five subunits assembled from a total of 19 subunits derived from a variety of different subunit classes (α_{1-6}, β_{1-3},

γ_{1-3}, δ, ε, θ, and π) (Barnard et al., 1988; Burt, 2003; Sieghart, 1995). $GABA_A$Rs are the most widespread inhibitory receptor in the central nervous system and are localized mainly on postsynaptic membranes of synaptic complexes. However, there is also evidence that they occur at presynaptic sites as well as at extrasynaptic sites on dendritic membranes (Kullmann et al., 2005). Several of the $GABA_A$R subunits are associated with the tubulin-linker protein gephyrin (Fritschy, Harvey, & Schwarz, 2008) that functions as a postsynaptic organizer molecule for major subtypes of $GABA_A$Rs and anchors them into postsynaptic membrane complexes (Moss & Smart, 2001; Tyagarajan & Fritschy, 2014). Additionally, other receptor-associated proteins, such as the TRAK family of proteins and the Huntington-associated protein, HAP1, are implicated in $GABA_A$R subunit trafficking (Stephenson, 2014). $GABA_A$Rs facilitate fast-response, inhibitory neurotransmission in the mammalian central nervous system, and the subunit configuration of each receptor complex determines the pharmacological and physiological responses of that individual receptor (Sieghart, 1995). Extensive research efforts have focused on the $GABA_A$Rs because of their highly complex heterogeneous subunit nature and many and varied roles in inhibitory mechanisms in the central nervous system. Dysfunction of $GABA_A$R responses and mutations in $GABA_A$R genes leads to various neurological disorders (Mohler, 2006b). A large number of drugs that act primarily via $GABA_A$Rs (e.g., benzodiazepines, barbiturates) are used to treat a variety of neurological conditions, for instance epilepsy, sleep disturbances, anxiety, schizophrenia, and others (Michels & Moss, 2007).

The detailed pharmacology and physiology related to the different configurations of subunits that produce the variety of $GABA_A$R subtypes (Mohler, 2006a, 2007; Wisden & Seeburg, 1992) are out of the scope of this chapter, but it provides the morphological basis for the complex $GABA_A$R pharmacology. This chapter reviews the morphological studies on the localization of $GABA_A$R subunits in the human basal ganglia which forms the basis for our understanding of the functions of $GABA_A$R subunits in this region of the brain.

1.1.1 Basal ganglia

The basal ganglia are a large group of nuclei located subcortically in the base of the forebrain which are comprised of predominantly GABAergic neurons and are intimately involved with the control of mood and movement and a variety of cognitive tasks (Alexander, Crutcher, & Delong, 1990). The nuclei belonging to the basal ganglia were originally considered to be the principal components of the "extrapyramidal system," and by convention, the term basal ganglia is now restricted mainly to: the striatum (comprised of the caudate nucleus (CN) and putamen), globus pallidus internal and external segments, the subthalamic nucleus, (STN), and the substantia nigra (Carpenter, 1976; Smith, Bevan, Shink, & Bolam, 1998). In the human brain, the striatum is divided into the two large nuclear masses: (i) the CN, which rostrally forms a head, more centrally a body, and posteriorly a tail region which extends dorsally over the thalamus adjacent to the wall of the lateral ventricle; and (ii) the putamen (P), which lies lateral to the CN and globus pallidus. The CN and P are separated by the condensed fibers of the internal capsule. The globus pallidus is divided into two parts: the external segment of the globus pallidus (GPe) and the internal segment of the globus pallidus (GPi) (also termed medial and lateral segments), which correspond to the globus pallidus and entopeduncular nucleus in rodent brains, respectively. The STN is located ventral to the thalamus, medial to the GPi, and rostrodorsal to the substantia nigra. The substantia nigra consists of two parts: the substantia nigra pars reticulata (SNr), which is located ventrally in the midbrain, and the substantia nigra pars compacta (SNc), which in humans and primates contains groups of pigmented cells lying in a crescent shape in the dorsal regions of the substantia nigra. Although the substantia nigra is located in the midbrain, it is considered part of the basal ganglia due to its close functional and connectional interrelationships with the striatum.

1.1.2 Neurochemical compartments

The mammalian striatum is further subdivided into two major interdigitating compartments comprised of the smaller neurochemically defined islands or patches of labeling termed "striosomes" within a larger surrounding extrastriosomal region termed the "matrix." These compartments were first identified by Graybiel and Ragsdale (1978) using acetylcholinesterase staining which labeled the matrix, leaving the smaller striosome compartment unstained. The striosome compartment is also identified with high concentrations of a variety of neurochemical markers such as neurotensin, LAMP, dopamine D2 receptors, substance P, and enkephalin, while the larger matrix compartment is characterized by high concentrations of other neurochemicals, AChE, the calcium-binding proteins calbindin and parvalbumin, tyrosine hydroxylase and the glutamatergic NMDA and AMPA receptors (Dure, Young, & Penney, 1992; Faull & Villiger, 1988;

Graybiel, 1990; Holt, Graybiel, & Saper, 1997; Parent, Fortin, Cote, & Cicchetti, 1996; Prensa, Gimenez-Amaya, & Parent, 1999; Waldvogel & Faull, 1993). These two neurochemical compartments, the striosomes and the matrix, are closely associated with either the limbic system or the sensory-motor system, respectively. The striosomes appear to be an extension of the ventral striatum and have close connections with the orbitofrontal and cingulate cortices and other components of the limbic system. The matrix component has strong connectional associations with the sensory-motor system particularly the frontal, motor, and sensory cortices (Alexander, DeLong, & Strick, 1986; Graybiel, 1990; Heimer & Wilson, 1975).

Within the striatum, the resident neurons fall principally into two major classes of neurons, first, the medium-sized spiny projection neurons and, second, the aspiny local circuit interneurons. These various classes of neurons are subclassified according to their size, neurochemistry, and connectional characteristics. The majority of these neurons (at least 95%) are the medium spiny projection neurons which use the inhibitory neurotransmitter GABA, and they project mainly to the GPe, GPi, and SNr. Those that project to the GPe contain the cotransmitter enkephalin, whereas those projecting to the GPi and SNr contain the cotransmitter substance P. The remaining neurons are a morphologically and neurochemically heterogeneous group of interneurons which modulate the function of the medium spiny output neurons. This heterogeneous group of interneurons modulates the activity of medium spiny neurons in a highly complex fashion. The majority of interneurons use GABA as their major neurotransmitter and can be labeled with glutamic acid decarboxylase (GAD), the enzyme which produces GABA, and these are subdivided into groups depending on the calcium-binding proteins they contain, principally parvalbumin- and calretinin-containing interneurons (Cicchetti & Parent, 1996; Cicchetti, Prensa, Wu, & Parent, 2000). The two other major types of interneurons in the human striatum are the large cholinergic interneurons, which also contain calretinin and the interneurons containing a mix of neuropeptide Y, somatostatin, NOS, and NADPH diaphorase.

The neurons of the globus pallidus and substantia nigra pars reticulata are large generally elongated neurons with long-branching dendrites and project to the thalamus; they are GABAergic and variably include the calcium-binding proteins, parvalbumin and calretinin (Waldvogel et al., 2008; Waldvogel, Kubota, Fritschy, Mohler, & Faull, 1999).

1.1.3 GABA$_A$Rs in the human basal ganglia

1.1.3.1 Regional localization

Autoradiographic studies using various ligands to the GABA receptor complex show that in the normal human brain, GABA$_A$Rs are detected at high concentrations throughout the basal ganglia. Studies using tritiated flunitrazepam ([^{3}H]FNZ) and tritiated Ro15-1788 ([^{3}H]Ro15-1788), ligands specific for GABAergic benzodiazepine receptors, show high levels of binding in the striatum but with lower levels in the globus pallidus with the external segment having higher levels than the internal segment (Faull & Villiger, 1988; Faull et al., 1993; Glass, Dragunow, & Faull, 2000; Zezula et al., 1988; Figs. 1, 6A and B, and 7A and B). In agreement with binding studies carried out in the human substantia nigra (Zezula et al., 1988), we have found high levels of binding with two benzodiazepine ligands, [^{3}H] FNZ and Ro15-1788, both showing higher levels of binding in the pars compacta than pars reticulata region (Waldvogel, Baer, & Faull, 2009; Fig. 5A).

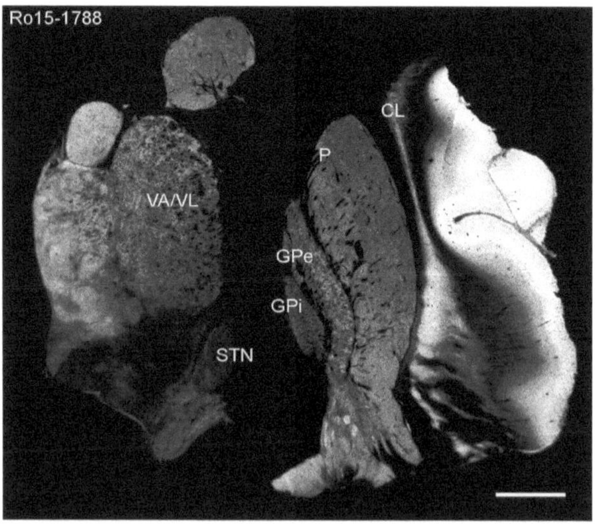

Figure 1 Binding of GABA$_A$/benzodiazepine receptor-specific label Ro15-1788 in the human basal ganglia; highest levels = bright levels or colors, lowest levels = darker levels or colors. Cl, claustrum; GPe, globus pallidus external segment; GPi, globus pallidus external segment; P, putamen STN, subthalamic nucleus; VA/VL, ventral anterior/ventral lateral nuclei of the thalamus. Scale bar = 0.5 cm.

1.2. $GABA_AR$ subunit localization

1.2.1 Regional distribution striatum (Fig. 2)

Our laboratory has investigated the regional and cellular immunohistochemical distribution patterns of the $GABA_AR$ subunits α_1, α_2, α_3, $\beta_{2,3}$, and γ_2 in the human striatum. As shown in Fig. 2, the regional labeling pattern for the five $GABA_AR$ subunits is markedly heterogeneous throughout the striatum. The reference label for the peptide enkephalin (Fig. 2F) is highly expressed in the striosomes and ventral striatum region and is used to delineate the striosome from the matrix compartments. The regional localization of α_1-subunit immunoreactivity produced a marked heterogeneous distribution throughout the striatum (Fig. 2A). The heterogeneity of labeling was evident as a very low level of labeling in the striosomes of various shapes and sizes in the dorsal striatum and putamen, and these regions corresponded closely to the regions of high enkephalin immunoreactivity in adjacent sections (arrow, Fig. 2A and F). These were surrounded by a background matrix of labeling of moderate α_1-subunit immunoreactivity in the striatal matrix. In the ventral striatum, the patches of low α_1-subunit immunoreactivity corresponded to the patches of low enkephalin immunoreactivity in adjacent sections (Fig. 2A). The α_2-subunit immunoreactivity (Fig. 2B) showed high intensities of labeling throughout the dorsal striatum, but there was especially intense labeling in patches which could be identified as striosomes on adjacent sections stained for enkephalin. Individual α_2-rich striosomes appear to be surrounded by an α_2-poor annular zone intervening between the intensely immunoreactive striosome and the moderately high intensity of labeling of the surrounding matrix compartment. In the ventral striatum, the low-density patches of α_2-subunit immunoreactivity (Fig. 2B, double arrow) corresponded to the low-density patches of enkephalin immunoreactivity. The α_3-subunit immunoreactivity (Fig. 2C) was moderately low throughout the striatum but did show some heterogeneity of staining. In the dorsal CN, patches of higher intensities of α_3-subunit labeling were evident and these appear to correspond with striosomes. In the ventral striatum, α_3-poor patches were scattered within regions of high α_3-subunit immunoreactivity. The $\beta_{2,3}$-subunit immunoreactivity (Fig. 2D) showed high intensities of labeling throughout both compartments of the striatum. In the dorsal striatum, patches of intense $\beta_{2,3}$-subunit labeling could be distinguished (arrows, Fig. 2D) which aligned with striosomes as identified by markers on adjacent sections. The γ_2-subunit immunoreactivity (Fig. 2E) showed moderate intensities throughout the striatum. The

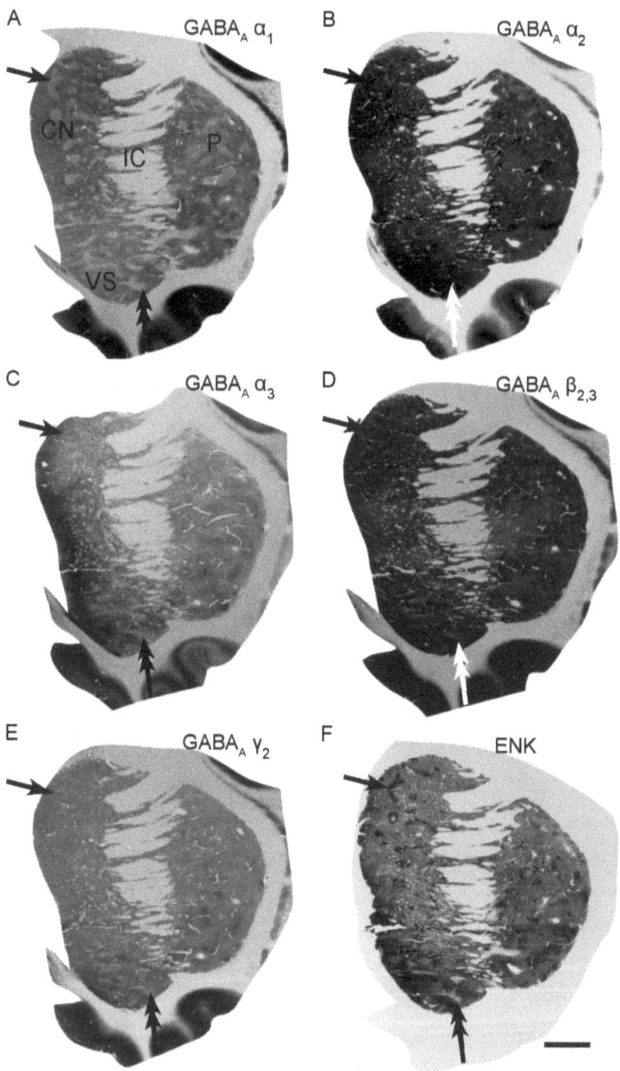

Figure 2 *Serial sections labeled for different GABA$_A$R subunits in the human caudate putamen.* (A) α_1 subunits; (B) α_2 subunits; (C) α_3 subunits; (D) $\beta_{2,3}$ subunits; (E) γ_2 subunits; (F) enkephalin staining is highly localized to the striosome compartment with lower levels in the striosomes; arrow in the dorsal compartment indicates a striosome, arrow in the ventral striatum show regions of low levels of subunit and enkephalin labeling. Scale bar = 0.5 cm. *Modified from Waldvogel et al. (1999), Wiley-Liss.*

CN and putamen showed relatively homogeneous labeling; however, small regions of higher intensity of γ_2-subunit labeling which corresponded to striosomes were evident in the dorsal region of the CN and putamen. In the ventral striatum, γ_2-subunit-poor patches were located within a more densely stained background matrix region. A variety of mRNA and immunohistochemical studies in the striatum of rodents and primates (Fritschy & Mohler, 1995; Goetz, Arslan, Wisden, & Wulff, 2007; Kultas-Ilinsky, Leontiev, & Whiting, 1998; Miralles, Li, Mehta, Khan, & De Blas, 1999; Pirker, Schwarzer, Wieselthaler, Sieghart, & Sperk, 2000; Wisden, Laurie, Monyer, & Seeburg, 1992) have found additional subunits exist in the striatum. For the $GABA_A$ α subunits in rodent studies, the α_2 and α_4 subunits have the highest concentrations in the striatum followed by weaker α_1 and α_3 subunits (Schwarzer et al., 2001). Although no mRNA was found for α_5 subunits (Wisden et al., 1992), the α_5 subunit protein was found at low levels (Fritschy & Mohler, 1995; Pirker et al., 2000; Schwarzer et al., 2001). For the β subunits, β_3 subunits showed the highest levels with moderate levels of β_2 subunits and low levels of β_1 subunits. The δ subunit was also found diffusely distributed to a moderate degree. Of the γ subunits, only γ_2 subunits were detected to any degree in the striatum. In the rat, the α and β subunits also showed patchy distribution reminiscent of the patch and matrix compartments (Schwarzer et al., 2001), although the α_1 subunit appears to be high in striosomes in rat, whereas as reported above, it is low in human striosomes. In summary, these findings suggest that the striosome compartment contains high densities of α_2, α_4, $\beta_{2,3}$, and γ_2 subunits, whereas the matrix shows moderate to high densities of α_1, α_2, α_4, $\beta_{2,3}$, γ_2, and δ subunits.

1.2.2 Regional distribution globus pallidus

The external and internal segments of the globus pallidus can be delineated by enkephalin and substance P staining, respectively (Allen, Waldvogel, Glass, & Faull, 2009; Deng et al., 2004). The α_1, α_3, $\beta_{2,3}$, and γ_2 subunits showed immunoreactive labeling of both segments of the globus pallidus and, as with the autoradiographic $GABA_AR$ ligand-binding studies, produced moderate levels of immunoreactivity in the external segment of the globus pallidus and lower levels in the internal segment. The α_1 subunit appeared to stain more intensely than α_3, $\beta_{2,3}$, or γ_2 subunits in both segments of the globus pallidus, the staining for the α_3 subunit was the least intense of the subunits, and there was barely detectable α_2-subunit immunoreactivity. The subunits observed in the human globus pallidus are also observed at high levels in rat, and additional subunits have also been reported. Several studies clearly show that the dominant α subunit is the α_1 subunit with low levels of all other α subunits and that the dominant β subunit is the β_2 subunit with little β_3 subunit being expressed. For the γ subunits, the rare γ_1 is found in the rat GP and γ_2 subunits are found in the rat and primate GP (Fritschy & Mohler, 1995; Kultas-Ilinsky et al., 1998; Pirker et al., 2000; Schwarzer et al., 2001; Somogyi, Fritschy, Benke, Roberts, & Sieghart, 1996). The α_1, $\beta_{2,3}$, and γ_2 subunits are found at the same synapses at the ultrastructural level indicating the existence of this well-recognized subunit combination in the globus pallidus (Somogyi et al., 1996). The δ subunit appears to be lacking in the rat globus pallidus but is found in the primate (Kultas-Ilinsky et al., 1998).

1.3. GABA$_A$ receptor subunit cellular distribution (Figs. 3 and 4)

1.3.1 Striatum

The cellular localization of the various GABA$_A$R subunits shows a heterogeneous distribution on the various neurochemically defined neuron types in the human striatum (Figs. 3A, C, E, G, and I). Various GABA$_A$R subunits are evident on the variety of striatal neuronal types.

1.3.1.1 Medium spiny neuron

The most common neuron type in the striatum is the medium spiny neuron. The full extent of the soma and spiny dendritic tree of the GABAergic medium-sized spiny projection neurons of the striatum is difficult to visualize using immunohistochemical stains, but the soma and proximal dendrites can be labeled with calbindin. In our double labelling studies on the human striatum, only the antibodies against the α_2 and $\beta_{2,3}$ subunits (Fig. 3C and G) colabeled with calbindin on the proximal dendrites of GABAergic medium-sized spiny projection neurons (e.g., Fig. 4A and B; Waldvogel et al., 1999). The high level of staining for both subunits in the neuropil (Figs. 2B and D and 3C and G) would also indicate the subunits α_2 and $\beta_{2,3}$ may be distributed on the spiny dendritic tree. Also medium spiny neurons are known to receive terminals from the parvalbumin-positive fast-spiking interneurons (Bennett & Bolam, 1994) and, coupled with the moderate level of α_1 subunit labeling in the matrix, indicate that α_1 subunit containing receptors may also be present on the medium-sized spiny neuron dendritic tree although to a lower degree. The α_1, $\beta_{2,3}$, and γ_2 subunits have been found together at striatal synapses on medium spiny neurons and interneurons (Fujiyama, Fritschy, Stephenson, & Bolam, 2000).

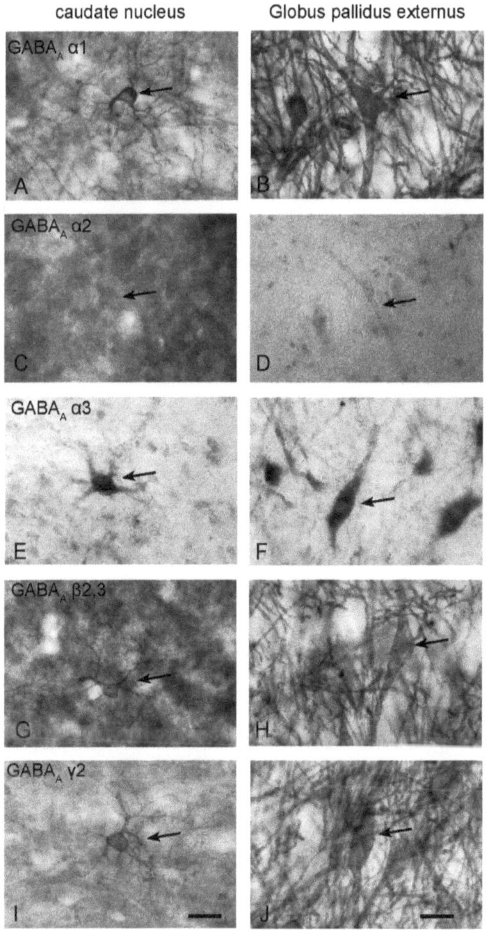

Figure 3 *Images of $GABA_A$ receptor subunit staining from human CN (A, C, E, G, I) and external segment of the globus pallidus (B, D, F, H, J).* The images show various neuronal staining patterns for the different subunits in these two regions of the basal ganglia. (A, B) α_1 subunits; (C, D) α_2 subunits; (E, F) α_3 subunits; (G, H) $\beta_{2,3}$ subunits; (I, J) γ_2 subunits. In the CN, the α_1, $\beta_{2,3}$, and γ_2 subunits produce similar cellular staining patterns on dendritic membranes, while α_2 and α_3 subunits produce different more intracellular staining patterns. Scale bars = 20 μm.

Further studies need to be carried out to determine the level of these and other subunits on medium spiny neurons.

1.3.1.2 Interneurons

Most GAD-positive interneurons in the human striatum were colocalized with α_1 and $\beta_{2,3}$ subunits (Fig. 4C and D). These subunits generally also

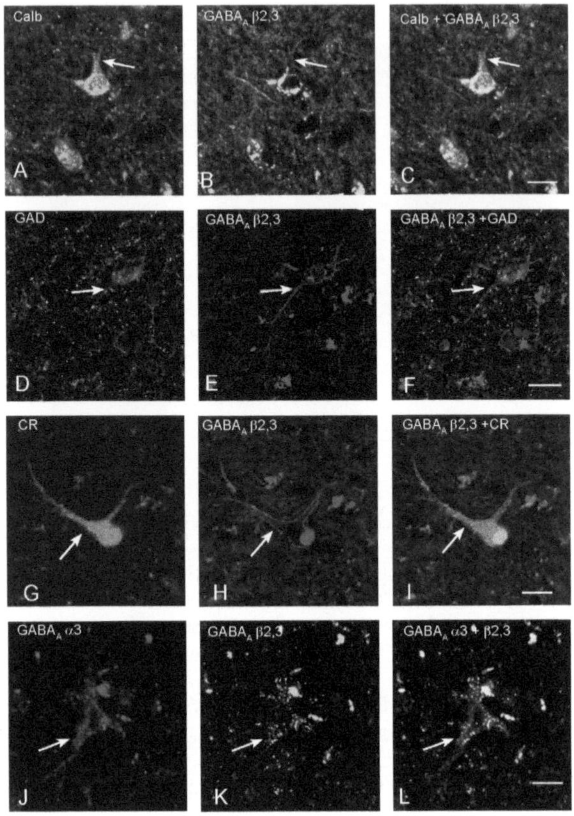

Figure 4 *Double labeling of GABA$_A$ receptor subunits and cellular markers in the human CN.* (A) A medium spiny calbindin-positive neuron, in (B) is double labeled with $\beta_{2,3}$ subunits on its proximal dendrites. C is the combined image. (D and E) A GAD-positive interneuron which is double labeled for $\beta_{2,3}$ subunits that completely outline its dendritic tree. F is the combined image. (G and H) A calretinin-positive neuron which is coated with $\beta_{2,3}$ subunits. This appears to be a unique cell type in the human striatum. I is the combined image. (J and K) The α_3 and $\beta_{2,3}$ subunits are found in the same cell, presumably a large cholinergic neuron due to its larger size neuron and morphology. L is the combined image. arrows indicate sites of receptor labelling. Scale bars = 20 μm. *Images A-C and G-I Reproduced from Waldvogel et al. (1999), Wiley-Liss. Images D-F and J-L kindly donated by V. Dieriks University of Auckland.*

colocalize with γ_2 subunits. The parvalbumin-positive neurons belong to a well-recognized group of striatal parvalbumin-positive fast-spiking interneurons which have been described in several different mammalian species (Bennett & Bolam, 1994; Bolam, Hanley, Booth, & Bevan, 2000; Kawaguchi, Wilson, Augood, & Emson, 1995; Waldvogel, Faull,

Williams, & Dragunow, 1991). These neurons have been associated with α_1, $\beta_{2,3}$, and presumably γ_2 subunit coating their dendritic membranes (Waldvogel et al., 1999).

In the human striatum, calretinin labels a diverse group of neurons (Cicchetti et al., 2000). These are grouped into the medium-sized most common type, a larger less common type 2 branched cell which appears to be a unique cell type to the human, and the large cholinergic interneurons which contain calretinin. The first two types colocalize with the α_1, $\beta_{2,3}$ subunits, whereas the large calretinin/ChAT cells do not contain the α_1 subunit.

The large cholinergic neurons have mainly the α_3 subunit (Waldvogel et al., 1999) as well as possibly the $\beta_{2,3}$ subunits (Fig. 4G and H). The group of interneurons containing Neuropeptide Y/somatostatin/NOS was not labeled by the five most common subunits and so far have not been shown to contain any GABA$_A$R subunits. In animal studies, additional subunits α_5 and both β_2 and β_3 were found on striatal neurons resembling interneurons, but the neurochemistry of these neurons was not further characterized (Schwarzer et al., 2001). What is interesting is that in the rat striatum, there is a dominant α_1 subunit neuron subtype (type 1) which is not parvalbumin positive and is not obvious in human and nonhuman primate (Hartig, Brauer, Fritschy, Bruckner, & Bigl, 1995; Waldvogel, Fritschy, Mohler, & Faull, 1998; Waldvogel et al., 1997) and so seems to be a distinct species difference.

1.3.2 Globus pallidus

The human globus pallidus contains populations of large GABAergic projection neurons which also contain calcium binding proteins, parvalbumin and calretinin, and a subpopulation of interneurons which are mainly calretinin positive. These neurons are associated with GABA$_A$R α_1, $\beta_{2,3}$, γ_2 subunits which coat the soma and the entire length of the dendritic tree (Fig. 3B, D, H, and J) and additionally have high levels of α_3 subunit on the soma and dendrites and in the cytoplasm of these pallidal neurons (Fig. 3F). However, they have only barely detectable levels α_2 subunits (Fig. 3D). Interestingly, higher levels of these subunits appear to occur in the external segment than the internal segment (Waldvogel et al., 1999). This generally agrees with what is observed in animal studies although additionally the γ_1 subunit is found on rat pallidal neurons giving rise to a possible α_1, β_2, γ_1 combination which has a different benzodiazepine-related pharmacology to those containing the γ_2 subunit (Ymer et al., 1990).

1.3.3 Substantia nigra (Fig. 5)

The two regions of the substantia nigra, the GABAergic substantia nigra pars reticulata (SNr) and the dopaminergic substantia nigra pars compacta (SNc), contain two different receptor combinations. Immunohistochemical studies revealed high levels of α_1, α_3, $\beta_{2,3,}$ and γ_2 subunits in the SNr which are localized on the large reticulata neurons (Fig. 5A, E, F, G, I, K, M, and N; Waldvogel et al., 2008). These subunits coated the membranes of the large reticulata neurons, and at high resolution, this coating was observed to often appear as "hot spots" on the cell soma and dendrites and also to have a lattice or web-like structure which outlines the surface membranes (Fig. 5K and N). In contrast, the pigmented dopaminergic neurons of the SNc contain mainly α_3 and γ_2 subunits and these appear to be cytoplasmic in these neurons (Fig. 5J and L; Petri et al., 2002; Waldvogel et al., 2008). An additional group of nondopaminergic neurons labeled for α_1, α_3, $\beta_{2,3}$, and γ_2 subunits which have been termed displaced reticulata cells were also scattered throughout the SNc (Fig. 5H). Also a small proportion of the SNc dopaminergic neurons that revealed the SNr-like combination (α_1, α_3, $\beta_{2,3,}$ γ_2) were shown to exist (Petri et al., 2002; Waldvogel et al., 2008). In rodent and primate studies, additional subunits have also been identified which may also occur in human substantia nigra. In particular, the α_4 and δ subunit and possibly the α_5 are found in compacta and reticulata with β_2 being the dominant β subunit in the SNr.

1.3.4 Subunit combinations

The distribution of the five $GABA_AR$ subunits α_1, α_2, α_3, $\beta_{2,3}$, and γ_2 throughout the basal ganglia (Table 1) suggests the most common $GABA_AR$ subunit combination which occurs in the $GABA_AR$ complexes on the different basal ganglia cell types is (α_1,$\beta_{2,3}$,γ_2). Other possible configurations which include α_2, α_3, and α_4 subunits are (α_1,α_3,$\beta_{2,3}$,γ_2), (α_2,α_3,$\beta_{2,3}$,γ_2), and (α_3,$\beta_{2,3}$), ($\alpha_3\gamma_2$) (α_4,$\beta_{2,3}$,γ_2). The GABAergic interneurons (stained with GAD, parvalbumin, calretinin) and the large projection neurons of the GPe and GPi and SNr are most likely to have receptor subunit combinations α_1,$\beta_{2,3}$,γ_2, or α_1,α_3,$\beta_{2,3}$,γ_2; the medium spiny neurons of the striatum the combinations including α_1,$\beta_{2,3}$ or α_2,$\beta_{2,3}$, the striatal cholinergic interneurons α_3,$\beta_{2,3}$, and the SNc dopaminergic neurons $\alpha_3\gamma_2$ with β_1 as a possible β subunit. No $GABA_AR$ subunits were detected on the striatal NPY/somatostatin/NOS neuronal cell type which as detailed later are resistant to degeneration in Huntington's disease.

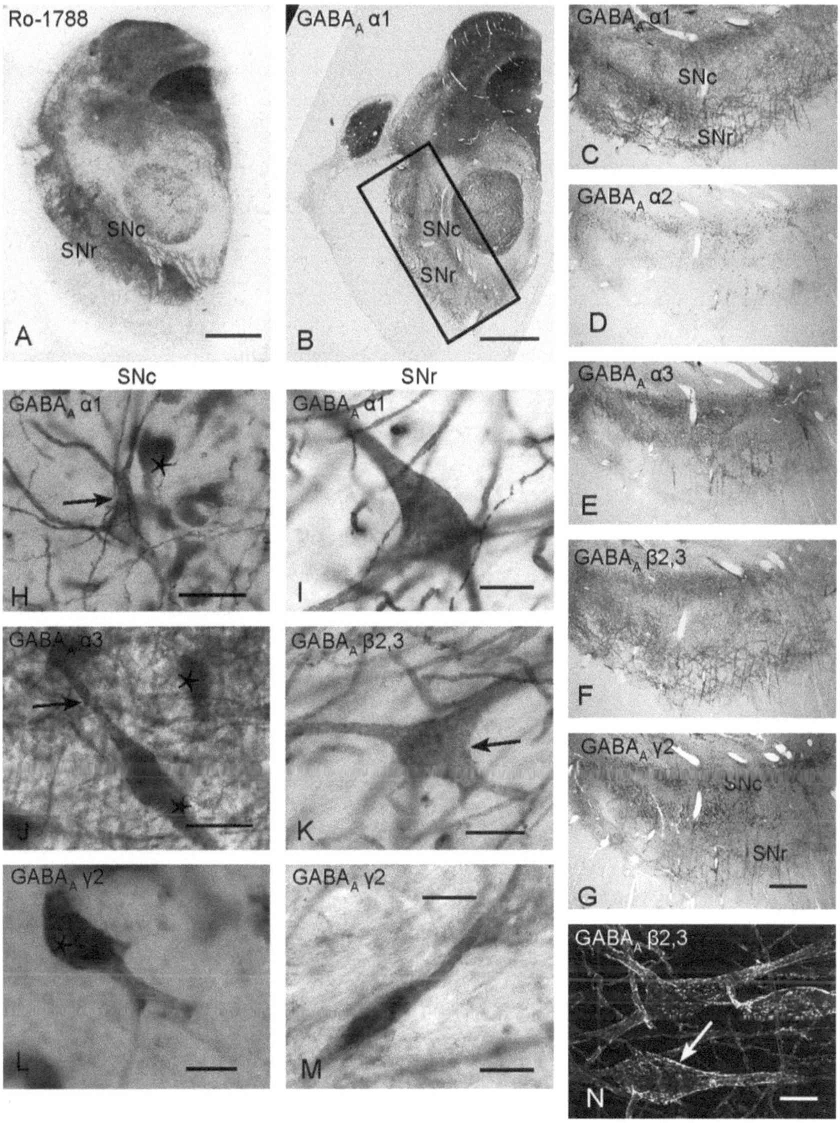

Figure 5 *$GABA_A$ receptor labeling in the human substantia nigra.* (A) A hemisection through the midbrain showing binding of the $GABA_A$/benzodiazepine receptor-specific label Ro15-1788 with very high levels of binding in the in the pars compacta region (SNc) and high levels in the pars reticulata region (SNr). (B) A similar section through the midbrain as in (A) labeled with the α_1 subunit. The boxed portion indicates the region illustrated in serial sections (C)–(G) for antibodies directed against the (C) α_1 subunits; (D) α_2 subunits; (E) α_3 subunits; (F) $\beta_{2,3}$ subunits; (G) γ_2 subunits. (H), (J), and (L) are images from the substantia nigra pars compacta region with neurons stained for: (H) α_1 subunits; (J) α_3 subunits; (L) γ_2 subunits. * indicates pigmented neurons (I), (K), and (M) illustrate neurons in the substantia nigra pars reticulata labeled for the subunits α_1, $\beta_{2,3}$, γ_2 subunits, respectively. (N) Confocal laser scanning microscopy image illustrating receptor clusters (arrow) on the surface membranes of neurons in the SNr. Scale bars (A) and (B) = 0.5 cm, (C)–(G) = 100 μm, (H)–(N) = 20 μm. *Images (B)–(G) sourced from Waldvogel et al. (2008), Wiley-Liss.*

Table 1 GABA$_A$ receptor subunits found in the various regions of the basal ganglia

Region	α_1	α_2	α_3	α_4	α_5	β_1	β_2	β_3	γ_1	γ_2	δ
Striatum	++	++++	++	++++	+	+	+	++++	–	++	++
GPe	+++	+	++	+	+	+	++++	+	++	++	++
GPi	++	–	+	+	+	+	+	+	+	++	++
SNc	+	+	+++	+	+	++	+	++	–	+++	+
SNr	++++	++	+++	+	+	+	++++	+++	+	++	+

++++, intense immunoreactivity (IR); +++, high IR; ++, moderate IR; +, weak IR; –, no detectable IR; results from human, rat, and primate studies.

These combinations compare well with other animal studies which have determined that the most common subunit combinations observed throughout the rat brain are (α_1,$\beta_{2,3}$,γ_2), (α_3,$\beta_{2,3}$,γ_2), (α_1,α_3,$\beta_{2,3}$,γ_2), ($\alpha_1\alpha_3\gamma_2$), and ($\alpha_3\gamma_2$) (Fritschy et al., 1992; Fritschy & Mohler, 1995; Mohler, Benke, Benson, Luscher, & Fritschy, 1995; Mohler et al., 1996). In the basal ganglia, the most common inferred combination of GABA$_A$R subunits is α_1,$\beta_{2,3}$,γ_2 on the cells of the striatum because the individual subunits α_1, $\beta_{2,3}$, and γ_2 showed intense labeling over their entire cell body and dendritic tree. Also a large neuronal type in the border region of the globus pallidus and ventral putamen contains the α_1, $\beta_{2,3}$, and γ_2 subunit combination. This type of neuron is thought to belong to a type of projection neuron (second type of striatonigral neuron) or an ectopic pallidal neuron (Bolam, Powell, Totterdell, & Smith, 1981).

The role of GABA$_A$Rs within the basal ganglia circuitry is highly complex. The basal ganglia are integrated into a circular interconnected forebrain loop which forms a cortical/basal ganglia/thalamus/cortical circuit (Nauta & Domesick, 1984). In this circuit, the cerebral cortex and thalamus provide a major excitatory glutamatergic input to the CN and putamen (Carpenter, 1976) that arises bilaterally from practically every cerebral cortical region (McGeorge & Faull, 1989) and an excitatory projection from the intralaminar nuclei and other nuclei of the thalamus (Sadikot, Parent, Smith, & Bolam, 1992). Other inputs include the major dopaminergic projection which arises from the substantia nigra pars compacta, a serotonergic input from the dorsal raphe nucleus (Graybiel & Ragsdale, 1979) and cholinergic and glutamatergic projections from the pedunculopontine nucleus in the midbrain. However, the main flow of cortical information from the striatum to the thalamus operates mainly through GABAergic projections.

These are termed the "direct" or "indirect" pathways (Albin, Young, & Penney, 1989; Parent & Hazrati, 1995; Smith et al., 1998). Understanding these pathways is critical to our understanding of HD pathophysiology. According to this model, cortical information which flows to the striatum is processed and transmitted through the basal ganglia via a direct GABAergic inhibitory pathway from the striatum to the GPi and the SNr (direct pathway) and includes the cotransmitter substance P and second, a GABAergic output from the striatum containing the cotransmitter peptide enkephalin that projects to the GPe, which, in turn, sends an inhibitory input to the STN and the STN sends an excitatory projection to the GPi (indirect pathway). Thus, the direct and indirect pathways converge on the inhibitory output projection neurons of the GPi which innervate the ventral anterior and ventral lateral (VA/VL) regions of the thalamus. The VA/VL thalamic nuclei which receive most of the input from the GPi and SNr send excitatory output to the frontal and premotor cortex (Kayahara & Nakano, 1996) which then influences the output of the motor cortex. This circuit thus converges on the output of the primary motor cortex and intimately controls the movement of muscles throughout the body. In addition, a recent pathway termed the hyperdirect pathway has been identified which is an excitatory link directly from the cerebral cortex to the STN, which can stimulate STN neurons to drive the GABAergic GPi output neurons to inhibit the thalamus, and in this way bypass the striatum (Nambu, Tokuno, & Takada, 2002).

Therefore, the main GABAergic inhibitory cellular targets of the basal ganglia operating through GABA$_A$Rs include the following: the medium spiny neurons modulated through a variety of GABAergic interneurons, as well as a direct feedback loop from the GPe (Bevan, Booth, Eaton, & Bolam, 1998), GABAergic axons terminating on the neurons of the GPe, GPi, and SNr, as well as the axons of the GPi neurons innervating VA/VL neurons of the thalamus. In addition, there is reciprocal inhibitory feedback between the GPe and GPi through collaterals of the pallidal neurons and there is also inhibitory input to the SNr from the globus pallidus (Bolam et al., 2000). From the receptor subunit configurations in the basal ganglia derived above, the major subtype consisting of $\alpha_1,\beta_{2,3},\gamma_2$ subunits corresponds to the principal type 1 benzodiazepine sensitive (now termed the A1a subtype) fast-acting GABA$_A$ receptor. In the striatum, this receptor subtype is found on most of the GABAergic interneurons. The population of striatal parvalbumin-positive neurons are also known as fast-spiking neurons and so this fast-acting receptor configuration would complement their function. The localization of this receptor type on the large pallidal and nigra pars reticulata neurons would suggest that this inhibitory input would act through the fast-acting receptors.

The final two groups of striatal interneurons are two important groups—the cholinergic and the neuropeptide Y/somatostatin/nitric oxide synthase interneurons. The cholinergic interneurons labeled intensely with the α_3 subunit; this is consistent with other studies in the rat brain which have shown that cholinergic neurons throughout the brain are usually associated with the α_3 subunit (Gao, Fritschy, Benke, & Mohler, 1993). The results of the studies in the human striatum show that the cholinergic neurons did not contain the γ_2 subunits which were relatively highly expressed in the striatum. However, as a general principal it is accepted that α, β, and γ subunits are necessary to constitute a functional $GABA_AR$ (Mohler et al., 1995; Pritchett et al., 1989), thus suggesting that either the γ_1 or γ_3 subunits in their $GABA_ARs$ would be included in this receptor complex on cholinergic neurons. However, molecular studies suggest a heterogeneous population of $GABA_AR$ subtypes on the cholinergic interneurons including one type which is deficient in γ subunits (Yan & Surmeier, 1997). Further studies are required to show more conclusively which subunits are found in these $GABA_ARs$ and help elucidate the complexities of $GABA_AR$ interactions on these cholinergic striatal interneurons.

The α_3-subunit labeling on the cholinergic interneurons and the neurons in the globus pallidus showed a greater degree of intracellular labeling and do not outline the profile of both of these cell types to the same extent as the subunits of the $\alpha_1,\beta_{2,3},\gamma_2$ combination (Waldvogel et al., 1999). The α_3 subunit was found on the cell membranes and within the cytoplasm of both the soma and proximal dendrites and but was not present on the membranes of the peripheral dendrites. The intracellular localization of the α_3 subunit suggests that as well as labeling receptors at the cell surface it may also be present in an intracellular pool of receptors not associated with the plasma membrane.

The localization of receptor subunits within the cytoplasm is consistent with their production in the endoplasmic reticulum of the cell and storage in Golgi followed by transport to the cell surface where they become functional receptors. Recent studies have identified several receptor-associated proteins of $GABA_AR$ subunits which influence the assembly and trafficking of $GABA_ARs$. These include GABA receptor-associated protein (GABARAP), Huntingtin-associated protein 1 (HAP1), and GRIF-1 which is now renamed TRAK2 which both belong to the family of kinesin adaptor proteins (Smith & Kittler, 2010). The kinesin adaptor proteins family of proteins are active transporter proteins also known to traffic mitochondria (Stephenson, 2014). Interestingly, expression studies in heterologous expression systems show that each of the three subunits (α, β, γ) is retained in the endoplasmic reticulum and gets degraded when expressed alone (Connolly et al., 1999; Pritchett et al., 1989).

Other subunits such as the α_1 and $\beta_{2,3}$ subunits labeled the cell membranes and dendrites even at their most distal branches. This may indicate that various $GABA_AR$ combinations with the α_1 and α_3 subunits may exist in combination with the $\beta_{2,3}$, and γ_2 subunits on individual neurons, and that receptors which have different subunit configurations may be localized in different regions of the cell as has been suggested for neurons in other regions of the brain (Fritschy et al., 1992; Fritschy & Mohler, 1995; Fritschy, Weinmann, Wenzel, & Benke, 1998; Gao et al., 1993).

The further detailed distribution and the precise localization of $GABA_AR$ subunits throughout the basal ganglia are therefore critical in determining their role in GABAergic transmission through this important forebrain complex.

2. NEUROPATHOLOGY OF THE BASAL GANGLIA IN HUNTINGTON'S DISEASE

2.1. Macroscopic changes

Huntington's disease is a hereditary neurodegenerative disorder caused by a mutation in the Huntingtin gene (*HTT*). This mutation is an expansion in the CAG repeat of the gene which is translated to the protein which interferes with many cellular processes and ultimately causes major cell death in the striatum and cerebral cortex. The main symptoms are involuntary choreiform movements as well as mood, cognitive and behavioral changes.

Gross examination of postmortem Huntington's disease (HD) human brain demonstrates a striking characteristic bilateral atrophy of the striatum (de la Monte, Vonsattel, & Richardson, 1988; Vonsattel & Difiglia, 1998; Vonsattel, Keller, & Del Pilar Amaya, 2008). This degeneration generally follows an ordered and topographical distribution and principally affects the striatal GABAergic medium spiny projection neurons. The tail and body of the CN show more degeneration than the head in the very early stages of the degenerative process. The pattern of degeneration in the CN and the putamen usually progresses from the tail of the CN to the head and body in the caudorostral and simultaneously in the dorsoventral and mediolateral directions (Vonsattel & Difiglia, 1998).

Macroscopically, the volume of the CN and putamen is reduced with subsequent alteration of their respective shapes indicating that in HD the putamen had an average 64% cross-sectional area loss compared with a 57% cross-sectional area loss in the CN (de la Monte et al., 1988). With progression of HD, the CN becomes progressively more atrophic changing from the characteristic normal convex shape defining the border of the lateral ventricle to a thinner and ultimately more concave shape with resultant enlargement of the lateral ventricles occurring in parallel. This gradually decreasing volume is due to the loss of especially the medium spiny neurons, and their dendritic arbors and heavily myelinated axonal projections. Combined with the neuronal degeneration, there is also marked gliosis by astrocytes and oligodendrocytes. The extent of both the macroscopic shape of the CN and associated ventricular enlargement and the microscopic striatal degeneration including the loss of striatal neurons and extent of gliosis provide the basis of the Vonsattel grading system (Vonsattel & Difiglia, 1998; Vonsattel et al., 1985).

2.2. Grading of striatal neuropathology

A neuropathological grading system for Huntington's disease was developed by Vonsattel based on the pattern of neurodegeneration in the HD striatum of a large number of HD brains (Vonsattel et al., 1985). The assignment of a grade of neuropathological severity from 0 to 4 is based on gross changes to the shape and shrinkage of the CN and the loss of neurons correlated with astrocytosis in the CN.

3. CELLULAR AND NEUROCHEMICAL CHANGES

3.1. Striatum

Various autoradiographic, *in situ* hybridization, and immunohistochemical studies have documented neuronal and glial changes in the striatum of HD and have reported loss of neurochemicals, neurotransmitters, and neurotransmitter-associated receptors in the HD striatum. The most affected neuronal populations in the HD striatum are the medium-sized spiny projection neurons (MSNs) that constitute ~90–95% of the total striatal neuronal population. In the human striatum, the cell bodies of these GABAergic inhibitory MSNs can be identified immunohistochemically as outlined above with calbindin and DARPP-32 (Deng et al., 2004; Guo et al., 2012; Holt et al., 1997). The loss of MSNs in the matrix compartment has been shown in immunohistochemical studies using the calcium-binding protein marker calbindin (Ferrante, Kowall, & Richardson, 1991; Seto-Ohshima, Lawson, Emson, Mountjoy, & Carrasco, 1988; Tippett et al., 2007), and a recent study showed a loss of 58–76% DARPP-32-positive neurons in the human HD putamen with increasing grade (Guo et al., 2012). The loss of DARPP32 neurons was correlated with the motor impairment score rather than chorea.

Enkephalin staining is dramatically lost in a grade-dependent manner in the GPe reflecting loss of the GABA-enkephalin-positive pathway to the GPe (indirect pathway). There is also a loss of substance P in the GPi and SNr reflecting the loss of the GABA-substance P-positive pathway projecting to these two striatal output nuclei (Albin et al., 1992; Allen et al., 2009; Deng et al., 2004; Reiner et al., 1988) and loss of cannabinoid receptors on the presynaptic terminals of both the direct and indirect pathways (Allen et al., 2009; Glass et al., 2000; Glass, Faull, & Dragunow, 1993). The loss of these pathways underlies the symptomatology of Huntington's disease. The indirect pathway is the most affected (Deng et al., 2004) and loss of the enkephalin containing MSNs of the indirect pathway especially early in the disease leads to less inhibition of the GPe and subsequent increased inhibition of the STN, decreased excitation of GPi, and subsequent decreased inhibition of the thalamus resulting in thalamic over excitation of the cerebral cortex leading to chorea.

In general, the striatal interneurons are less affected by the disease process than the medium spiny neurons, especially in lower grades (Ferrante et al., 1987); however in higher grades, the interneurons are also affected in a differential manner; that is, the fast-spiking neurons containing the calcium-binding protein, parvalbumin, are consistently lost in the HD striatum and are degenerated in a grade-dependent manner that they may be linked to patients developing dystonia (Reiner et al., 2013). The large-sized calretinin-positive neurons many of which may belong to the population of the large-sized cholinergic neurons are generally preserved until the higher grades (Cicchetti & Parent, 1996). By contrast, interneurons containing neuropeptide Y/somatostatin or NADPH diaphorase/NOS and the medium-sized calretinin-positive interneurons are largely spared even in relatively severe cases of striatal degeneration (Cicchetti & Parent, 1996; Ferrante et al., 1987). The reason for this is not yet clear, but it is postulated to be due to either the pattern of distribution of excitatory receptors on these neurons or possibly their genetic fingerprint and susceptibility to the toxic mutant *HTT* gene.

Due to either the cell autonomous or nonautonomous effects of the *HTT* gene causing major degeneration of the GABAergic neurons of the striatum, we have investigated the changes in GABA receptors that occur in this disease to find how the loss of GABAergic neurons affects the distribution of the $GABA_A$ receptor.

We found in the dorsal striatum a major dorsoventral gradient of loss of $GABA_AR$ binding using FNZ binding (Fig. 6A and B). Following on from this our immunohistochemical staining using antibodies to the $\beta_{2,3}$ of the GABA receptor we found when comparing the staining in the striatum from a range of different cases of HD, a very heterogeneous staining pattern across different cases, with some cases demonstrating a selective loss of the striosome compartment with a relative preservation of the matrix, while in other cases there was a higher preservation of the striosomes with a loss in the matrix compartment (Fig. 6C–F). The loss in the striosome compartment was found to be correlated with patients that had major problems with mood (Tippett et al., 2007). This finding correlated with previous studies, which found that gliosis in the striosomes as shown by GFAP staining, appears as the earliest observed change in low-grade HD cases (Hedreen & Folstein, 1995), indicating that the neurons in striosomes may be more vulnerable at an early stage of HD or lower grades of the disease than those in the matrix (Augood, Faull, Love, & Emson, 1996; Hedreen & Folstein, 1995). This also correlates with many cases of Huntington's disease where psychiatric problems, mood, and personality changes occur before the onset of the overt movement disorder. In other cases, generally with higher grades, there is a preferential loss of neurons and neurochemical markers in the matrix compartment with clear sparing of the striosomes (Faull et al., 1993; Ferrante et al., 1987; Seto-Ohshima et al., 1988). These findings detailing the heterogeneous pattern of compartmental striatal degeneration in different HD cases are interesting because studies in the rodent and primate brains show that the striosome and matrix compartments have different patterns of connectivity and suggest that the two compartments are functionally different such that the striosome compartment contains MSNs that are linked with the limbic system and also project to the dopamine-containing neurons in the substantia nigra pars compacta (Fujiyama et al., 2011; Gerfen, 1984; Tokuno, Chiken, Kametani, & Moriizumi, 2002). Therefore, the striosome compartment is thought to play a major "limbic" processing role in modulating mood and other related functions of the basal ganglia. In contrast, the matrix compartment receives

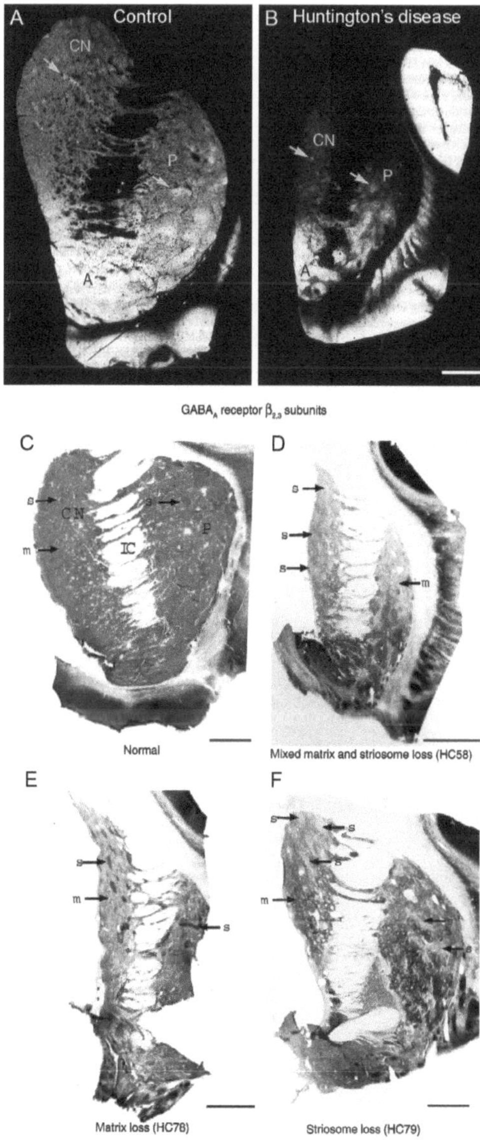

Figure 6 *Loss of $GABA_A$ receptors in the Huntington's disease caudate putamen.* (A and B) Autoradiograms of [^{3}H] flunitrazepam binding in the caudate-putamen from (A) a neurologically normal human brain and (B) a Huntington's disease human brain showing major loss of receptors in the Huntington's disease basal ganglia. arrows in A,B indicate striosomes. (C)–(F) Immunohistochemical labeling of the $\beta_{2,3}$ subunits from a normal (C) and three different cases (D)–(F) illustrating the high degree of variability in the loss of receptors in the various striatal compartments. S, striosomes; m, matrix. Scale bars = 0.5 cm. *Images (A) and (B) from Faull et al. (1993), Elsevier; images (C)–(D) from Tippett et al. (2007), Oxford University Press.*

topographically organized inputs from especially the sensori-motor and associative cortices, and hence, it is postulated to play a major role in the control of movement (Graham et al., 2009; Graybiel, 1990; Parent & Hazrati, 1995). That there is a higher level of $GABA_A$ receptor subunits in the striosomes compared to the matrix is also interesting, indicating a higher involvement of $GABA_ARs$ in the limbic region of the striatum. The changes in $GABA_ARs$ observed in the striatum of HD patients are principally thought to be an indicator of degeneration because of their dense distribution on the striatal MSNs and interneurons. It has not been established whether the changes in $GABA_ARs$ are in any way causative of cell death of MSNs, but the lack of inhibition in the striatum due to loss of these receptor subunits may be a factor in the observed excessive excitability of the striatum by glutamatergic afferents from the cerebral cortex in HD (Ferrante, Kowall and Richardson, 1991).

3.2. Globus pallidus

When investigating the changes that occur in $GABA_ARs$ in the globus pallidus, we and others have shown a marked upregulation of $GABA_AR$ binding, always with a higher level of upregulation in the external segment (Fig. 7). Previous studies have reported cell atrophy, cell loss, and volume loss ranging from approximately 40% to 50% in HD globus pallidus with the external segment always much more affected than the internal segment (Lange, Thorner, Hopf, & Schroder, 1976). The loss of neurons in the globus pallidus is in addition to the loss of the striatopallidal and striatonigral axons that release GABA into the GPe, GPi, and SNr. The resulting $GABA_AR$ response to these changes within the output nuclei is a major increase in postsynaptic $GABA_AR$ subunits. We have found that a majority of the $GABA_AR$ subunits in our studies (α_1, $\beta_{2,3}$, and γ_2) showed major increases in HD GP (Fig. 8), and while the α_3 subunit appeared to show some increase, it was not found to be significant, and the α_2 subunit was very weakly expressed in the human GP (Allen et al., 2009; Thompson-Vest, Waldvogel, Rees, & Faull, 2003). In addition, a marked increase of $GABA_B$ receptor R1 subunit was detected. The increase in $GABA_AR$ subunits is proposed to be a compensatory upregulatory response of $GABA_ARs$ on the postsynaptic membranes of the pallidal and nigral output neurons (Fig. 9). The exact mechanism of this response is still unclear. An increase or upregulation in $GABA_AR$ subunits can be replicated in a rat model where the striatonigral and striatopallidal pathways are reduced by placing

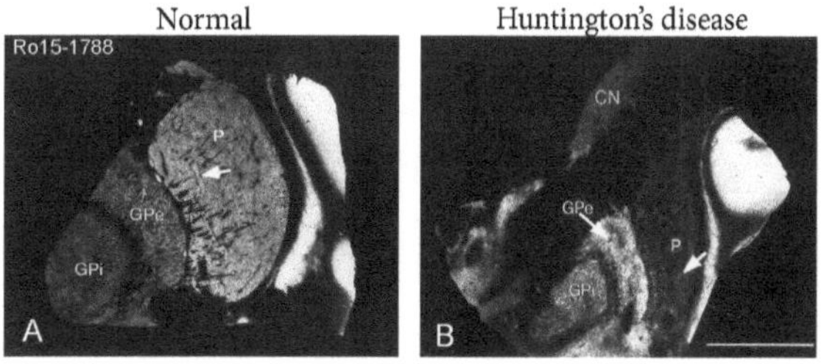

Figure 7 *Upregulation of GABA$_A$ receptors in Huntington's disease (HD) globus pallidus.* (A and B) Radioligand binding of GABA$_A$/benzodiazepine receptor-specific label [H^3] Ro15-1788 in; (A) control brain and (B) an HD brain showing loss of binding in the putamen (P) and increased binding in the globus pallidus external (GPe) and internal segments (GPi). Binding in control and Huntington's disease is higher in the external segment compared to the internal segment. Arrows indicate increased labeling of striosome-like patches in putamen. Scale bars (A,B) = 1 cm. *Images (A) and (B): Faull et al. (1993), Elsevier.*

excitotoxic lesions in the striatum (Faull et al., 1993; Nicholson, Faull, Waldvogel, & Dragunow, 1995). This response occurs within a few hours and remains upregulated. Whether there is also a genetic component related to the Huntington's gene remains unclear. It has been postulated that GABA release may be influenced by cannabinoid receptors, which are dramatically lost in HD GPe and SNr (Glass et al., 2000; Glass, Dragunow, & Faull, 1993). Transgenic HD mice models have not modeled this upregulation successfully. The HD transgenic models show a downregulation of CB1 and a downregulation of CB1-induced GABA release but no change in GABA$_A$Rs (Dowie et al., 2009). Interestingly, we also found no changes in gephyrin protein immunoreactivity in human HD brain both in our published and in unpublished studies using different antibodies. The gephyrin protein has been closely associated with the trafficking of the γ_2 subunit which changes dramatically in HD GP, thus suggesting that in the GP the scaffolding protein gephyrin does not increase its structure but remains a fixed entity bound to the microtubule structure of the cell.

3.3. Substantia nigra

Variable reports on loss of neurons have been published in the substantia nigra in HD. The SNc is thinner than normal with cell loss but less than that of the SNr (Oyanagi, Takeda, Takahashi, Ohama, & Ikuta, 1989; Richardson, 1990); however, other studies found no real loss of dopaminergic neurons in the SNc in rigid HD (Albin, Reiner, Anderson,

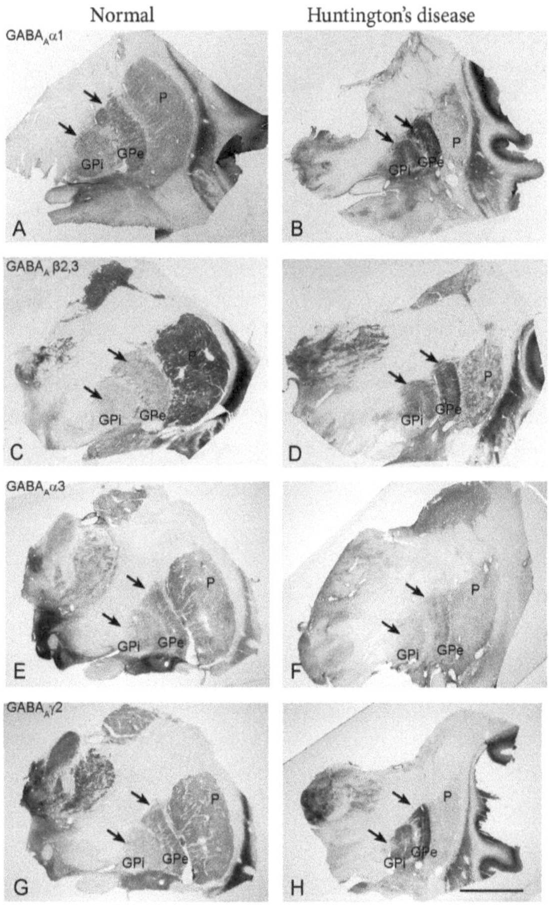

Figure 8 *Immunoreactivity observed in the globus pallidus external (GPe) and internal (GPi) segments for the α_1, $\beta_{2,3}$, α_3, and γ_2 subunits in normal (A, C, E, G) compared to HD (B, D, F, H).* (A and (B) Immunohistochemical labeling of the α_1 subunit in: (A) control and (B) HD brain showing increased immunoreactivity in the globus pallidus with an especially high increase in the external segment, with a loss of staining in the HD putamen. (C and D) Immunohistochemical labeling of the $\beta_{2,3}$ subunits in: (C) control and (D) HD brain, also showing increased immunoreactivity in the GPe compared to control with a lower increase in GPi but a patchy loss of immunoreactivity in the putamen. (E and F) There is no significant increase in α_3 subunits in the GPe or GPi in HD compared to normal. (G and H) There is a marked reduction in immunoreactivity of γ_2 subunits in the HD putamen and a marked upregulation in this subunit in HD GPe with a lower increase in GPi. Arrows (A)–(H) indicate immunoreactivity in GPe and GPi. Scale bar applies to (A)–(H) = 0.5 cm.

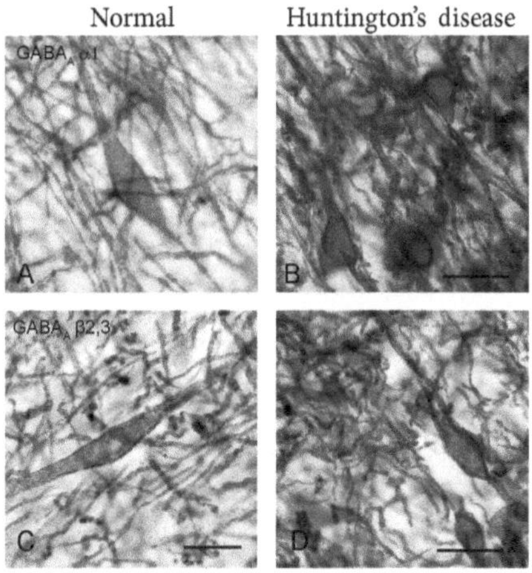

Figure 9 (A and B) High magnification images from the GPe showing increased immunoreactivity at synaptic sites along the soma and dendritic membranes of large pallidal neurons labeled for α_1 subunits: (A) is from a normal case with moderate immunoreactivity surrounding the cell body and dendritic membranes, (B) is from an HD case showing highly increased labeling over the soma and dendrites. (C) $\beta_{2,3}$ subunits on soma and dendrites in control GPe and (D) in HD GPe showing increased labeling on soma and dendrites which show a distorted morphology. Scale bars (A)–(D) = 20 μm.

Penney, & Young, 1990). There is loss of TH protein and mRNA from the SNc (Yohrling et al., 2003), but a loss of TH in the matrix of the striatum has not been conclusively shown (Ferrante & Kowall, 1987). However, neurons of the SNc have a major dopaminergic projection to the full extent of the striatal matrix and these studies suggest that a loss of dopamine in the striatum may contribute to the symptoms of HD (Yohrling et al., 2003).

In the SNr in postmortem brain, no GABA$_A$R subunit changes have been reported. Our preliminary unpublished studies observe increases in GABA$_A$ α_1 and $\beta_{2,3}$ subunits in SNr but not to the same extent as in the GP. Excitotoxic quinolinic acid lesions of the rat striatum do produce a dramatic upregulation of GABA$_A$R subunits and the increase in receptor staining was shown to be at the remaining synapses and not the degenerating ones (Nicholson et al., 1995). As the GP and SNr show many similarities in their morphology, similar mechanisms may be in play in both the substantia nigra and globus pallidus regarding upregulation of GABA$_A$R numbers in normal and Huntington's disease brain. Additionally, the SNc receives GABAergic input from both the striatum and the SNr but due to a different receptor configuration which is most likely to be α_3, β_1, γ_2 in the SNc, the responses to GABA by these receptor complexes are likely to be different to those on neurons of the

SNr. The plasticity of $GABA_A$ receptor subunits on the dopaminergic neurons of the SNc still needs to be elucidated.

3.4. Parkinson's disease

Parkinson's disease (PD) is also a basal ganglia-related disease, however, in both its symptoms, and pathogenesis is the "mirror" image of HD with its key feature being bradykinesia rather than hyperkinesia as in HD. Also one of the key pathological features of PD is the loss of dopaminergic nigrostriatal neurons in the SNc, rather than the striatonigral and striatopallidal fibers in HD, and the consequent reduction of dopamine release in the striatum which reduces the activation of D1 and D2 dopamine receptors on the striatal MSNs. This leads to a decreased activation of the GPe and a hyperactive indirect pathway with increased activity in the SNr and GPi resulting in an increased inhibition of the thalamic neurons and decreased excitation of the cerebral cortex (Albin et al., 1989; Delong, 1990). Most studies which have examined the GABAergic system in PD, although producing variable results, indicate that generally only little change is seen in GABA levels in the striatum. Slightly higher levels of GABA levels were found in the striatum of postmortem PD brain (Gerlach et al., 1996; Kish et al., 1986), and in animal models, elevated striatal mRNA for GAD was found, especially in striatopallidal neurons (Levy et al., 1995). In animal models, no increases were seen in the levels of mRNA for GAD by *in situ* hybridization or in GABA levels in microdialysis studies in the GPe (Robertson, Graham, Sambrook, & Crossman, 1991; Soghomonian & Laprade, 1997). On the other hand, increases in GAD mRNA in the GPi and SNr were found in animal models (Kincaid, Albin, Newman, Penney, & Young, 1992). $GABA_AR$-binding studies found no significant changes in $[H^3]$ FNZ levels in the caudate and putamen of postmortem human or MPTP-treated monkeys (Griffiths, Sambrook, Perry, & Crossman, 1990; Robertson, Clarke, Boyce, Sambrook, & Crossman, 1990). In contrast, in the GPe, increased or decreased levels of GABA have been reported, whereas in the GPi and SNr increased levels of GABA were measured (Calon et al., 1999; Griffiths et al., 1990; Robertson et al., 1990). No studies on $GABA_AR$ subunit changes have been published on human PD basal ganglia, but animal models such as the 6-OHDA rat model of PD show decreases in α_1, α_2, β_2, and γ_2 subunits in the globus pallidus (Chadha, Dawson, Jenner, & Duty, 2000; Galvan & Wichmann, 2007) and increases in SNr and entopeducular nucleus. The

SNr and GPi therefore show similar likely compensatory changes to HD, most likely due to the decreased striatonigral GABA output (Katz, Nielsen, & Soghomonian, 2005; Yu, Wang, Liu, & Yin, 2001).

In the later stages of Huntington's disease, the chorea is reduced and the patients become hypoactive and rigid similar to the late stages in PD. This outcome for late-stage HD is thought to be related to the loss of striatal neurons in the direct pathway reducing the inhibition of GPi and therefore excessively inhibiting the thalamus thereby reducing movement.

3.5. Subventricular zone and neurogenesis in Huntington's disease

The subventricular zone (SVZ) which lies along the margin of the CN adjacent to the lateral ventricle has become a region of intense interest with the discovery of adult neural stem cells in this region. In the normal human SVZ, neural precursors have been identified (Curtis et al., 2005; Kam et al., 2009) using PCNA as a marker for proliferating cells coupled with neuronal stem cell markers to identify their neuronal phenotype. In Huntington's disease brain, an increase in cell proliferation was found in the SVZ with evidence for increased neurogenesis and increasing thickness of the SVZ with increasing grade (Curtis, Connor, & Faull, 2003); that is, with increased cell death in the CN, there is increased neurogenesis in the SVZ. This raises the exciting possibility of stimulating the production of new neurons from precursors in the SVZ and subsequent migration directly or via the rostral migratory stream (Curtis et al., 2007) into the cell-depleted HD striatum as a possible therapy for HD. The subsequent integration of these newly formed neurons into the basal ganglia circuitry will be critical. Studies in the rat indicate that stem cells have the ability to migrate into the quinolinic acid lesioned rat striatum (Tattersfield et al., 2004), but further studies are needed to determine if this is a potentially viable therapy in humans.

It is well known that GABA plays a very important role in development and has both inhibitory and excitatory functions which allows correct interneuronal connections and the correct configurations of synaptic complexes to be assembled in the developing brain (Sernagor, Chabrol, Bony, & Cancedda, 2010). The role of GABA in the proliferation of cells in the human SVZ is still unclear, but in animal studies, GABA and GABA$_A$Rs are known to regulate the movement of neuroblasts within the SVZ (Ge, Pradhan, Ming, & Song, 2007; Platel, Dave, & Bordey, 2008). The production of new neurons in the adult SVZ begins when a neural stem cell which is a GFAP-positive astrocyte-like cell (B cell) divides to produce a transit-amplifying cell (C cell). The transit-amplifying cells divide to produce neuroblasts which migrate toward the olfactory bulb. Neurogenesis is closely regulated through a nonsynaptic GABAergic feedback mechanism whereby neuroblasts generated from GFAP-positive cells release GABA that decreases the proliferation of GFAP-positive progenitors through tonically activating $GABA_A$ nonsynaptic receptors on these cells (Platel et al., 2008). Therefore, lower levels of GABA in the environment reactivate proliferation of new neural cells moving toward the olfactory bulb (Carleton, Petreanu, Lansford, Alvarez-Buylla, & Lledo, 2003; Kriegstein, 2005; Liu, Wang, Haydar, & Bordey, 2005). In light of the large number of species differences that appear in the composition of the SVZ, we investigated the $GABA_AR$ subunit localization in the human SVZ to determine whether similar mechanisms may operate in human (Dieriks, Waldvogel, Monzo, Faull, & Curtis, 2013). The organization of the SVZ is quite different to that of the normal underlying CN and so the localization of $GABA_AR$ subunits in the human SVZ were oriented differently to those in the CN. $GABA_AR$ subunits were present in the SVZ in varying proportions and were expressed with variable amounts on most cell types within the SVZ. By far, the most widespread distribution was that of the $GABA_AR$ γ_2 subunit which was only expressed on dendritic-like processes mainly orientated perpendicular to the ventricular endothelial cell layer of the SVZ including those staining for GFAP delta and PSA-NCAM indicating a role in proliferating cells. These processes also had a very similar morphology to those of radial glia. These processes protruded into the CN, but the number of processes decreased with increasing distance from the SVZ. Further studies are required to reveal whether γ_2 subunits colocalize with radial glia markers. However, in the subgranular zone of the hippocampus, radial glia express γ_2 subunits and, in the adult mouse hippocampus, nestin-expressing radial glia-like quiescent neural stem cells respond tonically to the neurotransmitter GABA through activation of γ_2 subunit containing $GABA_ARs$. These receptors control the proliferation of quiescent adult neural stem cells in response to neuronal activity and experience (Song, Zhong, Bonaguidi, Sun, Hsu., 2012). Therefore, if this mechanism is also active in the human SVZ, GABA released from terminals from within the SVZ or from external sources may influence proliferation of neural precursor stem cells. Interestingly in the human, the most common subunit in the brain, the $GABA_AR$ α_1 subunit was expressed at low levels within the SVZ and was much more abundant in the CN just below the myelin layer. $GABA_AR$ $\beta_{2,3}$ subunits were only found on the cell processes of occasional cells in the SVZ; however, a large number of cells in the

SVZ expressed $GABA_AR$ $\beta_{2,3}$ subunits in their cell bodies. $GABA_AR$ α_2 subunit was present in cell bodies in the SVZ and appeared as small cells evenly distributed throughout the SVZ and $GABA_AR$ α_3 subunit was also present in cell bodies in the SVZ and CN. Colocalization studies found that various combinations of these subunits were found in proliferating cells which were GABAergic. We found $GABA_AR$ γ_2, α_3 and α_2 subunits to colabel with $GABA_AR$ $\beta_{2,3}$-expressing cells. Based on these data, we can extrapolate $\alpha_2\beta_{2,3}\gamma_2$ and $\alpha_3\beta_{2,3}\gamma_2$ as potential subunit compositions of possible functional $GABA_ARs$ in the SVZ (Dieriks et al., 2013). Several types of subunit combinations have been identified throughout the brain using the techniques of immunohistochemical colocalization and coimmunoprecipitation studies (Olsen & Sieghart, 2008). Additionally, negative allosteric modulation of $GABA_ARs$ in the SVZ has been shown by the diazepam-binding inhibitor (DBI) which is able to promote neurogenesis in the SVZ (Alfonso, Le Magueresse, Zuccotti, Khodosevich, & Monyer, 2012). DBI thereby counteracted the effect of the GABAergic brake on proliferation. The potential of using stem cells for replacement therapy is an exciting prospect but still requires more investigation regarding promoting their growth into target areas. The integration of new neurons into damaged tissue is crucial to this procedure as excessive growth can also be detrimental. The true numbers of stem cells in the SVZ have been difficult to estimate particularly in humans as this is also different across different animal species (Low, Faull, Bennet, Gunn, & Curtis, 2013), with estimates of 10,000–30,000 neurons migrating to the olfactory bulb each day (Lois & Alvarezbuylla, 1994) and approximately 100,000 new proliferating cells are thought occur in the Rostral Migratory Stream in an adult human (Curtis, Low, & Faull, 2012) at any time. Whether these can be successfully stimulated to integrate into damaged human brain such as into the striatum in Huntington's disease is still under investigation, although there are promising results from animal studies (Tattersfield et al., 2004) which may have an application in the future for HD therapeutics.

3.6. Huntington's disease-related proteins association with $GABA_A$ receptor subunits

One very interesting association with the Huntington's disease gene and $GABA_ARs$ is the association that the Huntington's-associated protein (HAP-1) has with $GABA_ARs$. It was shown that HAP-1 binds more tightly to mutant Huntingtin with higher CAG repeat number (Li et al., 1995). HAP-1 binds $GABA_ARs$ to KIF5 (kinesin motor protein 5) and this complex is transported to synapses. Disrupting this complex in experimental conditions decreases $GABA_AR$ numbers and reduces GABA currents (Twelvetrees et al., 2010). The KIF5 complex has been shown to be associated with the $GABA_AR$ β_{1-3} and γ_2 subunits. Therefore, disruption of $GABA_AR$ trafficking in Huntington's disease could directly contribute to reduced inhibition leading to excessive excitation which is one of the pathological features noted in Huntington's disease. Overactive excitability has been shown in cortical neurons and is a possible mechanism of cell death in the striatum of Huntington's disease (Gu et al., 2005, 2007; Spampanato, Gu, Yang, & Mody, 2008). This may therefore be one of the multiple contributors to cell death in Huntington's disease.

3.7. Overall distribution and function of the $GABA_A$ receptor in the human basal ganglia

In summary, regarding the distribution of α subunits in the basal ganglia, the α_1 subunit is most highly expressed on the membranes of interneurons throughout the striatum, as well as on the output neurons of the basal ganglia located in the globus pallidus and substantia nigra pars reticulata. The receptors form "hot spots" of label along dendritic membranes which often form a web-like distribution pattern on each neuron when viewed with high-resolution confocal laser microscopy. The α_2 subunit is very highly expressed throughout the striatum but is low in the output nuclei. The α_3 subunit is restricted mainly to cholinergic interneurons in the striatum and to dopaminergic neurons of the substantia nigra pars compacta as well as at lower levels in the output nuclei. The α_3 subunit is commonly associated with monaminergic and cholinergic neurons (Fritschy & Mohler, 1995; Gao et al., 1993; Rudolph & Antkowiak, 2004) throughout the brain and has been implicated in muscle relaxant activity and thalamic oscillations via the reticular nucleus (Sohal, Keist, Rudolph, & Huguenard, 2003).

The $\beta_{2,3}$ subunits are highly distributed throughout all the regions and are often associated with α_1, α_2, and α_3 subunits with the exception of α_3 subunits localized in the dopaminergic neurons of the SNc where the $\beta_{2,3}$ subunits were not detected. The γ_2 subunit distribution is relatively highly correlated with the pattern of all of the above subunits although the expression level is usually lower.

Furthermore, as detailed above, HD specifically targets the GABAergic system of the basal ganglia. The primary neurons that degenerate are the GABAergic medium spiny projection neurons of the striatum which causes a marked upregulation of $GABA_AR$ subunits ($\alpha_1\beta_2\gamma_2$) in the postsynaptic neurons of the globus pallidus and substantia nigra. The GABA$_A$R subunit changes thought to occur throughout the life of HD and PD patients may naturally slow the progression of the motor and mood symptomatology by compensating for the loss of striatal GABAergic neurons in HD and dopaminergic neurons in PD. Importantly, the death of the striatal projection neurons results in a reactive proliferative neurogenesis of the progenitor neurons in the SVZ lining the lateral ventricle. GABA$_A$Rs comprised from a range of subunits located on the variety of proliferating cells in the SVZ are thought to play an important role in the extent of neuronal repair in the diseased human basal ganglia.

CONFLICT OF INTEREST

The authors declare no financial or other conflict of interest.

ACKNOWLEDGMENTS

We have been very fortunate to have developed collaborative research studies with Professor Hanns Möhler in 1989 when he so generously provided samples of his bd17 and bd24 antibodies to the α_1 and $\beta_{2,3}$ subunits, respectively, of the GABA$_A$ receptor to trial on our postmortem human brain. This collaboration for our broad spectrum programe on human neurodegenerative diseases enabled us to especially concentrate on the regional, cellular, and subcellular localization of GABA$_A$ receptor subunits in the human brain and spinal cord which has blossomed and flourished over 25 years and resulted in over 25 research and review papers on GABA$_A$Rs and related neurochemicals in the normal and Huntington's disease human central nervous system. Hanns and his colleague Jean-Marc Fritschy have been an inspiration and wonderful collaborators to our brain research and have been instrumental in enabling us to make pioneering contributions to our GABA research in the human brain and spinal cord. Hanns has also provided leadership, wisdom, and advice as an inaugural founding member of the international scientific advisory board of the Centre for Brain Research at the University of Auckland, and we look forward to his continuing guidance over the years ahead. We salute Professor Hanns Möhler on his 75th birthday as a very special friend and esteemed colleague and thank him for his continuing collaborations between our laboratories in Switzerland and New Zealand.

REFERENCES

Albin, R. L., Makowiec, R. L., Hollingsworth, Z. R., Dure, L. S., Penney, J. B., & Young, A. B. (1992). Excitatory amino acid binding sites in the basal ganglia of the rat: A quantitative autoradiographic study. *Neuroscience*, *46*, 35–48.

Albin, R. L., Reiner, A., Anderson, K. D., Penney, J. B., & Young, A. B. (1990). Striatal and nigral neuron subpopulations in rigid Huntingtons-disease—Implications for the functional-anatomy of chorea and rigidity-akinesia. *Annals of Neurology*, *27*(4), 357–365.

Albin, R. L., Young, A. B., & Penney, J. B. (1989). The functional-anatomy of basal ganglia disorders. *Trends in Neurosciences*, *12*(10), 366–375.

Alexander, G. E., Crutcher, M. D., & Delong, M. R. (1990). Basal ganglia-thalamocortical circuits—Parallel substrates for motor, oculomotor, prefrontal and limbic functions. *Progress in Brain Research*, *85*, 119–146.

Alexander, G. E., DeLong, M. R., & Strick, P. L. (1986). Parallel organisation of functionally segregated circuits linking basal ganglia and cortex. *Annual Review of Neuroscience*, *9*, 357–381.

Alfonso, J., Le Magueresse, C., Zuccotti, A., Khodosevich, K., & Monyer, H. (2012). Diazepam binding inhibitor promotes progenitor proliferation in the postnatal SVZ by reducing GABA signaling. *Cell Stem Cell*, *10*(1), 76–87.

Allen, K. L., Waldvogel, H. J., Glass, M., & Faull, R. L. (2009). Cannabinoid (CB(1)), $GABA_A$ and $GABA_B$ receptor subunit changes in the globus pallidus in Huntington's disease. *Journal of Chemical Neuroanatomy*, *37*(4), 266–281.

Augood, S. J., Faull, R. L., Love, D. R., & Emson, P. C. (1996). Reduction in enkephalin and substance P messenger RNA in the striatum of early grade Huntington's disease: A detailed cellular in situ hybridization study. *Neuroscience*, *72*(4), 1023–1036.

Barnard, E. A., Darlison, M. G., Fujita, N., Glencorse, T. A., Levitan, E. S., Reale, V., et al. (1988). Molecular biology of the $GABA_A$ receptor. *Advances in Experimental Medicine & Biology*, *236*, 31–45.

Bennett, B. D., & Bolam, J. P. (1994). Synaptic input and output of parvalbumin-immunoreactive neurons in the neostriatum of the rat. *Neuroscience*, *62*, 707–719.

Bevan, M. D., Booth, P. A. C., Eaton, S. A., & Bolam, J. P. (1998). Selective innervation of neostriatal interneurons by a subclass of neuron in the globus pallidus of the rat. *Journal of Neuroscience*, *18*(22), 9438–9452.

Bolam, J. P., Hanley, J. J., Booth, P. A., & Bevan, M. D. (2000). Synaptic organisation of the basal ganglia. *Journal of Anatomy*, *196*(Pt 4), 527–542.

Bolam, J. P., Powell, J. F., Totterdell, S., & Smith, A. D. (1981). The proportion of neurons in the rat neostriatum that project to the substantia nigra demonstrated using horseradish peroxidase conjugated with wheatgerm agglutinin. *Brain Research*, *220*, 339–343.

Burt, D. R. (2003). Reducing GABA receptors. *Life Sciences*, *73*(14), 1741–1758.

Calon, F., Morissette, M., Goulet, M., Grondin, R., Blanchet, P. J., Bedard, P. J., et al. (1999). Chronic D1 and D2 dopaminomimetic treatment of MPTP-denervated monkeys: Effects on basal ganglia $GABA_A$/benzodiazepine receptor complex and GABA content. *Neurochemistry International*, *35*(1), 81–91.

Carleton, A., Petreanu, L. T., Lansford, R., Alvarez-Buylla, A., & Lledo, P. M. (2003). Becoming a new neuron in the adult olfactory bulb. *Nature Neuroscience*, *6*(5), 507–518.

Carpenter, M. B. (1976). Anatomy of the basal ganglia and related nuclei: A review. *Advances in Neurology*, *14*, 7–48.

Chadha, A., Dawson, L. G., Jenner, P. G., & Duty, S. (2000). Effect of unilateral 6-hydroxydopamine lesions of the nigrostriatal pathway on $GABA_A$ receptor subunit gene expression in the rodent basal ganglia and thalamus. *Neuroscience*, *95*(1), 119–126.

Cicchetti, F., & Parent, A. (1996). Striatal interneurons in Huntington's disease: Selective increase in the density of calretinin-immunoreactive medium-sized neurons. *Movement Disorders*, *11*(6), 619–626.

Cicchetti, F., Prensa, L., Wu, Y., & Parent, A. (2000). Chemical anatomy of striatal interneurons in normal individuals and in patients with Huntington's disease. *Brain Research Reviews*, *34*(1–2), 80–101.

Connolly, C. N., Uren, J. M., Thomas, P., Gorrie, G. H., Gibson, A., Smart, T. G., et al. (1999). Subcellular localization and endocytosis of homomeric gamma 2 subunit splice variants of gamma-aminobutyric acid type A receptors. *Molecular and Cellular Neuroscience*, *13*(4), 259–271.

Curtis, M. A., Connor, B., & Faull, R. L. (2003). Neurogenesis in the diseased adult human brain—New therapeutic strategies for neurodegenerative diseases. *Cell Cycle*, *2*(5), 428–430.

Curtis, M. A., Kam, M., Nannmark, U., Anderson, M. F., Axell, M. Z., Wikkelso, C., et al. (2007). Human neuroblasts migrate to the olfactory bulb via a lateral ventricular extension. *Science*, *315*(5816), 1243–1249.

Curtis, M. A., Low, V. F., & Faull, R. L. M. (2012). Neurogenesis and progenitor cells in the adult human brain: A comparison between hippocampal and subventricular progenitor proliferation. *Developmental Neurobiology*, *72*(7), 990–1005.

Curtis, M. A., Penney, E. B., Pearson, J., Dragunow, M., Connor, B., & Faull, R. L. (2005). The distribution of progenitor cells in the subependymal layer of the lateral ventricle in the normal and Huntington's disease human brain. *Neuroscience*, *132*(3), 777–788.

de la Monte, S. M., Vonsattel, J. P., & Richardson, E. P., Jr. (1988). Morphometric demonstration of atrophic changes in the cerebral cortex, white matter, and neostriatum in Huntington's disease. *Journal of Neuropathology and Experimental Neurology*, *47*(5), 516–525.

Delong, M. R. (1990). Primate models of movement disorders of basal ganglia origin. *Trends in Neurosciences*, *13*(7), 281–285.

Deng, Y. P., Albin, R. L., Penney, J. B., Young, A. B., Anderson, K. D., & Reiner, A. (2004). Differential loss of striatal projection systems in Huntington's disease: A quantitative immunohistochemical study. *Journal of Chemical Neuroanatomy*, *27*(3), 143–164.

Dieriks, B. V., Waldvogel, H. J., Monzo, H. J., Faull, R. L. M., & Curtis, M. A. (2013). GABA$_A$ receptor characterization and subunit localization in the human sub ventricular zone. *Journal of Chemical Neuroanatomy*, *52*, 58–68.

Dowie, M. J., Bradshaw, H. B., Howard, M. L., Nicholson, L. F. B., Faull, R. L. M., Hannan, A. J., et al. (2009). Altered Cb1 receptor and endocannabinoid levels precede motor symptom onset in a transgenic mouse model of Huntington's disease. *Neuroscience*, *163*(1), 456–465.

Dure, L. S., Young, A. B., & Penney, J. B. (1992). Compartmentalization of excitatory amino-acid receptors in human striatum. *Proceedings of the National Academy of Sciences of the United States of America*, *89*(16), 7688–7692.

Faull, R. L. M., & Villiger, J. W. (1988). Multiple benzodiazepine receptors in the human basal ganglia: A detailed pharmacological and anatomical study. *Neuroscience*, *24*(2), 433–451.

Faull, R. L. M., Waldvogel, H. J., Nicholson, L. F., & Synek, B. J. (1993). The distribution of GABA$_A$-benzodiazepine receptors in the basal ganglia in Huntington's disease and in the quinolinic acid-lesioned rat. *Progress in Brain Research*, *99*, 105–123.

Faull, R. L. M., Waldvogel, H. J., Nicholson, L. F. B., Williams, M. N., & Dragunow, M. (1994). GABA$_A$ receptor changes in the basal ganglia in Huntington's disease in the human brain and in the quinolinic acid lesioned rat model of the disease following fetal neuron transplants. In D. Tracey, J. Stone, & G. Paxinos (Eds.), *Neurotransmitters in the human brain*. New York: Plenum Press.

Ferrante, R. J., & Kowall, N. W. (1987). Tyrosine hydroxylase-like immunoreactivity is distributed in the matrix compartment of normal human and Huntington's disease striatum. *Brain Research*, *416*(1), 141–146.

Ferrante, R. J., Kowall, N. W., Beal, M. F., Martin, J. B., Bird, E. D., & Richardson, E. P., Jr. (1987). Morphologic and histochemical characteristics of a spared subset of striatal neurons in Huntington's disease. *Journal of Neuropathology and Experimental Neurology*, *46*(1), 12–27.

Ferrante, R. J., Kowall, N. W., & Richardson, E. P. (1991). Proliferative and degenerative changes in striatal spiny neurons in Huntington's disease—A combined study using the section Golgi method and calbindin-D28k immunocytochemistry. *Journal of Neuroscience*, *11*(12), 3877–3887.

Fritschy, J. M., Benke, D., Mertens, S., Oertel, W. H., Bachi, T., & Mohler, H. (1992). Five subtypes of type A gamma-aminobutyric acid receptors identified in neurons by double and triple immunofluorescence staining with subunit-specific antibodies. *Proceedings of the National Academy of Sciences of the United States of America, 89*(15), 6726–6730.
Fritschy, J. M., Harvey, R. J., & Schwarz, G. (2008). Gephyrin: where do we stand, where do we go? *Trends in Neurosciences, 31*(5), 257–264.
Fritschy, J. M., & Mohler, H. (1995). Gaba(a)-receptor heterogeneity in the adult rat brain—Differential regional and cellular distribution of seven major subunits. *Journal of Comparative Neurology, 359*(1), 154–194.
Fritschy, J. M., Weinmann, O., Wenzel, A., & Benke, D. (1998). Synapse-specific localization of NMDA and GABA(a) receptor subunits revealed by antigen-retrieval immunohistochemistry. *Journal of Comparative Neurology, 390*(2), 194–210.
Fujiyama, F., Fritschy, J. M., Stephenson, F. A., & Bolam, J. P. (2000). Synaptic localization of $GABA_A$ receptor subunits in the striatum of the rat. *Journal of Comparative Neurology, 416*(2), 158–172.
Fujiyama, F., Sohn, J., Nakano, T., Furuta, T., Nakamura, K. C., Matsuda, W., et al. (2011). Exclusive and common targets of neostriatofugal projections of rat striosome neurons: A single neuron-tracing study using a viral vector. *European Journal of Neuroscience, 33*(4), 668–677.
Galvan, A., & Wichmann, T. (2007). GABA and the Basal Ganglia: From Molecules to Systems. GABAergic circuits in the basal ganglia and movement disorders. *Progress in Brain Research, 160*, 287–312.
Gao, B., Fritschy, J. M., Benke, D., & Mohler, H. (1993). Neuron-specific expression of $GABA_A$-receptor subtypes: Differential association of the a1 and a3-subunits with serotonergic and GABAergic neurons. *Neuroscience, 54*, 881–892.
Ge, S. Y., Pradhan, D. A., Ming, G. L., & Song, H. J. (2007). GABA sets the tempo for activity-dependent adult neurogenesis. *Trends in Neurosciences, 30*(1), 1–8.
Gerfen, C. R. (1984). The neostriatal mosaic—Compartmentalization of corticostriatal input and striatonigral output systems. *Nature, 311*(5985), 461–464.
Gerlach, M., Gsell, W., Kornhuber, J., Jellinger, K., Krieger, V., Pantucek, F., et al. (1996). A post mortem study on neurochemical markers of dopaminergic, GABA-ergic and glutamatergic neurons in basal ganglia thalamocortical circuits in Parkinson syndrome. *Brain Research, 741*(1–2), 142–152.
Glass, M., Dragunow, M., & Faull, R. L. M. (2000). The pattern of neurodegeneration in Huntington's disease: A comparative study of cannabinoid, dopamine, adenosine and $GABA_A$ receptor alterations in the human basal ganglia in Huntington's disease. *Neuroscience, 97*(3), 505–519.
Glass, M., Faull, R. L., & Dragunow, M. (1993). Loss of cannabinoid receptors in the substantia nigra in Huntington's disease. *Neuroscience, 56*(3), 523–527.
Goetz, T., Arslan, A., Wisden, W., & Wulff, P. (2007). GABA and the Basal Ganglia: From Molecules to Systems. $GABA_A$ receptors: Structure and function in the basal ganglia. *Progress in Brain Research, 160*, 21–41.
Graham, R. K., Pouladi, M. A., Joshi, P., Lu, G., Deng, Y., Wu, N. P., et al. (2009). Differential susceptibility to excitotoxic stress in YAC128 mouse models of Huntington disease between initiation and progression of disease. *Journal of Neuroscience, 29*(7), 2193–2204.
Graybiel, A. M. (1990). Neurotransmitters and neuromodulators in the basal ganglia. *Trends in Neurosciences, 13*(7), 244–254.
Graybiel, A. M., & Ragsdale, C. W., Jr. (1978). Histochemically distinct compartments in the striatum of human, monkeys, and cat demonstrated by acetylthiocholinesterase staining. *Proceedings of the National Academy of Sciences of the United States of America, 75*(11), 5723–5726.

Graybiel, A. M., & Ragsdale, C. W., Jr. (1979). Fiber connections of the basal ganglia. *Progress in Brain Research*, *51*, 237–283.

Griffiths, P. D., Sambrook, M. A., Perry, R., & Crossman, A. R. (1990). Changes in benzodiazepine and acetylcholine receptors in the globus pallidus in Parkinson's disease. *Journal of the Neurological Sciences*, *100*(1–2), 131–136.

Gu, X., Andre, V. M., Cepeda, C., Li, S. H., Li, X. J., Levine, M. S., et al. (2007). Pathological cell-cell interactions are necessary for striatal pathogenesis in a conditional mouse model of Huntington's disease. *Molecular Neurodegeneration*, *2*, 8.

Gu, X., Li, C., Wei, W., Lo, V., Gong, S., Li, S. H., et al. (2005). Pathological cell-cell interactions elicited by a neuropathogenic form of mutant Huntingtin contribute to cortical pathogenesis in HD mice. *Neuron*, *46*(3), 433–444.

Guo, Z. H., Rudow, G., Pletnikova, O., Codispoti, K. E., Orr, B. A., Crain, B. J., et al. (2012). Striatal neuronal loss correlates with clinical motor impairment in Huntington's disease. *Movement Disorders*, *27*(11), 1379–1386.

Hartig, W., Brauer, K., Fritschy, J. M., Bruckner, G., & Bigl, V. (1995). Regional and cellular expression sites of the alpha(1) subunit of gaba(a) receptors in the rat basal forebrain—A cytochemical study with glutamic-acid decarboxylase, choline-acetyltransferase, calcium-binding proteins and nitric-oxide synthase as 2nd markers. *Brain Research*, *692*(1–2), 215–226.

Hedreen, J. C., & Folstein, S. E. (1995). Early loss of neostriatal striosome neurons in Huntington's disease. *Journal of Neuropathology and Experimental Neurology*, *54*(1), 105–120.

Heimer, L., & Wilson, R. (1975). The subcortical projections of the allocortex: Similarities in the neural associations of the hippocampus, the piriform cortex, and the neocortex. In M. Santini (Ed.), *Golgi Centennial symposium* (pp. 177–193). New York: Raven Press.

Holt, D. J., Graybiel, A. M., & Saper, C. B. (1997). Neurochemical architecture of the human striatum. *Journal of Comparative Neurology*, *384*, 1–25.

Kam, M., Curtis, M. A., McGlashan, S. R., Connor, B., Nannmark, U., & Faull, R. L. M. (2009). The cellular composition and morphological organization of the rostral migratory stream in the adult human brain. *Journal of Chemical Neuroanatomy*, *37*(3), 196–205.

Katz, J., Nielsen, K. M., & Soghomonian, J. J. (2005). Comparative effects of acute or chronic administration of levodopa to 6-hydroxydopamine-lesioned rats on the expression of glutamic acid decarboxylase in the neostriatum and $GABA_A$ receptors subunits in the substantia nigra, pars reticulata. *Neuroscience*, *132*(3), 833–842.

Kawaguchi, Y., Wilson, C. J., Augood, S. J., & Emson, P. C. (1995). Striatal interneurones—Chemical, physiological and morphological characterization. *Trends in Neurosciences*, *18*(12), 527–535.

Kayahara, T., & Nakano, K. (1996). Pallido-thalamo-motor cortical connections—An electron microscopic study in the macaque monkey. *Brain Research*, *706*(2), 337–342.

Kincaid, A. E., Albin, R. L., Newman, S. W., Penney, J. B., & Young, A. B. (1992). 6-Hydroxydopamine lesions of the nigrostriatal pathway alter the expression of glutamate-decarboxylase messenger-RNA in rat globus-pallidus projection neurons. *Neuroscience*, *51*(3), 705–718.

Kish, S. J., Rajput, A., Gilbert, J., Rozdilsky, B., Chang, L. J., Shannak, K., et al. (1986). Elevated gamma-aminobutyric-acid level in striatal but not extrastriatal brain-regions in Parkinson's disease—Correlation with striatal dopamine loss. *Annals of Neurology*, *20*(1), 26–31.

Kriegstein, A. R. (2005). GABA puts the brake on stem cells. *Nature Neuroscience*, *8*(9), 1132–1133.

Kullmann, D. M., Ruiz, A., Rusakov, D. M., Scott, R., Semyanov, A., & Walker, M. C. (2005). Presynaptic, extrasynaptic and axonal $GABA_A$ receptors in the CNS: Where and why? *Progress in Biophysics and Molecular Biology*, *87*(1), 33–46.

Kultas-Ilinsky, K., Leontiev, V., & Whiting, P. J. (1998). Expression of 10 $GABA_A$ receptor subunit messenger RNAs in the motor-related thalamic nuclei and basal ganglia of Macaca mulatta studied with in situ hybridization histochemistry. *Neuroscience*, *85*(1), 179–204.

Lange, H., Thorner, G., Hopf, A., & Schroder, K. F. (1976). Morphometric studies of neuropathological changes in choreatic diseases. *Journal of the Neurological Sciences*, *28*(4), 401–425.

Levy, R., Herrero, M. T., Ruberg, M., Villares, J., Faucheux, B., Guridi, J., et al. (1995). Effects of nigrostriatal denervation and L-Dopa therapy on the gabaergic neurons of the striatum in MPTP-treated monkeys and Parkinson's disease—An in situ hybridization study of Gad(67) messenger-Rna. *European Journal of Neuroscience*, 7(6), 1199–1209.

Li, X. J., Li, S. H., Sharp, A. H., Nucifora, F. C., Schilling, G., Lanahan, A., et al. (1995). A Huntingtin-associated protein enriched in brain with implications for pathology. *Nature*, *378*(6555), 398–402.

Liu, X. X., Wang, Q., Haydar, T. F., & Bordey, A. (2005). Nonsynaptic GABA signaling in postnatal subventricular zone controls proliferation of GFAP-expressing progenitors. *Nature Neuroscience*, *8*(9), 1179–1187.

Lois, C., & Alvarezbuylla, A. (1994). Long-distance neuronal migration in the adult mammalian brain. *Science*, *264*(5162), 1145–1148.

Loup, F., Wieser, H. G., Yonekawa, Y., Aguzzi, A., & Fritschy, J. M. (2000). Selective alterations in $GABA_A$ receptor subtypes in human temporal lobe epilepsy. *Journal of Neuroscience*, *20*(14), 5401–5419.

Low, V. F., Faull, R. L. M., Bennet, L., Gunn, A. J., & Curtis, M. A. (2013). Neurogenesis and progenitor cell distribution in the subgranular zone and subventricular zone of the adult sheep brain. *Neuroscience*, *244*, 173–187.

McGeorge, A. J., & Faull, R. L. M. (1989). The organization of the projection from the cerebral cortex to the striatum in the rat. *Neuroscience*, *29*, 503–537.

Michels, G., & Moss, S. J. (2007). $GABA_A$ receptors: Properties and trafficking. *Critical Reviews in Biochemistry and Molecular Biology*, *42*(1), 3–14.

Miralles, C. P., Li, M., Mehta, A. K., Khan, Z. U., & De Blas, A. L. (1999). Immunocytochemical localization of the beta(3) subunit of the gamma-aminobutyric acid(A) receptor in the rat brain. *Journal of Comparative Neurology*, *413*(4), 535–548.

Mohler, H. (2006a). $GABA_A$ receptor diversity and pharmacology. *Cell and Tissue Research*, *326*(2), 505–516.

Mohler, H. (2006b). $GABA_A$ receptors in central nervous system disease: Anxiety, epilepsy, and insomnia. *Journal of Receptor and Signal Transduction Research*, *26*(5–6), 731–740.

Mohler, H. (2007). Molecular regulation of cognitive functions and developmental plasticity: Impact of $GABA_A$ receptors. *Journal of Neurochemistry*, *102*(1), 1–12.

Mohler, H., Benke, D., Benson, J., Luscher, B., & Fritschy, J.-M. (1995). $GABA_A$-receptor subtypes in vivo: Cellular localisation, pharmacology and regulation. In F. Biggio, A. Sanna, & E. Costa (Eds.), *$GABA_A$ receptors and anxiety: From neurobiology to treatment.* New York: Raven Press.

Mohler, H., Fritschy, J.-M., Luscher, B., Rudolph, U., Benson, J., & Benke, D. (1996). The $GABA_A$ receptors from subunits to diverse functions. In T Narahashi, (Ed.), *Vol. 4, Ion channels*, New York: Plenum Press.

Moss, S. J., & Smart, T. G. (2001). Constructing inhibitory synapses. *Nature Reviews. Neuroscience*, *2*(4), 240–250.

Nambu, A., Tokuno, H., & Takada, M. (2002). Functional significance of the cortico-subthalamo-pallidal 'hyperdirect' pathway. *Neuroscience Research*, *43*(2), 111–117.

Nauta, W. J. H., & Domesick, V. B. (1984). Afferent and efferent relationships of the basal ganglia. In E. V. Evarts (Ed.), *Functions of the basal ganglia* (pp. 3–29). London: Pitman.

Nicholson, L. F., Faull, R. L., Waldvogel, H. J., & Dragunow, M. (1995). GABA and $GABA_A$ receptor changes in the substantia nigra of the rat following quinolinic acid lesions in the striatum closely resemble Huntington's disease. *Neuroscience*, *66*(3), 507–521.

Olsen, R. W., & Sieghart, W. (2008). International Union of Pharmacology. LXX. Subtypes of gamma-aminobutyric acid(A) receptors: Classification on the basis of subunit composition, pharmacology, and function. Update. *Pharmacological Reviews*, *60*(3), 243–260.

Oyanagi, K., Takeda, S., Takahashi, H., Ohama, E., & Ikuta, F. (1989). A quantitative investigation of the substantia nigra in Huntington's disease. *Annals of Neurology*, *26*(1), 13–19.

Parent, A., Fortin, M., Cote, P. Y., & Cicchetti, F. (1996). Calcium-binding proteins in primate basal ganglia. *Neuroscience Research*, *25*(4), 309–334.

Parent, A., & Hazrati, L. N. (1995). Functional-anatomy of the basal ganglia. 1. The cortico-basal ganglia-thalamo-cortical loop. *Brain Research Reviews*, *20*(1), 91–127.

Petri, S., Krampel, K., Dengler, R., Bufler, J., Weindl, A., & Arzberger, T. (2002). Human $GABA_A$ receptors on dopaminergic neurons in the pars compacta of the substantia nigra. *Journal of Comparative Neurology*, *452*(4), 360–366.

Pirker, S., Schwarzer, C., Wieselthaler, A., Sieghart, W., & Sperk, G. (2000). $GABA_A$ receptors: Immunocytochemical distribution of 13 subunits in the adult rat brain. *Neuroscience*, *101*(4), 815–850.

Platel, J. C., Dave, K. A., & Bordey, A. (2008). Control of neuroblast production and migration by converging GABA and glutamate signals in the postnatal forebrain. *Journal of Physiology—London*, *586*(16), 3739–3743.

Prensa, L., Gimenez-Amaya, J. M., & Parent, A. (1999). Chemical heterogeneity of the striosomal compartment in the human striatum. *Journal of Comparative Neurology*, *413*(4), 603–618.

Pritchett, D. B., Sontheimer, H., Shivers, B. D., Ymer, S., Kettenmann, H., Schofield, P. R., et al. (1989). Importance of a novel $GABA_A$ receptor subunit for benzodiazepine pharmacology. *Nature*, *338*(6216), 582–585.

Reiner, A., Albin, R. L., Anderson, K. D., D'Amato, C. J., Penney, J. B., & Young, A. B. (1988). Differential loss of striatal projection neurons in Huntington disease. *Proceedings of the National Academy of Sciences of the United States of America*, *85*(15), 5733–5737.

Reiner, A., Shelby, E., Wang, H. B., DeMarch, Z., Deng, Y. P., Guley, N. H., et al. (2013). Striatal parvalbuminergic neurons are lost in Huntington's disease: Implications for dystonia. *Movement Disorders*, *28*(12), 1691–1699.

Richardson, E. P. (1990). Huntington's disease—Some recent neuropathological studies. *Neuropathology and Applied Neurobiology*, *16*(6), 451–460.

Robertson, R. G., Clarke, C. A., Boyce, S., Sambrook, M. A., & Crossman, A. R. (1990). The role of striatopallidal neurons utilizing gamma-aminobutyric-acid in the pathophysiology of Mptp-induced Parkinsonism in the primate—Evidence from [H-3] flunitrazepam autoradiography. *Brain Research*, *531*(1–2), 95–104.

Robertson, R. G., Graham, W. C., Sambrook, M. A., & Crossman, A. R. (1991). Further investigations into the pathophysiology of Mptp-induced parkinsonism in the primate—An intracerebral microdialysis study of gamma-aminobutyric-acid in the lateral segment of the globus-pallidus. *Brain Research*, *563*(1–2), 278–280.

Rudolph, U., & Antkowiak, B. (2004). Molecular and neuronal substrates for general anaesthetics. *Nature Reviews. Neuroscience*, *5*(9), 709–720.

Sadikot, A. F., Parent, A., Smith, Y., & Bolam, J. P. (1992). Efferent connections of the centromedian and parafascicular thalamic nuclei in the squirrel-monkey—A light and electron-microscopic study of the thalamostriatal projection in relation to striatal heterogeneity. *Journal of Comparative Neurology*, *320*(2), 228–242.

Schwarzer, C., Berresheim, U., Pirker, S., Wieselthaler, A., Fuchs, K., Sieghart, W., et al. (2001). Distribution of the major gamma-aminobutyric acid(A) receptor subunits in the basal ganglia and associated limbic brain areas of the adult rat. *Journal of Comparative Neurology*, *433*(4), 526–549.

Sernagor, E., Chabrol, F., Bony, G., & Cancedda, L. (2010). GABAergic control of neurite outgrowth and remodeling during development and adult neurogenesis: General rules and differences in diverse systems. *Frontiers in Cellular Neuroscience*, *4*, 1–11.

Seto-Ohshima, A., Lawson, E., Emson, P. C., Mountjoy, C. Q., & Carrasco, L. H. (1988). Loss of matrix calcium-binding protein-containing neurons in Huntington's disease. *The Lancet*, *331*(8597), 1252–1255.

Sieghart, W. (1995). Structure and pharmacology of gamma-aminobutyric acidA receptor subtypes. *Pharmacological Reviews*, *47*(2), 181–234.

Smith, Y., Bevan, M. D., Shink, E., & Bolam, J. P. (1998). Microcircuitry of the direct and indirect pathways of the basal ganglia. *Neuroscience*, *86*(2), 353–387.

Smith, K. R., & Kittler, J. T. (2010). The cell biology of synaptic inhibition in health and disease. *Current Opinion in Neurobiology*, *20*(5), 550–556.

Soghomonian, J. J., & Laprade, N. (1997). Glutamate decarboxylase (GAD67 and GAD65) gene expression is increased in a subpopulation of neurons in the putamen of parkinsonian monkeys. *Synapse*, *27*(2), 122–132.

Sohal, V. S., Keist, R., Rudolph, U., & Huguenard, J. R. (2003). Dynamic $GABA_A$ receptor subtype-specific modulation of the synchrony and duration of thalamic oscillations. *Journal of Neuroscience*, *23*(9), 3649–3657.

Somogyi, P., Fritschy, J. M., Benke, D., Roberts, J. D. B., & Sieghart, W. (1996). The gamma-2 subunit of the Gaba(a) receptor is concentrated in synaptic junctions containing the alpha-1 and beta-2/3 subunits in hippocampus, cerebellum and globus pallidus. *Neuropharmacology*, *35*(9–10), 1425–1444.

Song, J., Zhong, C., Bonaguidi, M. A., Sun, G. J., Hsu, D., Gu, Y., et al. (2012). Neuronal circuitry mechanism regulating adult quiescent neural stem-cell fate decision. *Nature*, *489*, 150–154.

Spampanato, J., Gu, X., Yang, X. W., & Mody, I. (2008). Progressive synaptic pathology of motor cortical neurons in a BAC transgenic mouse model of Huntington's disease. *Neuroscience*, *157*(3), 606–620.

Stephenson, F. A. (2014). Revisiting the TRAK family of proteins as mediators of $GABA_A$ receptor trafficking. *Neurochemical Research*, *39*(6), 992–996.

Tattersfield, A. S., Croon, R. J., Liu, Y. W., Kells, A. P., Faull, R. L. M., & Connor, B. (2004). Neurogenesis in the striatum of the quinolinic acid lesion model of Huntington's disease. *Neuroscience*, *127*(2), 319–332.

Thompson-Vest, N. M., Waldvogel, H. J., Rees, M. I., & Faull, R. L. M. (2003). $GABA_A$ receptor subunit and gephyrin protein changes differ in the globus pallidus in Huntington's diseased brain. *Brain Research*, *994*(2), 265–270.

Tippett, L. J., Waldvogel, H. J., Thomas, S. J., Hogg, V. M., van Roon-Mom, W., Synek, B. J., et al. (2007). Striosomes and mood dysfunction in Huntington's disease. *Brain*, *130*(Pt 1), 206–221.

Tokuno, H., Chiken, S., Kametani, K., & Moriizumi, T. (2002). Efferent projections from the striatal patch compartment: Anterograde degeneration after selective ablation of neurons expressing mu-opioid receptor in rats. *Neuroscience Letters*, *332*(1), 5–8.

Twelvetrees, A. E., Yuen, E. Y., Arancibia-Carcamo, I. L., MacAskill, A. F., Rostaing, P., Lumb, M. J., et al. (2010). Delivery of $GABA_A$ Rs to synapses is mediated by HAP1-KIF5 and disrupted by mutant huntingtin. *Neuron*, *65*(1), 53–65.

Tyagarajan, S. K., & Fritschy, J. M. (2014). Gephyrin: A master regulator of neuronal function? *Nature Reviews. Neuroscience*, *15*(3), 141–156.

Vonsattel, J. P., & Difiglia, M. (1998). Huntington disease. *Journal of Neuropathology and Experimental Neurology*, *57*(5), 369–384.

Vonsattel, J. P., Keller, C., & Del Pilar Amaya, M. (2008). Neuropathology of Huntington's disease. *Handbook of Clinical Neurology*, *89*, 599–618.

Vonsattel, J. P., Myers, R. H., Stevens, T. J., Ferrante, R. J., Bird, E. D., & Richardson, E. P. (1985). Neuropathological classification of Huntington's disease. *Journal of Neuropathology and Experimental Neurology*, *44*, 559–577.

Waldvogel, H. J., Baer, K., & Faull, R. L. M. (2009). The localization of inhibitory neurotransmitter receptors on dopaminergic neurons of the human substantia Nigra. *Journal of Neural Transmission. Supplementum*, *73*, 59–70.

Waldvogel, H. J., Baer, K., Gai, W. P., Gilbert, R. T., Rees, M. I., Mohler, H., et al. (2008). Differential localization of $GABA_A$ receptor subunits within the substantia nigra of the human brain: An immunohistochemical study. *Journal of Comparative Neurology*, *506*(6), 912–929.

Waldvogel, H. J., & Faull, R. L. M. (1993). Compartmentalization of parvalbumin immunoreactivity in the human striatum. *Brain Research*, *610*, 311–316.

Waldvogel, H. J., Faull, R. L., Jansen, K. L., Dragunow, M., Richards, J. G., Mohler, H., et al. (1990). GABA, GABA receptors and benzodiazepine receptors in the human spinal cord: An autoradiographic and immunohistochemical study at the light and electron microscopic levels. *Neuroscience*, *39*(2), 361–385.

Waldvogel, H. J., Faull, R. L., Williams, M. N., & Dragunow, M. (1991). Differential sensitivity of calbindin and parvalbumin immunoreactive cells in the striatum to excitotoxins. *Brain Research*, *546*(2), 329–335.

Waldvogel, H. J., Fritschy, J. M., Mohler, H., & Faull, R. L. M. (1998). Gaba(a) receptors in the primate basal ganglia—An autoradiographic and a light and electron microscopic immunohistochemical study of the alpha(1) and beta(2,3) subunits in the baboon brain. *Journal of Comparative Neurology*, *397*(3), 297–325.

Waldvogel, H. J., Kubota, Y., Fritschy, J. M., Mohler, H., & Faull, R. L. M. (1999). Regional and cellular localisation of $GABA_A$ receptor subunits in the human basal ganglia: An autoradiographic and immunohistochemical study. *Journal of Comparative Neurology*, *413*(3), 313–340.

Waldvogel, H. J., Kubota, Y., Trevallyan, S. C., Kawaguchi, Y., Fritschy, J. M., Mohler, H., et al. (1997). The morphological and chemical characteristics of striatal neurons immunoreactive for the alpha(1)-subunit of the $GABA_A$ receptor in the rat. *Neuroscience*, *80*(3), 775–792.

Whitehouse, P. J., Trifiletti, R. R., Jones, B. E., Folstein, S., Price, D. L., Snyder, S. H., et al. (1985). Neurotransmitter receptor alterations in Huntington's disease: Autoradiographic and homogenate studies with special reference to benzodiazepine receptor complexes. *Annals of Neurology*, *18*, 202–210.

Wisden, W., Laurie, D. J., Monyer, H., & Seeburg, P. H. (1992). The distribution of 13 $GABA_A$ receptor subunit mRNAs in the rat brain. I. Telencephalon, diencephalon, mesencephalon. *Journal of Neuroscience*, *12*(3), 1040–1062.

Wisden, W., & Seeburg, P. H. (1992). GABA(a) receptor channels: From subunits to functional entities. *Current Opinion in Neurobiology*, *2*, 263–269.

Yan, Z., & Surmeier, D. J. (1997). D5 dopamine receptors enhance Zn^{2+}-sensitive $GABA_A$ currents in striatal cholinergic interneurons through a PKA/PP1 cascade. *Neuron*, *19*(5), 1115–1126.

Ymer, S., Draguhn, A., Wisden, W., Werner, P., Keinanen, K., Schofield, P. R., et al. (1990). Structural and functional-characterization of the gamma-1 subunit of gaba-a benzodiazepine receptors. *The EMBO Journal*, *9*(10), 3261–3267.

Yohrling, G. J., Jiang, G. C. T., DeJohn, M. M., Miller, D. W., Young, A. B., Vrana, K. E., et al. (2003). Analysis of cellular, transgenic and human models of Huntington's disease reveals tyrosine hydroxylase alterations and substantia nigra neuropathology. *Molecular Brain Research*, *119*(1), 28–36.

Yu, T. S., Wang, S. D., Liu, J. C., & Yin, H. S. (2001). Changes in the gene expression of $GABA_A$ receptor alpha(1) and alpha(2) subunits and metabotropic glutamate receptor 5 in the basal ganglia of the rats with unilateral 6-hydroxydopamine lesion and embryonic mesencephalic grafts. *Experimental Neurology, 168*(2), 231–241.

Zezula, J., Cortes, R., Probst, A., & Palacios, J. M. (1988). Benzodiazepine receptor sites in the human brain: Autoradiographic mapping. *Neuroscience, 25*(3), 771–795.

INDEX

Note: Page numbers followed by "*f*" indicate figures and "*t*" indicate tables.

H

FSC
www.fsc.org
MIX
Papier | Fördert gute Waldnutzung
FSC® C083411

Kolibri 360 GmbH
Ferdinand-Jühlke-Straße 7
99095 Erfurt, Deutschland
produktsicherheit@kolibri360.de